NASA SP—4022

ASTRONAUTICS AND AERONAUTICS, 1977

A Chronology

Eleanor H. Ritchie

The NASA History Series

Scientific and Technical Information Branch   1986
National Aeronautics and Space Administration
Washington, DC

Four spacecraft launched by NASA in 1977: left to right, top, ESA's Geos 1 *and NASA's* Heao 1; *bottom, ESA's* Isee 2 *on NASA's* Isee 1, *and Italy's* Sirio. *(NASA 77-H-157, 77-H-56, 77-H-642, 77-H-484)*

Contents

Preface v
January 1
February 21
March 47
April 61
May 77
June 101
July 127
August 143
September 165
October 185
November 201
December 217
Appendixes
A. Satellites, Space Probes, and Manned Space Flights, 1977 237
B. Major NASA Launches, 1977 261
C. Manned Space Flights, 1977 265
D. NASA Sounding Rocket Launches, 1977 267
E. Abbreviations of References 273
Index 279

Illustrations

Four spacecraft launched by NASA, 1977 ii
Space Shuttle orbiter Enterprise en route to testing 17
Artist's concept, world's largest wind-turbine blade 34
Artist's concept, solar sail rendezvous with comet 49
Landsat images, Sierra Nevada snowcover, 1975 and 1977 81
Dr. Wernher von Braun with Saturn V, 1969 116
"Sounds of Earth" recording goes on *Voyager 2* spacecraft 144
Space Shuttle orbiter Enterprise on first free flight 151

NASA maintains an internal history program for two principal reasons: (1) Sponsorship of research in NASA-related history is one way in which NASA responds to the provision of the National Aeronautics and Space Act of 1958 that requires NASA to "provide for the widest practicable and appropriate dissemination of information concerning its activities and the results thereof." (2) Thoughtful study of NASA history can help agency managers accomplish the missions assigned to the agency. Understanding NASA's past aids in understanding its present situation and illuminates possible future directions. The opinions and conclusions set forth in this book are those of the author; no official of the agency necessarily endorses those opinions or conclusions.

Preface

Astronautics and Aeronautics for 1977 is the 17th volume in a series of annual chronological digests of principal events in the fields of astronautics and aeronautics. Initiated in 1961 with a compendium of events from 1915–1960, this series is designed to serve as a reference for policymakers, reference librarians, and researchers. Each entry is followed by one or more citations of sources to which users may turn for further information on the particular event or development being chronicled.

The *Astronautics and Aeronautics* series is one way in which the NASA History Office has attempted to implement its charter to disseminate information about the activities of the National Aeronautics and Space Administration, and to provide support for historians and other researchers in the area of United States aeronautical and space programs. As a result of staff and budgetary limitations, the NASA History Office has been unable to maintain the annual currency of this series. We expect, however, to bring the series up to date with a forthcoming compendium volume covering principal events over a six-year period.

This volume was written by a member of the NASA History Office staff, Eleanor H. Ritchie. Michael A. Dennis, Stuart Rosenbaum, and Charles Scott also worked on the volume.

Reliability is essential to the usefulness of any reference work of this scope. As with previous volumes, readers can contribute by notifying our office of any errors so that we can publish them in subsequent volumes.

Sylvia D. Fries
Director, NASA History Office

December 1985

January

January 2: Leaders of 28 major U.S. scientific societies had endorsed tests to determine the value of a "science court"—an impartial quasi-judicial body—in evaluating conflicting claims on controversial national issues such as nuclear-reactor safety, food additives, pesticides, and offshore oil exploration, and in determining the facts, according to *New York Times* science reporter John Noble Wilford. The concept, debated for several yr, had received the endorsements in the last months of 1976. Although it had the support of President Ford's scientific advisers, proponents were not sure how President-elect Carter's advisers stood. (*NYT,* Jan 2/77, 28)

January 3: NASA announced award of what *Aviation Week* called "the largest NASA contract ever issued for an unmanned space program" to Western Union Space Communications Co., TRW Systems, and Harris Electronics Systems Div. for development of the tracking and data-relay satellite system (TDRSS) aimed at replacing present ground stations by offering orbital-communications coverage for 85 to 100% of spacecraft orbit, compared to 15% coverage now available from ground-based facilities. The system would consist of two specialized relay spacecraft in synchronous earth orbit plus a ground terminal at White Sands, N.M., supporting 20 or more satellites orbiting at up to 5000km altitude. The 10yr contract at $79.6 million per yr would include work on spacecraft and ground stations, as well as all operating costs.

TRW would be prime manufacturer under subcontract with WU to build six spacecraft plus components for a seventh, using technology developed for the FltSatCom spacecraft and hardware used in WU's Westar domestic comsats; Harris Electronics would build the White Sands station and provide antenna components. Missions using the TDRSS would include scientific, applications, and manned satellites such as Space Shuttle, Spacelab, and Shuttle-launched spacecraft scheduled for the 1980s. (*AvWk,* Jan 3/77, 14; NASA Release 76-213)

• Seven European nations planning to use data from the U.S. Seasat A, due for launch in May 1978, were forming an organization to coordinate their activities, *Aviation Week* magazine reported. Denmark, Finland, France, West Germany, Britain, Norway, and Spain would use Seasat measurement of wave heights to forecast floods in the North Sea area and to design ships or offshore structures such as oil rigs, or in harbor construction. The group planned to have five laser-ranging stations and up to 10 Doppler tracking stations ready by the time Seasat would be operational. (*AvWk*, Jan 3/77, 20)

January 4: Kennedy Space Center announced it would hold a symposium Jan. 15 on minority opportunities at the Spaceport, to include talks on KSC procurement, future construction plans, and explanations of the KSC affirmative action plan and equal employment opportunity program. KSC had invited representatives of minority business firms and the general public to attend the symposium. (KSC Release 2-77)

January 5: The U.S. Air Force Systems Command reported it had worked out a way to suppress noise of jet-aircraft engines during maintenance operations on the ground, as a means of reducing hearing loss in maintenance personnel and noise impact on the community. Fifty-eight Air Force bases worldwide had received the suppressor units, consisting of primary and secondary air intakes and enclosures, augmenter tube, exhaust muffler, and support structure to house controls and water pumps. After backing an aircraft into the secondary enclosure, workers would enclose the air intakes at the front in the primary-intake mufflers; the secondary enclosure, covering the aft fuselage and engine exhaust, would include openings to be sealed against the fuselage and stabilizers to keep noise inside. Inside the augmenter tube, secondary cooling air would mix with the exhaust, and spray bars in the tube would spray water at 800gal per min during afterburner operation. (OIP Release 284.76)

• The International Telecommunications Satellite Organization (INTELSAT) announced it would award to a British company, AERE Hartwell, a 15mo $71 540 fixed-price contract to develop low-pressure, nickel-hydrogen power cells that would use a lanthanum-alloy member both as hydrogen reservoir and as negative electrode. Use of lanthanum would lower pressure in the cells, reduce the heat dissipated during discharge, and reduce the volume of cell density. If successfully developed, the cells would operate at a maximum pressure of about 6 atmospheres and would have a minimum life of 500 charge-discharge cycles when used in synchronous-satellite applications. (INTELSAT Release 77-1-M)

• INTELSAT announced award of a 15mo $72 536 fixed-price contract to EIC Corp. of Newton, Mass., for development of an improved high-pressure nickel-hydrogen power cell that would increase reliability and attain a minimum 10yr cyclical life in synchronous-satellite applications.

Such an advance would permit about 4000 daily charge-discharge cycles, or about 10yr of maintenance-free operations. (INTELSAT Release 77-2-M)

January 6: NASA formally adjudged the Viking 75 mission, launched toward Mars on Aug. 20 and Sept. 9, 1975, successful, according to the post-landing mission operations report. All Viking instruments except the seismometer on the first lander (which failed to uncage itself) operated successfully and were continuing to send data to earth. The prelaunch mission objectives (advancing knowledge of Mars by orbital observation and by direct measurements in the atmosphere and on the planet's surface) had been realized, the report said. (MOR S-815-75-01/02)

• NASA announced that a test model of *Pioneer 10*, first spacecraft to reached Jupiter, would be on display at the Natl. Air and Space Museum in Washington, D.C., beginning Jan. 11. The actual spacecraft, launched March 3, 1972, had reached Jupiter in Dec. 1973 and was now about 1.6 billion km (about a billion miles) from earth headed for the orbit of Uranus, which it should reach in Sept. 1979, by the time its sister spacecraft *Pioneer 11* would reach Saturn.

Headed generally toward Aldebaran in the constellation Taurus, *Pioneer 10* would reach the orbit of Pluto (about 4 billion miles out) in 1987, but radio-signal reception would be doubtful beyond the orbit of Uranus. The model of the first spacecraft scheduled to leave the earth's solar system would appear in the Milestones of Flight hall with the Wright Flyer, the Spirit of St. Louis, and the *Apollo 11*. (NASA Release 77-1; ARC Release 77-1)

• NASA Hq announced appointment of Curtis M. Graves as director of the Community Affairs Division, effective Jan. 3. Graves came to NASA from the Natl. Civil Service League, where he was managing associate for continuing education from 1973 to 1977. For 6yr he was a member of the Texas legislature, representing a Houston district beginning in 1966 as the first black elected to that body since 1891. Born in New Orleans, he received a bachelor's degree from Texas Southern University and an honorary doctorate from Union Baptist Bible College in Houston. He also served as branch manager of a savings and loan association and as advertising manager and editor of two Houston newspapers. In his new assignment with NASA he would supervise the agency's relations with communities where NASA centers and installations are located, promoting understanding of NASA's economic and social impact on these areas among local citizens and NASA personnel nationwide. (NASA Release 77-3)

- India and the Soviet Union had reached agreement on coordinating their satellite systems, said the Indian news agency Samachar, as reported by the Foreign Broadcast Information Service. The agreement resulted from a recent visit to Moscow by a team of experts from India's Ministry of Communications and Department of Space. The problem arose because of possible interference between the proposed India domestic comsat Insat, to be placed in geostationary orbit over the Indian Ocean, and existing or proposed satellite systems of the USSR. The aim of the talks was to ensure avoidance of satellite interference through discussion of the technical aspects of the systems, including the distance between satellites. Satellites already in operation in that area included those of the Intl. Telecommunications Satellite Organization, Indonesia, and others. (FBIS, Delhi *Patriot,* Dec 26/76, 1)

January 7: NASA announced that four customers had signed up for "getaway special" payloads to fly on the Space Shuttle on a space-available basis beginning in 1980. The small self-contained payloads must weigh no more than 90kg (200lb) each, measure no more than $0.15m^3$ ($5ft^3$), and require no services such as electrical power or deployment into space.

Owners of the research and development packages were R. Gilbert Moore of Utah, a private citizen who offered half his payload to Utah State Univ., which in turn would make the space available to high school students who submitted proposals to fly their own experiments; Dr. L.R. Megill, chairman of the space science experiment committee at Utah State, who funded a payload as a follow-on to Moore and also planned to invite student proposals; and Reiner Klett, representing independent West German researchers, who signed for two payloads, one biological and one for space processing. (NASA Release 77-4)

- Radio signals between the four Viking spacecraft on Mars and earth-based antennas had provided the most accurate confirmation so far of Einstein's theory of relativity in an experiment conducted Nov. 25, 1976, when Mars moved behind the sun relative to the earth, causing a total blackout of communications with the Viking spacecraft. Radio signals transmitted to the Viking spacecraft from antennas at Goldstone, Calif., and Canberra, Australia, had produced "incredibly precise" measurement of the radio transmission. According to Adam I. Braufman of the Mass. Institute of Technology, the "uncertainty" over a space of some 322 million km was less than 2m, an accuracy of five parts in 10 million millionths.

The Viking radio science team was trying to confirm Einstein's prediction that the sun's gravitational force would bend and delay radio signals (or any form of radiation) traveling close to a massive body such as the sun. Previous tests using the spacecraft communications systems tended to confirm the theory, but the Viking test proved much more accurate than the previous ones. (*NYT,* Jan 7/77, A-8; *W Post,* Jan 7/77, A-1)

• NASA announced it would participate in a joint agro-environmental monitoring system that would improve crop management in the state of Virginia through knowledge gained from the automated data-collection and processing systems developed for satellite and space-exploration programs. Funded by Wallops Flight Center at Wallops Island, Va., and by the Virginia Polytechnic Institute and State Univ. at Blacksburg, a network of data collection stations would automatically track "degree-day" units necessary for the emerging of certain insects or diseases; when a certain total was reached, the system would alert a farmer to spray his crop. Savings would result from reduced use of insecticides during less effective periods, reduced possibility of pollution, and fewer requirements for labor and equipment. Loyd C. Parker, Wallops project manager, said design work was complete and by early spring the five monitoring stations and central computer should be ready for use. (NASA Release 77-6)

• Marshall Space Flight Center announced it would accept proposals for a study contract to define possibilities of receiving electrical energy from the sun in large quantities for use on earth, through a concept called Satellite Power Systems, consisting of a number of spacecraft in stationary orbit using photovoltaic or solar-thermal power instrumentation to capture solar energy, convert it to electrical energy, convert that to microwave energy, transmit it to earth stations for conversion back to electrical energy, and distribute the power to users. Other NASA studies had established the technical and economic feasibility of satellite power systems, but an agreement between NASA and the Energy Research and Development Administration would result in additional data on key issues and critical areas, and the ability of the concept to deal with these issues and areas. MSFC would accept proposals for the study through Jan. 24. (MSFC Release 77-1)

January 8: James B. Irwin, eighth astronaut to walk on the moon, remained in satisfactory condition after triple bypass surgery at the Texas Heart Institute. "He is still in intensive care but is doing very well," said a spokesman at St. Lukes Episcopal Hospital in Houston. The operation consisted of replacing three vein segments with veins removed from Irwin's legs. Irwin, 46, lunar-module pilot on the *Apollo 15* mission July 26, 1971, had resigned from the astronaut corps in May 1972. (*NYT,* Jan 8/77, 17)

January 10: AvWk reported that the seismometer carried on the Viking 2 lander might have recorded the first seismic event since the spacecraft touched down Sept. 3, 1976, on the Utopia Planitia area of the planet Mars. Beginning about 7pm local time for the lander, on Sol 60 (Nov. 7, 1976), the activity continued for about 30min. Viking project observers said the data seemed to have the "proper signature" of a seismic event, similar to data recorded on the lunar surface during seismic events. Donald Anderson, leader of the Viking seismology team, said rough estimates put the center of the activity several thousand km from the lander, but cautioned against labeling the activity a "marsquake" until further studies were done. (*AvWk,* Jan 10/77, 21)

• MSFC reported that the number of visitors to the center in 1976 had increased 11% over the previous yr to a total of 95 529, representing each of the 50 United States and several foreign countries. The guided bus tours of the center, cosponsored by MSFC and the Alabama Space and Rocket Center, last year added new items and increased activities at some of the former Saturn test sites. The tour stop presenting the story of the Space Shuttle and its payload had added a full-scale mockup of the Shuttle main engine. Tourists could also enter a full-scale mockup of Skylab to see how the astronauts lived in orbit; and the east and west test areas being converted for Shuttle testing were now available for viewing. The Redstone test stand already on the tour route had been refurbished and was added to the Natl. Register of Historic Places. (MSFC Release 77-3)

• The U.S. Dept. of Defense had begun development of plans to wage war in outer space, the *Washington Post* reported, and had contracted with six manufacturers to study ways of letting its satellites sound an alarm if approached, sound another alarm if attacked, and defend itself from an enemy satellite that approached too close.

Quoting "knowledgeable sources . . . close to the Pentagon," the *Post* said the USSR's resumption last year of hunter-killer satellite testing after a 5yr hiatus had caused growing concern over the threat of war in space. The Pentagon presently had at least a dozen comsats in orbit linking U.S. military bases, ships, and aircraft around the world, and another dozen navigation satellites to guide Navy surface ships and submarines. A killer satellite launched by an enemy could knock out the entire U.S. defense communications system within a wk, according to the *Post* account, and the Pentagon had considered putting an entire array of comsats in orbit and turning off their radios until needed: "With their radios off," said a source, "the opposition couldn't track their radio beacons. And if they can't track them they can't find them." No one had forecast a war in space within the next 10yr, the *Post* added, but the Pentagon planned to be prepared after that. Asked if the U.S. could fight a war on earth today if it had lost its satellites in space, a source replied: "Not successfully." (*W Post,* Jan 10/77, A-26)

January 17: NASA announced plans to launch NATO III-B, second in a new series of comsats to serve the North Atlantic Treaty Organization, from the Eastern Test Range at Cape Canaveral, Fla., on Jan. 27 as NASA's first launch of 1977. The three spacecraft in the Phase III series, funded entirely by NATO, were built by Ford Aerospace; the third would be launched in 1978. The Delta rocket provided by Goddard Space Flight Center that would put NATO III-B in orbit had successfully launched three previous NATO satellites: *Nato II-A* and *-B,* in March 1970 and Feb. 1971, and *Nato III-A* in April 1976. (NASA Release 77-5)

• Rockwell Intl.'s B-1 Division announced award of contracts for more than $3 million to two Phoenix, Ariz., firms—Sperry Rand Flight Systems Division and AiResearch Manufacturing Co.—for work on the new USAF B-1 strategic bomber. The B-1 Division was system contractor to the Air Force for the program, which received DOD production approval on Dec. 2, 1976. Sperry would receive a $1 840 500 contract to build a gyrostabilization subsystem and a vertical situation display for a B-1 prototype that would enter flight testing in 1979; AiResearch would receive a $1 218 102 contract to produce a secondary power subsystem for the same prototype. The gyrostabilization subsystem would provide the B-1 crew with the aircraft's course heading and flight angle with relation to the horizon; the secondary power subsystem would provide extra power for engine start and operation of support systems on the B-1. (Rockwell Release LA-1)

January 18: MSFC announced that the state of Georgia would be first in the nation to have its own Landsat digital-analysis system, under a contract between the Georgia Institute of Technology and MSFC enabling state planners to assess land-resources usage and to monitor changes, and to set up a standard data base for a statewide land-use program. The two Landsat satellites now in near-polar orbit would provide images and recorded magnetic tape transmitted to a ground receiving station. A data-retrieval system developed by Johnson Space Center would help in establishing a state organization structure to distribute the data to regional, county, and local users. A local official could find out how a tract as small as 10 acres was being used, merely by dialing a central computer at Ga. Tech and getting a printout by telephone into a local terminal. (MSFC Release 77-8)

January 19: NASA announced it would share with the city of Hampton, Va., the cost of building a refuse-burning, steam-producing plant to serve the city and four federal agencies in the Tidewater area. First of this type of community facility to be jointly sponsored by a city government and federal agencies, the project was the result of studies at Langley Research Center on better use of potential energy sources and improvement of the environment. Cost of the plant, about $8 million, would be shared by Langley Air Force Base, the City of Hampton, and LaRC, and should be amortized within 20yr. Beginning in 1979, it would consume about 70% of the city's waste and all refuse from LaRC, the air force base, the Army's Fort Monroe, and the VA hospital in Hampton, and would produce about 306 million lb of steam per yr. Built on NASA-owned land, the plant would remain the property of NASA though leased to and operated by the city for at least 20yr. Ash produced through special incineration at up to 1800°F would be sterile, nonorganic, and nonpolluting, amounting to a seventh of the refuse burned. (NASA Release 77-8)

- A Trident missile successfully launched from Cape Canaveral had passed its first flight test and met all major test objectives, according to a *NY Times* account. The U.S. Navy reported that the missile, carrying an inert payload, flew nearly its 9500km range to an area near Ascension Is. in the South Atlantic. Like its predecessors, the Polaris and Poseidon strategic missile systems, Trident would become the nation's primary undersea intercontinental weapon. At a cost of $21.5 billion, the Trident missile would be operative well into the 1980s. New Trident class submarines would carry 24 of the new missiles, whereas the Polaris and Poseidon had carried only 16. *Ohio,* first Trident class submarine under construction, and costing $882 million, would be ready for patrol duty in 1979 with a full complement of 24 Trident missiles, the report said. (*NYT,* Jan 19/77, A-14; *W Post,* Jan 19/77, A-16)

- The Dept. of Defense announced plans to begin full-scale development of cruise missiles made by Boeing Co. and General Dynamics Corp., with the first to be available in 1979, the *Wall St. Journal* reported. Cruise missiles (small low-flying pilotless jet planes designed to evade enemy radar) could carry either nuclear or conventional warheads and could be launched from ships, planes, or trucks, or submerged submarines. Boeing and General Dynamics had each developed a cruise missile: Boeing the Air Force version, and General Dynamics, the Navy version called the Tomahawk, designed for launch from submarines and surface vessels. Although the two versions used different airframes, they used the same engines, guidance systems, and warheads. After reviewing both versions, the DOD decided to continue both, and established a joint program office under the direction of Navy Capt. Walter Locke, whose orders required maximum cooperation with the Air Force. In addition to its "short" cruise missile, 1083cm long with a range of 1127km, the Air Force supported development of a stretch version 1470cm long with a range of 2254km for more distant targets. The DOD also expressed support for advanced technology to refine missile accuracy and for a supersonic version to exceed present speed of 885kph. It had asked Congress for $402.7 million for the programs in FY 1978, up from $198.7 million in 1977. (*WSJ*, Jan 19/77, 38)

January 20: The Netherlands had "finally decided" at the end of 1976 to join the U.S. and ESA in financing a second astronomy satellite (IRAS, the infrared astronomy satellite), but not all its citizens were enthusiastic over the anticipated expense, said a report in *Nature* magazine. Starting with a preliminary definition study in Jan. 1975, followed by 13 proposals for experiments in response to a NASA announcement, a U.S. study team of 11 astronomers had begun discussion with Dutch colleagues on the scientific program and had reached agreement at the beginning of 1976.

NASA had expected signing of a memorandum of understanding between the U.S. and the Netherlands by July 1976; the Dutch government had agreed in principle in July, but it had taken 6mo to find the money for the project. The U.S. and the Netherlands would each pay about 110 million guilders, and the United Kingdom Science Research Council, which had taken active part in the U.S.-Dutch negotiations, would contribute 10 million guilders. Plans were to launch IRAS from the Western Test Range in the spring of 1981, to trace and map about 10 million infrared sources. Dutch associations of scientific workers had protested in 1974 against further space activity, especially in behalf of underdeveloped countries lacking means to take advantage of such effort. (*Nature*, Jan 20/77, 202)

• The Max Planck Institute of Astronomy announced that, for the first time, astronomers had observed the presence of water outside earth's galaxy, indicating the possibility of life in outer space, the *NY Times* reported. Astronomers from the U.S., France, and West Germany, using the 32.91m Effelsberg radiotelescope, largest in the world, found rotating molecules of water at the edge of nebula IC 133, 2.2 million light-years from earth. The water molecules, when struck by light, apparently emit excess energy and (like laser beams) vibrate in unison, giving off radio signals. The discovery meant that other solar systems with the same physical conditions as earth's might exist, along with planets and stars also formed by condensation of dust and gas, with the same type of lifespan.

"What is decisive," according to Otto Hachenberg, director of the

Bonn Institute for Radio Astronomy, "is that we find the same conditions of physical matter one billion light years away from earth as on earth." Discovery of water vapor in another galaxy was expected to inspire scientists to look for such vapor in other galaxies. (*NYT,* Jan 20/77, 23)

January 21: The United States planned to orbit in 1983 a 9080kg 645cm telescope that would observe images almost to the edge of space and the beginning of time, Thomas O'Toole reported in the *Washington Post.* "Without the blurring of the atmosphere," said Dr. John Bahcall of Princeton Univ.'s Institute for Advanced Study, "we would be able to see things that are 10 times smaller than we have seen, or 10 times farther away. It would be like reading the writing on a quarter that we could only tell was round before."

If approved by Congress, the $435 million Space Telescope would be carried from earth by astronauts manning the Space Shuttle. The telescope would investigate quasars radiating thousands of times the energy generated by ordinary stars of the same size, and would observe the birth of stars, map nearby planets and galaxies, and compute distances to the nearest stars.

In addition to optics half the size of Mount Palomar's, the Space Telescope would carry lenses and instruments to scan ultraviolet and infrared spectra not accessible to earthbound telescopes because of atmospheric blockage. The telescope would relay its views to earth using geostationary satellites over the U.S. NASA would send Shuttle astronauts to the telescope every 5yr to replace wornout or damaged parts. (*W Post,* Jan 21/77, C-24)

• The Communications Satellite Corporation announced that its earth stations in Andover, Maine; Etam, W.Va.; and Jamesburg, Calif., had operated with two TV channels in each of four Intelsat satellites to transmit coverage of President Jimmy Carter's inauguration to about 70 countries around the world. Total transmit-receive time for the satellite TV coverage was 61-1/6 half-channel hr. The European Broadcasting Union (EBU), the U.S. Information Agency (USIA), the Armed Forces Radio and TV Service, and other organizations had reached audiences in both western and eastern Europe through the ComSat network. Countries in Africa, the Middle East, Latin America, Asia, and Australia had seen inaugural events transmitted either live or on film and videotape news services transmitted by the satellites. (ComSat Release 77-8)

• The Soviet Union had radioed commands to its orbiting *Salyut 5* space station to position it for rendezvous with a forthcoming Soyuz launch, according to a story in the *Washington Post.* Ideal time for a Soyuz launch would be within the next 5 days, the report said, and two Soviet tracking ships were already on station in the Atlantic Ocean and Caribbean Sea to keep in touch with the cosmonauts. (*W Post,* Jan 21/77, C8)

January 24: NASA Hq announced appointment of R.D. Ginter as assistant administrator for energy programs, effective Jan. 2. Ginter had been acting in that position since Sept. 1975, following the departure of astronaut (now U.S. Senator) Harrison H. Schmitt (R-NM).

Ginter had joined NASA in 1960 after 8yr as a missile project officer with the U.S. Navy and served in a variety of positions, including chief of the Scout launch vehicle program, director of the Technology Applications Division, and director of the Energy Systems Division. He received NASA's Exceptional Service Medal in 1966 for his work as program manager for the Atlas-Centaur launch vehicle. (NASA Release 77-9)

January 25: MSFC announced selection of a Huntsville, Ala., firm for award of its largest contract to a minority business under the Small Business Act. Robinson Printing Co. would receive $998 474 to provide reproduction, documentation, and photographic services for the Michoud Assembly Facility at New Orleans; the cost-plus-fixed-fee contract would be for 1yr with two 1yr renewal options. Robinson would be one of four minority firms working at Michoud under MSFC contract. (MSFC Release 77-12)

• MSFC announced that four of its engineers had received awards conferred by the Soviet Union for participating in the Apollo-Soyuz test project of 1975. Dr. William R. Lucas, director of MSFC, presented the S. Korolev medal to Harold Ledford (now of the Shuttle projects office) for "outstanding management and technical proficiency as chief engineer for science and engineering" from beginning to end of the ASTP.

The USSR also sent Yuri Gagarin diplomas for "exemplary leadership" to three engineers who had chaired stress-corrosion review teams after prelaunch inspection of the long stored Saturn 1B launch vehicle revealed a cracked fin. The teams had investigated all launch stages and associated hardware, independently assessing use of corrosion-susceptible materials. Jewel W. Moody's review team and a Chrysler Corp. team assessed flightworthiness of the Saturn 1B; Charles E. Cataldo's team, with personnel from McDonnell Douglas and the Rocketdyne Division of Rockwell, investigated the S-IVB stage and rocket engines used on the various launch stages to assure they were flightworthy; and Paul H. Schuerer's team, with personnel from IBM and the Bendix Corp., assessed flightworthiness of the Saturn instrument unit and inertial-guidance platform. (MSFC Release 77-13)

• The Natl. Oceanic and Atmospheric Administration (NOAA) announced plans to use a new tool—satellite pictures—in establishing hunting regulations for Canada geese and other arctic-nesting game birds. Wildlife managers in Canada and the U.S. would use imagery from the *Noaa 5* polar-orbiting satellite, showing snow and ice conditions in traditional nesting areas too remote for on-site inspection, to monitor habitat conditions.

Identification of probable areas of castastrophic or outstanding goose production would enable managers to impose restrictions to prevent overkill of birds or to bring the populations to levels suitable for hunting. Factors affecting the nesting of arctic geese would include the timely disappearance of snow and ice, and availability of melt water to allow rearing of young; late seasons or adverse weather could prevent nesting or reduce clutch or brood sizes. (Dept. of Commerce/NOAA Release 77-13)

• Timothy Leary, once described by the media as "high priest of LSD," visited Washington, D.C., on behalf of his latest enthusiasm, space migration, according to a story in the *Washington Star*. Leary, at his first press conference in 7yr, described "a raging epidemic of hope in this country" and said it was time to find a new frontier. He wanted his support taken seriously, and cited the studies by Gerard K. O'Neill of Princeton Univ. and those of the L-5 Society. "Soon people will find it cheaper to build a new world than to fight over an old one," he added. When would space migration become a reality? "Ten years after people stop laughing about it," Leary replied. (*W Star*, Jan 25/77, D-3)

January 26: Lewis Research Center announced it would participate in programs to demonstrate the feasibility of using an airborne infrared scanner to pinpoint residential heat loss. The cities of Cleveland, Ohio, and Springfield, Ill., would reimburse the center for its services. A NASA C-47 aircraft, flown over areas designated by Cleveland's city council as eligible for low-interest winterizing and rehabilitation loans from the U.S. Dept. of Housing and Urban Development (HUD), would record on magnetic tape the images of rooftops showing hot areas as white, warm areas as gray, and cool areas as black. Homeowners in the target areas could view the results at community centers and apply for block grant money from HUD as necessary. The city of Springfield had requested similar assistance for a HUD loan program. LeRC had used a thermal infrared technique in 1975–76 to scan NASA centers for energy loss, saving the agency about $350 000 in the first yr of the program. (NASA Release 77-13; *Lewis News,* March 11/77, 3)

• A television relay satellite 22 000 miles in space was the focus of an experiment Jan. 25 in using TV transmissions to conduct judicial proceedings, reducing the travel time and cost of witnesses and lawyers, and to increase the efficiency of the courts, the *New York Times* reported.

The project, result of a yr's planning between the Univ. of Maryland law school in Baltimore and that of Ohio Northern in Lima, was only partly successful because the yr-old communications technology satellite *Cts* orbiting over the equator south of Denver repeatedly sent to Baltimore the jagged horizontal lines of TV "snow" during the closing arguments of the mock hearing at Lima. Also, the voice-activated TV cameras, two at each location to cover the two pairs of lawyers and judges, would take over the projection screen whenever anyone wearing a lapel microphone coughed or cleared his throat.

The Maryland Center for Public Broadcasting, which sponsored the experiment, had paid nothing for use of the system, and the *Cts* project leader (Herbert Nunelly of Westinghouse Electric Corp.) operating the teleconference for NASA said he could not compute the cost of an hour's or day's use of the satellite. The American Bar Assn. had approved the mock appeals hearing before Judge Joseph F. Weis, Jr., of the U.S Court of Appeals for the Third Circuit, and three lawyers before whom law school students were arguing the case. Judge Weis's ABA project director, Howard Primer of Chicago, cited several actual trials conducted with TV assistance. (*NYT*, Jan 26/77, B-1)

- First appraisals of a Soviet MiG-25 aircraft flown by a Soviet pilot to Japan on Sept. 6, 1976, had been revised, according to U.S. Air Force specialists quoted by the *NY Times*. Further analysis had revealed that the aircraft did not lag behind the advanced technology of U.S aircraft, as previously stated, but contained some surprises. After 7wk of study, officials reported the MiG-25's radar to be more powerful and less vulnerable to enemy jamming than U.S. radars, although operated on vacuum-tube technology. The Soviet aircraft's radar also lacked the "look-down" capability of U.S. aircraft, so that Soviet MiG-25's could not detect low-flying enemy aircraft that blended in with the earth's surface. Similarly, Air Force specialists described the MiG-25's Tumansky jet engines as 15yrs outmoded: the flight computer—called impressive even though it too was based on vacuum-tube technology—required extra space, maintenance, and cooling mechanisms. The USSR apparently was reluctant to entrust its pilots with total control of their aircraft, which explained the "very sophisticated data base" which could not only handle fire control and sensor systems but could also return the craft to any of four predesignated landing fields. "The biggest advantage," said one U.S. Air Force officer, "is that we allow, we train, our men to think, to adapt. This does not occur in many cases with the Soviet air force." Designed in 1960 to combat the U.S. B-70 bomber, which never went into production, the MiG-25 had been altered later for use against the highflying Lockheed SR-71 reconnaissance plane. (*NYT*, Jan 26/77, A-11)

- The *New York Times* reported that France had inaugurated its first operating solar power plant, in what officials predicted would be a race with the U.S. to sell solar power to Arab and third world countries. The plant, located at Odeillo in the Pyrenees mountains of southern France, would contribute about 64 kilowatts to the national electricity grid (enough to run about 60 household electric irons, the *NYT* said). Jean-Claude Colli, the government's director for new energy sources, noted that the French could claim operational equipment, whereas "the Americans are presenting futuristic projects." France expected to sign a major solar energy development contract with Saudi Arabia within the week and to host a sales meeting at the end of the week with 26 countries from the Persian Gulf and Mediterrranean regions. The new plant was said to receive about 180 days of sun per yr, and to use collector mirrors for reflecting the rays to a boiler heating steam to drive a turboalternator that would produce the electricity. (*NYT,* Jan 26/77, A-9)

January 27: NASA announced publication of a "space photo album": an atlas-size book containing 400 images of earth's natural and cultural features taken by the Landsat earth resources satellite, most of the pictures in full color and near full-page size. Authors were three scientists from GSFC (Drs. Nicholas M. Short, Paul D. Lowman, and Stanley C. Freden) and Dr. William A. Finch, Jr., from San Diego State Univ. Designed as a reference, a textbook, a teachers' guide, and a supplemental tool for research specialists, the book devoted 40% of its content to images of the 50 United States but included images from all over the world, and repeated some standard views to show seasonal changes. The pictures had a grid-number system for quick location of geological and geographical features described in the captions. The book included sections telling the story of the Landsat program, its systems, applications, and accomplishments, and a supplemental teacher's guide with glossary and suggestions for classroom use. (NASA Release 77-14)

January 28: NASA launched from the Eastern Test Range on a Delta vehicle the second in a series of three communications satellites for use by the North Atlantic Treaty Organization, into an orbit with 35,797km preliminary apogee, 35 777km perigee, 1436.2min period, and 2.6° orbital inclination. The cylindrical *Nato IIIB* spacecraft had a covering of solar cells for power generation and was gyrostabilized, providing stability for the antenna system which would counter-rotate to point to the earth at all times. All active communications components were redundant, having an identical backup part on board. After fifth apogee, about 48hr after launch, firing of an apogee motor would put *Nato IIIB* into a circular synchronous orbit at about 35 900km altitude, where its onboard reaction-control system would take it to a predetermined location. The U.S. Air Force's satellite-control facility would operate these activities on behalf of NATO, which had funded three Phase III comsats built by Ford Aerospace to replace the *Nato II* satellites in the NATO integrated communications systems. (NASA Release 77-5; *Spacewarn* SPX-280; *Sf Satellite Digest,* July-Aug 77, 298)

• NASA announced resumption of communication with the four Viking-mission vehicles—two landers and two orbiters—at the end of the "conjunction" period during which contact was degraded and temporarily lost because of the changing position of Mars, behind the sun with relation to the earth since Nov. 1976.

The Viking mission extended into the post-conjunction period would use the functioning spacecraft to obtain seasonal-variation data, do long duration sampling for statistically important experiments, and gather additional information not possible during the primary mission because of time or observational constraints. No known limitations would preclude completion of at least a year of extended mission and probably more, the announcement said, as the orbiters had made minimal use of their consumables during the early mission and had performed a minimal number of battery-recharge cycles. Lander B's location at Mars 44°N would encounter temperatures below the qualification limits in the winter season, but the effect of this was unpredictable because lander tests had not gone beyond qualification limits.

Extended-mission data collection had actually begun last Nov. when controllers had used the spacecraft transponders passively to measure ranging, tracking, signal-time delay, and solar-corona-induced perturbation. The extended mission would continue through May 1978, the announcement said, and data analysis would continue through Sept. 1977; requirements for data analysis beyond FY 1978 were under study. (MOR-S-815-75-01/02)

• NASA announced plans to join with the Energy Research and Development Administration (ERDA) in a program to use solar energy for heat and hot water supply at 6 space centers from Va. to Calif. NASA would invest $500 000 and ERDA $1.25 million in up to 10 separate projects that had been scheduled for completion at NASA facilities by the end of the fiscal year, Oct. 1977. The programs could reduce utility bills by up to a quarter million dollars/yr.

Five of the solar projects, all at MSFC, would heat an office building; heat liquid nitrogen to a gas, for experiments and manufacturing at 2 MSFC facilities; supply hot water for the cafeteria and restrooms in a 10-story building; and heat a warehouse used for storing hydraulic equipment. Other projects would supply hot water for cafeterias at Dryden Flight Research Center in Calif. and KSC in Fla.; hot water for photo labs at Ames Research Center in Calif. and LaRC in Va.; and heat and dehumidification for areas containing sensitive computer and other electronic equipment in the mission control center at JSC in Tex. NASA would also provide funds for 5 smaller solar heat projects at ARC, KSC, and MSFC, to be operational by the end of Aug. The program should demonstrate the possibility of offsetting the comparatively high initial cost of solar energy systems through substantial savings on oil, gas, and electrical energy. (NASA Release 77-15; MSFC Release 77-15; ERDA Release 77-18)

• The European Space Agency (ESA) announced that its 1977 budget for scientific and general activities had been unanimously approved by the ESA Council meeting in Paris. Individual budgets for projects Aerosat, Ariane, Marots, Meteosat, OTS, and Spacelab had already been approved by their respective program boards. The Council had reduced by 3 million accounting units (MAU) the total 481.8 MAU proposed for all ESA work in 1977, approving the equivalent of $526.68 million in U.S. dollars. (In 1977, one AU was equal to 1.1 U.S. dollars.) (ESA news release, Jan 28/77; ESA newsletter, Jan 77, 1)

January 29: H. Julian Allen, originator of the concept of bluntness as an aerodynamic technique for reducing heating of spacecraft reentering earth's atmosphere, died of a heart attack at Stanford Univ. hospital at the age of 66. A major figure in modern aerospace technology, Allen's work had made possible the safe return to earth of U.S. astronauts and had revolutionized the design of ballistic missiles. After earning bachelor's and engineering degrees from Stanford, Allen in 1936 had joined the Natl. Advisory Committee for Aeronautics (NACA) at its Langley laboratory in Hampton, Va., and had returned to Calif. in 1941 when the Ames installation came into being. He had worked there ever since, serving as ARC director from 1965 to 1969.

His scientific contributions included a general theory of subsonic airfoils which led to use of low-drag airfoils in World War II on aircraft such as the Mustang fighter. He led in developing high-speed wind tunnels at ARC, now recognized as primary national resources for advancing aircraft design and national defense. He had received national and international honors, including NASA's distinguished service medal and medal for exceptional scientific achievement. He was a fellow of many societies, including the Royal Aeronautical Society of London. (ARC Release 77-06)

January 30: The possibility of teleportation-transfer of objects from one place to another by radio waves came closer with the announcement of a new way to create identical copies of a three-dimensional object by using a sculpture-photography system based on the same principles of chemistry and electronics now used in making printing plates. As reported in the *Chicago Tribune*, Wyn Kelly Swainson, a young man with a degree in English literature and some training in science, had obtained his first patent on a device being developed with the help of the Battelle Memorial Institute in Columbus, Ohio.

Swainson began with the idea of reproducing great works of sculpture without painstaking handwork: a plastic material hardened by light rays would be targeted by intersecting laser beams, controlled by a computer programmed with the image of the object to be copied. Swainson said IBM had already produced an instrument to record computerized shapes, convert the electronic data into light rays, and feed them through the lasers scanning a container of plastic. Where the beams intersected, they would cause a chemical reaction hardening the plastic; washing off the unhardened material would leave an exact copy of the object previously computerized. In theory, a three-dimensional hologram of an object even of complex metals could be analyzed for its precise composition, converted to electronic data, transmitted, and reproduced as often as desired. (*C Trib*, Jan 30/77, 1-5)

Space Shuttle orbiter Enterprise, with tailcone covering engine-nozzle location, moves on a 90-wheel rig January 31 from the Rockwell plant to DFRC 35mi away in preparation for approach and landing tests. (NASA 77-H-77)

January 30—February 1: Enterprise, the first Space Shuttle, towed atop a 90-wheel trailer from Rockwell Intl.'s plant in Palmdale, Calif., made its first trip to Edwards AFB 57km away. Escorted by a convoy of security and patrol cars, the Enterprise moved at 8kph and drew hundreds of spectators to view its configuration (24m wingspan and 21m vertical tailfin). Tests of the 38m-long space vehicle would begin Feb. 18 with a series of "captive" flights; first manned flight to an unpowered landing would occur July 22. (*W Star*, Jan 30/77, A-7; *NYT*, Jan 31/77, 12; *W Post*, Feb 1/77, A-3)

January 31: MSFC announced that on Jan. 28 NASA had invited industry to submit proposals for supplying design, development, and manufacture of Space Telescope support systems module and optical telescope assembly. The telescope should permit observation of about 350 times the volume of space now acessible to ground-based telescopes. MSFC said it expected award of contracts by Oct. or Nov. 1977 if Congress authorized the mission and appropriated the necessary funds.

The 2.4m telescope, included in NASA's FY 1978 budget recently submitted to Congress, would weigh about 10 tons. The Space Shuttle would carry it as a payload in 1983, launching it to orbit at about 500km altitude at an inclination of 28.8° to the equator. Once there, it could serve for a decade as an in-space observatory operated remotely from the ground, but would be designed for maintenance and servicing by a space-suited astronaut or for retrieval and return by Shuttle for overhaul and subsequent relaunch. NASA would announce to the scientific community in Feb. an opportunity to submit proposals for scientific instruments to be carried on the Space Telescope for its initial launch. (MSFC Release 77-16; NASA Release 77-19)

• LaRC announced that Alex Haley, author of *Roots*, would be the speaker Feb. 7 at an LaRC colloquium and at an "Our Future in the Cosmos" public lecture. Demand for tickets to the lecture, part of a series sponsored by LaRC and the College of William and Mary, was so great that LaRC had shifted the Haley lecture from the Hampton high school to the Hampton Coliseum. Title of Haley's talk was "A Saga of Black History." (LaRC anno Jan 31/77)

• The U.S Air Force Systems Command announced plans to experiment with dispersal of cloudcover by sowing minute amounts of silver iodide into supercooled clouds over a sparsely populated area of Michigan, using two small civilian aircraft, one to dispense the chemical and the second to photograph effects of the treatment. North American Weather Consultants of Goleta, Calif., contractor to the AF Geophysics Laboratory, would conduct the experiments to disperse cloudcover that could interfere with military or civilian flight operations.

The experiments, aimed at clearing specific areas of cloud during landings, would occur only on days that were overcast with little or no precipitation, said Bruce A. Kunkel, project scientist for AFGL. Clouds sought for the tests would be supercooled (consisting of water drops remaining liquid though below freezing temperature) so that they would form ice crystals when treated with the chemical and would fall from the sky, clearing up to a few square miles for up to an hour. Some snow might reach the ground, but most should evaporate after falling from the cloud, and the amount of silver iodide needed would be extremely small notwithstanding its visually dramatic effect. (OIP Release 004.77)

• Canadians monitoring the mysterious radio signals emanating from the Soviet Union, said to have disrupted worldwide communications [cf. A&A76, Oct. 29 and Nov. 10-11], suggested that the signals might correspond to transmission of electrical energy without the use of wires, through a process developed by Nikola Tesla, Yugoslav inventor who died in 1943.

According to a story in the *Wash. Star,* the Canadian Department of Communications received a request some months ago to have its nine listening posts analyze the radio signals in an effort to identify electrical-energy problems in eastern Canada. By this time, the signals had caused complaints from European governments and were becoming shorter in duration. Although the signals were high frequency (3 to 30megahertz) and Tesla's experiments had been on low frequencies (6 to 10k cycles per second), the Canadians theorized that the transmissions might have been harmonics of VLF transmissions. Tesla had discovered in 1900 that the earth itself could serve as a conductor of electricity, being "responsive as a tuning fork" to electrical vibrations of a certain pitch; he had succeeded in lighting 200 electric lamps from a distance of 25mi without wires. One scientist working with the Canadians suggested that the signals might aid in modifying weather by affecting electrically charged particles in the upper atmosphere. (*W Star,* Jan 31/77, A-5)

During January: The Natl. Aeronautic Assn. reported that 1975 was one of the safest years experienced by scheduled U.S. airlines, with a passenger-fatality rate of 0.07 per 100 million passenger miles, beating the railroads' record of 0.08 fatalities per 100 million passenger miles. In 1975, over 205 million passengers on scheduled U.S. airlines (almost as many as live in the entire country, the newsletter noted) had flown more than 162 million passenger miles; only 124 persons had died in 3 U.S.air-carrier accidents, whereas highway accidents had killed 44 690. On the basis of the 1975 accident rate, the newsletter calculated a passenger's chances of safe arrival at destination to be 99.99998%. (NAA newsletter, Jan 77, 3)

• MSFC employees Alfred G. Orillion, Advanced Projects Office, and James E. Downey, III, deputy director for program development, received special awards for the Alabama section of the American Inst. of Aeronautics and Astronautics at the AIAA honors banquet in Washington, D.C., Jan 13. The awards were for a 1975 space industrialization symposium at MSFC of which Downey was general chairman; Orillion was chairman of AIAA's Alabama section. (MSFC Release 77-9)

• *AvWk* reported that James S. Martin, Jr., manager of the Viking project, would receive the Goddard award for "brilliant leadership of the Viking project to land an instrumented, automated spacecraft on the planet Mars." (*AvWk,* Jan 10/77, 9)

• Cosmonaut trainees from Czechoslovakia, Poland, and East Germany had arrived at Zvezdnoy Gorodok, USSR, late in Dec. to begin training for flight positions on Soyuz/Salyut missions. Flight candidates from Bulgaria, Hungary, Cuba, Mongolia, and Romania were to arrive for training at the center during 1977. (*AvWk,* Jan 10/77, 9)

February

February 1: NASA announced it had received 1147 applications for its Space Shuttle astronaut candidate program, with openings for at least 15 pilot and 15 mission specialist candidates. Deadline for applications would be June 30, 1977. Those selected would report July 1, 1978, to Johnson Space Center at Houston for 2yr of training and evaluation; appointment as astronaut would depend on satisfactory completion of the course.

NASA had mailed 11 822 applications to persons outside the agency, and another 350 to employees at NASA centers. All the applications received so far had been from civilians; the astronaut selection board would receive applications from members of the military services just before the deadline. Most of the applications were for the mission specialist category: 922 for that group, 225 for pilot, and 118 for both.

Pilot astronauts would control the Shuttle during launch, orbit, and landing, and would be responsible for vehicle systems. Mission specialists would be responsible for coordinating Shuttle operations with the commander and pilot in areas of consumables usage, experiment operations, and crew activities. Crews would number up to seven: commander, pilot, mission specialist, and as many as four payload specialists (who need not be NASA employees and would be nominated by sponsors of the payload being flown). Payload specialists would operate specific payload equipment needing special skills. NASA would seek especially to recruit qualified minorities and women as Shuttle astronauts. (NASA Release 77-18)

• Kennedy Space Center reported that NASA had awarded the Univ. of Ariz. a $50K contract extension to continue studying thunderstorm electrical fields and other phenomena in the Spaceport area. Total value of the contract, awarded in April 1974 and now extended to Jan. 15, 1978, would be $116 000.

Under the direction of Dr. Phillip Krider of UA's Inst. of Atmospheric Physics, scientists working with UA graduate and undergraduate students would use KSC's sophisticated instrumentation during summer thunderstorms to study electrical-field changes and recovery curves created by lightning under varied conditions, and would create computer programs to determine and display the location of lightning discharges and thunderstorm characteristics. Techniques resulting from the study would help make Space Shuttle vehicle-processing, launch, and landing at KSC safe and efficient even in marginal weather.

The U.S. studies would supplement a program known as TRIP (thunderstorm research international program), cosponsored by the Am. Geophysical Union and the Am. Meteorological Society, under which teams of atmospheric scientists would conduct research at KSC between 1976 and 1978. (KSC Release 33-77)

- At least 7 defense contractors had performed studies on space defense systems, *Aerospace Daily* reported, predicting that spending on this activity would double in the upcoming federal budget (from $61 million to $126 million in FY 1978, and double again to $265 million in FY 1979).

  In its concern for detection of hostile satellites, especially in the infrared frequencies, the Dept. of Defense had officially confirmed four USSR intercept launches last yr. DOD considered comsats and early-warning satellites in synchronous orbits most vulnerable; the USAF was also concerned about its reconnaissance satellites, especially the Big Bird, only one of which is normally in orbit at a time.

  Contractors working on the problem were Science Applications, Inc., doing a $60 000 study of charge-coupled device (CCD) sensors for a terminal optical-warning system; Rockwell Intl., with a $50 000 study of potential maneuvers for the NavStar global-positioning satellite to be used for guidance aid in tactical-weapons delivery; TRW, Inc., a $50 000 study to identify maneuvering possibilities for the AfSatCom (USAF satcom) system; and three other firms studying standby survival launch capabilities applicable to all USAF missions (Hughes, $250 000; Boeing, $250 000; and Lockheed, $100 000). The Air Force also reported that Westinghouse had made a $50 000 study of an impact sensor to detect attacks with non-nuclear pellet warheads against a spacecraft in orbit. (*ASD*, Feb 1/77, 158)

- Western Union Corp., first of the 1000 largest U.S. industrial companies required to publish replacement-cost data called for this year by the Securities and Exchange Commission, claimed that its 1976 earnings of $34 million would be erased had it based its depreciation charges on current costs of replacing its entire physical plant. *New York Times* reporter Robert D. Hershey, Jr., cited the argument of many companies that releasing the figures required by the SEC would depress prices of their stocks. WU said its plant and equipment would have cost $1.61 billion to replace at the end of 1976, not the $1.11 billion its books showed as historical cost. The SEC ruled that, for fiscal yrs beginning after Dec. 1975, the annual reports of companies with at least $100 million in physical assets representing at least 10% of total assets must give actual replacement-cost data, to let stockholders see the effect of inflation on financial statements. (*NYT*, Feb. 1/77, 37)

• The Institute of Space and Aeronautical Science at the Univ. of Tokyo announced postponement (until Feb. 15 or later) of the launch of a satellite-carrying MU-3H rocket, because of mechanical trouble. An inspection had revealed problems with the rocket's attitude-control system, the announcement said. (FBIS, Tokyo Kyodo in English, Feb 1/77)

February 2: The new administration of President Jimmy Carter was reevaluating the NASA budget submitted by President Ford, and would study the possibility of increases, however marginal, in that budget, *Defense/Space Daily* reported. NASA's energy program, budgeted at $4.5 million in FY 1958 ($1 million to study satellite solar power and $3.5 million for energy technology that could be transferred to ERDA), would probably be recommended for increase within the next few weeks, the report said. (*D/SD*, Feb 2/77, 185)

February 3: Tass reported that Soviet space station *Salyut 4*, launched Dec. 26, 1974, to house cosmonauts working in space up to 90 days, disintegrated over the Pacific Ocean when it moved on ground command into "a descent trajectory" and burned up. Tass noted that the station had flown for 3mo joined to the unpiloted space capsule *Soyuz 20*. The longest, most successful career of a Salyut in the 16yr-old USSR space program had included what Tass called "an extensive program of medical and biological research" as well as study of the sun, stars, and planets in the electromagnetic spectrum; views of Soviet territory in the middle and southern latitudes; and a great amount of data on physical processes in earth atmosphere and in space.

Cosmonauts Georgy Grechko and Aleksey Gubarev had docked *Soyuz 17* with *Salyut 4* for 29 days between Jan. 10 and Feb. 9, 1975; Vasily Lazarev and Oleg Makarov failed to dock with *Salyut 4* on Apr. 5 of that yr; and Pyotr Klimuk and Vitaly Sevastyanov had docked *Soyuz 18* with *Salyut 4* May 25 and remained for 63 days, during which they talked with their compatriots on the ASTP mission and returned July 24 with the Soviet duration record for a manned mission. *Salyut 4* had completed 12,188 orbits around the earth before being destroyed, Tass said. (*Today*, Feb 4/77, 10A; FBIS, Dom Svc Rusn, Feb 3/77)

February 4: NASA announced a yr-long schedule of test runs and captive-flight tests on the first-built Space Shuttle orbiter, Enterprise, that would take place at the Dryden Flight Research Center, Edwards, Calif. Called approach and landing tests (ALT), the series of flights with a modified Boeing 747 serving as a ferry aircraft and airborne launch platform would begin with several taxiing tests with the orbiter atop the carrier; following would be six inert-captive flights in which the jumbo jet would carry the unmanned orbiter to a 7620m (25 000ft) altitude. Unmanned flights would verify performance of the two vehicles in mated flight. Then would come a series of captive-active flights with the orbiter's systems powered up and the Enterprise manned by 2 astronauts.

The first orbiter off the assembly line arrived at DFRC Jan. 31 from Rockwell Intl.'s Palmdale plant; upon completion of ALT, the carrier would take the prototype to Marshall Space Flight Center for ground-vibration tests. After these tests, NASA said, the Enterprise would return to Palmdale to be prepared for orbital flight in the early 1980s. The orbiter now under construction (OV-102) would be the first used in actual orbital-flight tests, now scheduled to begin in mid-1979; six test flights would demonstrate the orbiter's capabilities in earth orbit before the start of operational flights, scheduled to begin in 1980. (NASA Release 77-16)

• Two giant crawler-transporters that had ferried Saturn V rockets to the launch pad at KSC and were scheduled for similar work in the Space Shuttle program had been designated National Historic Mechanical Engineering Landmarks, *Today* newspaper reported. At ceremonies held Feb. 3 beneath one of the vehicles, more than 100 persons watched the unveiling of a plaque noting the creation of the 18th national landmark identified since the Am. Society of Mechanical Engineers began singling them out in 1973.

Present at the ceremony were KSC employees who had helped build the pair of 2.7-million-kg platforms; ASME officials and members from across the U.S. on their way to a national meeting in southern Fla.; and former astronaut Donn F. Eisele, who had been command-module pilot on the 1968 *Apollo 7* flight that resumed moonshots more than a yr after the fatal fire in an Apollo training exercise. Eisele had since become a manager for Marion Power Shovel Co., the Ohio firm that built the transporters at a cost of $14 million. Ray Clark, KSC director of design engineering, accepted the plaque from Dr. Stothe P. Kezios, president-elect of ASME, with the comment that the transporters were "barely broken in" with mileage on each at just over 800km; they might accumulate 10 times that amount carrying Shuttle equipment during their second yr of operation, he added. (*Today*, Feb 4/77, 10A)

• JSC announced award of a cost-plus-fixed-fee contract worth $9 083 303 to Northrop Services, Inc., for operation and maintenance at JSC of life sciences and engineering laboratories and the lunar curatorial facility, effective Feb. 1, 1977, and expiring Jan. 31, 1978. (JSC Release 77-08)

February 7: NASA revealed the contents of the space and aeronautics exhibit that would appear in the U.S. pavilion at the 32nd Paris Air Show, under the theme "The Continuous Challenge: 50 Years Since Lindbergh." NASA planned to display concepts of space communities and their construction; living, farming, and manufacturing in space; and solar-power stations in space, together with a 3-screen presentation of the 50yr of flight since Lindbergh. A 2.4m Shuttle model would include the European Space Agency's Spacelab, part of a major international cooperative space program, in its cargo bay.

Also on display would be the Galileo 2 flying laboratory, its interior configured to simulate the reusable Spacelab with facilities for space experiments in medicine, manufacturing, astronomy, and production of pharmaceuticals. A life-size model of Mars landscape viewed by the Viking lander would demonstrate space exploration, and Landsat stereo photographs of earth would show the increasingly routine use of space to survey crops and natural resources, monitor pollution and weather, and collect global weather data. Other models and presentations would show the scope of NASA programs such as government-industry work to halve fuel consumption by transport aircraft and to develop safer, more efficient, and more environmentally acceptable civil aircraft. (NASA Release 77-20)

• MSFC announced award of a 15-mo contract worth $99 663 to Desert Research Institute—part of the Univ. of Nevada system—for development and fabrication of a prototype particle generator for possible use in the Atmospheric Cloud Physics Laboratory (ACPL), a payload scheduled for an early Shuttle or Spacelab flight. The generator would form nuclei with diameters of a hundredth to a thousandth of a micron (one thousandth of a mm) upon which water would condense to form clouds. Researchers would use the zero-gravity environment on Shuttle flights to perform cloud-physics experiments leading to an understanding of microphysical processes in the atmosphere, and to eventual prediction, alteration, and control of weather. (MSFC Release 77-19)

• MSFC announced that a special hammer developed in the 1960s for Saturn fabrication had solved a major problem of manufacturing external tanks for the Space Shuttle. Martin Marietta Aerospace, prime contractor for the tanks, in welding four large sections of a liquid-oxygen container at the Michoud assembly facility, had found that peaking of the weld joints left some areas as much as 5° out of tolerance. The special tool MSFC had created for removing distortions in component welds would not impact a surface in the usual sense but, when in contact with the metal, would send a hydrodynamic force uniformly into the thickness of the metal when an electrical charge dumped into the hammer's coil created an expanding magnetic field. The field would exert an evenly distributed three-dimensional force that would remove the distortions without leaving hammer marks on the target.

After Robert M. Avery of MSFC's Materials and Processes

Laboratory showed how to use the tool at the Michoud facility, the contractor was able to remove most of the distortions during the week of demonstration. Two working hammers would remain at the plant for future use. (MSFC Release 77-21)

• U.S. aviation experts had changed their minds about the USSR's MiG-25 "Foxbat" interceptor plane and now considered it superior to western planes of its type, the *Chicago Tribune* reported. *New Scientist* magazine had quoted aviation experts who had examined the Foxbat flown to Japan last Sept. by defecting pilot Viktor Belenko; their conclusions were completely opposed to earlier assessments that described the Foxbat as crude and outdated.

Shortly after its landing in northern Japan, the Foxbat had been described as a "manned rocket" less advanced than previously thought. The magazine said that U.S. officials now considered it unsurpassed in ease of cockpit maintenance and a masterpiece of standardization unmatched in the U.S. The experts said they were impressed not only with the speed and rate of climb, but also with the electronics on the Foxbat, now believed to have the most powerful airborne fire-control radar in the world as well as a highly sophisticated computerized flight-communications system that could perform most of the missions without the pilot [see Jan. 26]. U.S. and Japanese experts had examined the Foxbat for 2mo before returning it to the USSR; Belenko, now in the U.S., was still being debriefed. (*C Trib*, Feb 7/77, 1-2)

February 7-25: The Soviet Union launched a manned Soyuz spacecraft from its Baykonur space center at 9:12 local time (1612 GMT) to try for a rendezvous with the *Salyut 5* space station, following an unsuccessful docking attempt last Oct. *Soyuz 24* carried as commander Col. Viktor V. Gorbatko, who had flown on *Soyuz 7* in Oct. 1969 as part of a 7-man mission using 3 Soyuz spacecraft, and rookie engineer Lt. Col. Yuri Glazkov, who became the 39th cosmonaut launched into space. Lt. Gen. Vladimir Shatalov, chief of cosmonaut training, described the *Soyuz 24* mission as routine, adding that the scientific and economic possibilities for using the Soyuz-Salyut combination were "inexhaustible."

Orbit parameters reported Feb. 8 were: 281km apogee, 218km perigee, 89.2-min period, 51.6° inclination. This first manned launch of the year was part of a USSR program of earth-orbital research using the Soyuz capsules on a one-mission basis, according to *NY Times* reporter Christopher S. Wren, who noted that the USSR had continued manned flights while the U.S. was waiting until the Space Shuttle became available in 1979 or 1980. Thomas O'Toole reported in the *W Post* that the cosmonauts might stay in orbit as long as 2mo; both pilots had been trained for space walks, and the choice of an all-Air Force crew indicated that *Soyuz 24* might be a military mission.

Tass reported successful docking and occupation of the *Salyut 5;* it later described housekeeping and biological experiments, plus communications with ocean-going vessels, studies of crystal growth and surface tension, and use of an infrared radiometer for long-distance nighttime observation of earth's surface and temperatures. The cosmonauts continued exercising with special equipment to measure body functions. On Feb. 21, they tried a new system for recharging the space station atmosphere: Tass said the air was satisfactory, but the test would verify the system for future prolonged flights. A multifunctional system would supply compressed air to control stabilization or to compensate for leaks in the compartments. Tass announced Feb. 23 that the crew had begun preparations for leaving the station.

(The unexpectedly early termination of the *Soyuz 24* mission recalled reports that the *Soyuz 21* mission ended ahead of time because the *Salyut 5* environmental-control system had produced an "acrid odor" that "became unbearable" [*A&A 76,* Oct 18].)

A commentary recorded by Gen. Shatalov for broadcast Feb. 24 on the Moscow Domestic Service noted "a whole series of cyclones and strong winds with snow" over the landing site in Kazakhstan. However, the cosmonauts landed safely at dusk Feb. 25 on a farm near Arkalyk on the Kustanay steppe, and were brought to the local airport by helicopter and flown from there to Baykonur. They held a press conference at Baykonur Feb. 24, in which Glazkov called the *Salyut 5* spacecraft "a fine old work horse" that had brought them home safely. (*NYT,* Feb 9/77, 11, B6; *W Post,* Feb 8/77, A16; Feb 9/77, A12; *C Trib,* Feb 9/77, 1:12; FBIS, Tass in English, Feb 7-26/77; Moscow Dom Svc Rusn, Feb 10-25/77; Moscow Intl Svc Rusn, Feb 9-25/77; *W Star,* Feb 25/77, A4)

February 9: Fifteen yr almost to the day after the U.S. placed its first man in earth orbit, NASA would begin flight tests of the Enterprise—first Space Shuttle orbiter off the assembly line—to be carried piggyback on a Boeing 747 on Feb. 11, 1977, for its first trip through earth's atmosphere. The flight of Marine Lt. Col. John H. Glenn on Feb. 20, 1962, lasted less than 5hr; since that flight, the U.S. had accumulated 22 504 man-hours in space by using 43 astronauts on 31 separate manned missions that included 9 trips around the moon, 6 moon landings, and 3mo earth-orbital missions on Skylab.

Glenn's Mercury spacecraft *Friendship 7* weighed only 1315kg, with barely enough room for the pilot and a few instruments, and had parachuted into the Atlantic after its one-time use. The orbiter would weigh 67 500kg empty, could carry up to 7 crew members, and was designed to land on a runway and be prepared for another flight within weeks. In contrast to Glenn's 3-orbit flight, Spacelab missions carried by the Shuttle might last from 7 to 30 days. First of 6 orbital tests of the Shuttle was set for 1979, and operational flight for 1980. (NASA Release 77-21; NASA pre-ALT rept Feb 11/77)

• NASA announced it had certified the General Aviation Design and Analysis Center at Ohio State Univ. airport in Columbus to make analyses of single-element airfoils, a first step toward providing a complete service for the general aviation community. Developed under a 3yr contract with LaRC and OSU's aeronautical and astronautical research laboratory, the ADAC would provide directly to aircraft designers and manufacturers services such as analysis and design of 2-dimensional airfoil shapes, investigations of high-lift devices and aerodynamic controls, compilation of airfoil aerodynamic characteristics, and technical assistance (including consultation and data interpretation) for wind-tunnel and flight testing of airfoils. ADAC would provide the services on a fee basis. ADAC personnel under Dr. G.M. Gregorek would work with NASA on refining present codes used in computation; as NASA developed new techniques and improved codes, ADAC would use the improvements to solve more difficult design problems. (NASA Release 77-22)

• JSC and the Lunar Science Institute announced plans for the 8th annual lunar science conference in Houston, March 14–18. More than 600 scientists from around the world would discuss new data about the moon, derived from studies of lunar samples and from Apollo surface and orbital experiments, as well as new interpretations of previous models of lunar origin and history. Several Soviet scientists would attend, to give NASA materials from the *Luna 24* mission for analysis. Dr. Michael G. Duke, acting chief of JSC's Lunar and Planetary Sciences Division, and Dr. Robert Pepin, outgoing director of the Lunar Science Institute, would cochair the conference.

The conference would also be the second meeting on "planetary" science, continuing the practice of the 1976 conference, where planetary interests were discussed for the first time as a result of the accumulation of data from NASA and USSR missions to Venus, Mercury, Mars, and the outer planets. The assembled data on planetary origin and evolution had proved of primary interest to lunar scientists. (JSC Release 77-09)

February 10: KSC announced it would reduce its dependency on fossil fuels as early as Oct., when 3 of its buildings would begin heating water with solar energy: the Visitors Center cafeteria and the Banana River repeating station, which should be operating on solar by late May, and the KSC headquarters building. Up to 70% of hot water for the cafeteria would be heated in a flat-plate collector on the ground next to the cafeteria, with a conventional water heater as backup for long periods of cloudy weather.

A larger flat-plate collector would provide up to 70% of the energy needed by the Banana River station, an electronic-communications relay near the eastern shore of Merritt Is., for hot water used in heating and air conditioning reheating. All three systems would be able to heat water to about 370K, although flow through the collectors would move the hot water into storage tanks at a maximum 333K. The flat-plate collectors would face south at a 28° angle for maximum efficiency. ERDA would fund the headquarters conversion jointly with NASA, which would fund the other projects alone. Energy savings over an 8yr period would be about $185K. (KSC Release 56-77)

• Soviet news agency Tass reported the death at age 82 of Sergei V. Ilyushin, designer of more than 50 planes from dive bombers of World War II to modern passenger jets. The report did not give the date or cause of death.

Mobilized into the czarist army in 1914, Ilyushin had begun as a sweeper in an airplane hangar, became a mechanic, and graduated from flying school in 1917. Shortly after the Russian revolution, he entered the Soviet army and rose to be a lieutenant general. After graduating from the air force engineering academy, where he had designed and built gliders, he was put in charge of airplane construction and after 1931 devoted himself exclusively to aircraft design. His most famous design was the Il-2 attack plane called the Stormovik, one of the first to prove the effectiveness of small single-engine craft operating near ground level. The Il-2, a two-seater carrying a rear gunner behind the pilot, was the first close-support plane to fire rockets to any degree and was known to the German forces in WWII as the "flying death." The Il-62, a 4-jet passenger plane, had begun Aeroflot passenger service from Moscow to New York in July 1968.

Ilyushin was fourth of the "big six" Soviet aircraft designers to die: A.A. Mikoyan, who with M.I. Gurevich designed the MiG series of fighter planes, died in 1970; A.N. Tupolev, father of the Tu-144 supersonic transport, in 1972; Gurevich in Nov. 1976. Survivors were O.K. Antonov, designer of An-12 and An-22 cargo planes, and Ilyushin's close associate A.S. Yakovlev, whose Yak fighter usually teamed with the Il-2. Ilyushin's son Vladimir, a test pilot, might have beaten Yuri A. Gagarin into space in 1961 by several days; officials denied this. (*NY Times*, Feb 11/77, A24; *W Post*, Feb 11/77, C6)

• Japan's Natl. Space Development Agency announced that launch of a three-stage "N" rocket carrying an engineering test satellite had been postponed to Feb. 14 because of trouble in the control system of the rocket's first stage. Launch had been scheduled originally for Feb. 6 from the space center at Tanegashima Is. (FBIS, Hong Kong AFP in English, Feb 10/77)

February 11: Langley Research Center announced award of a cost-plus-fixed-fee contract worth about $1.8 million to Vought Corp. for design, development, and qualification of a Scout launch-vehicle guidance system. Teledyne would do the designing under subcontract and Vought would integrate the system into the Scout. Work would proceed at the Vought plant in Dallas and Teledyne's plant in Northridge, Calif., under LaRC direction. (LaRC Release 77-3)

• NASA announced it had adjudged successful the mission of Gravity Probe A (GP-A), launched June 18, 1976, on a Scout vehicle from Wallops Flight Center. Trajectory parameters computed for GP-A were within preflight predictions, and the experiment and support systems had operated normally. Principal investigator Dr. R.F.C. Vessot had begun data reduction and reported achieving an accuracy of 150 parts per million, surpassing the prelaunch accuracy objective of 200ppm. Associate Administrator for Space Science Dr. Noel W. Hinners noted that the payload had impacted in the Atlantic Ocean as planned and was not recovered. (MOR S-879-76-01 [postlaunch], Feb 14/77)

• Analysis of telemetry data and flight films from the SPAR III mission launched Dec. 14 at White Sands Missile Range indicated successful performance of the electromagnetic levitation device designed to provide the first "containerless" way to process materials in space, MSFC announced.

Containerless processing, in which materials could be suspended, melted, and resolidified without touching a container, was thought to permit a degree of purity never before achieved in high-temperature processing, as the material in a conventional container would always exhibit contamination in some degree from the container itself. Apparatus used on SPAR III to process beryllium satisfied all operational requirements, like two other items in the payload: thermal migration of bubbles, and liquid mixing. An apparatus anomaly resulted in only "probable" success of an experiment on epitaxial growth; a fifth experiment, on coalescing of viscous bodies, apparently failed but would be carried again on a later flight. SPAR III rocket flights could offer 5min of near-weightlessness in suborbital coast. (MSFC Release 77-23)

• *W Post* staff writer Thomas O'Toole quoted administration sources as saying President Carter had chosen Dr. Frank Press, head of the department of earth and planetary sciences at the Mass. Inst. of Technology, to be White House science adviser. Press, considered one of the world's leading seismologists, had been one of a team of geophysicists who first described Antarctica as a continent rather than as an island of ice floating on earth's crust. He had also helped design and build the seismometers left on the moon by six crews of Apollo astronauts. A member of the Natl. Science Board, Press had been consultant to 7 federal agencies including DOD, the State Dept., and the Arms Control and Disarmament Agency. Offered the job of science adviser to the president, Press apparently had not accepted it at once, and said by telephone from MIT that "nothing is settled." (*W Post*, Feb 11/77, A-2)

February 13: NASA announced it had commanded the *Viking 1* lander on Mars to dig a ft-deep trench in the surface and bring up soil samples to use in finding out more about the complex chemistry of the planet. The sequence, which would continue through March 15, would include pictures taken by the lander to show mission controllers how the digging progressed. The Jet Propulsion Laboratory had sent commands to the *Viking 1* lander Feb. 10 and would start the *Viking 2* lander's digging sequence about Feb. 16. *Viking 1*'s orbiter would make a series of close passes near Mars's satellite Phobos to take pictures and measure its temperature, while observing the effect of its gravity on the course of the spacecraft. Afterward, mission controllers would drop the orbiter to about 300km altitude over Mars from the current close-approach altitude of 1500km. (NASA Release 77-24; JPL Release Feb 10/77)

February 15: A project known as DUMAND (deep underwater muon and neutrino detection) would use a large area of the Pacific Ocean near Hawaii to detect the arrival of extremely high-energy particles from space, Walter Sullivan reported in the *NY Times*.

Neutrinos, considered the clues to several astronomical mysteries, had been hard to detect because of their lack of mass or electrical charge; moving at the speed of light, they apparently could pass entirely through the earth with no trace. A new science, neutrino astronomy, would use an array of microphones laid out on the sea bottom northeast of the island of Maui, in an area shown to be "pool-table flat" by survey ships from the Hawaiian Inst. of Geophysics, to listen for sound impulses generated by high-energy neutrinos colliding with atoms in the sea water. Collision would produce particle showers that should generate sound pulses as well as flashes of light known as Cerenkov radiation, resulting from the passage of particles through the water at speeds faster than the speed at which light proceeds through water. The original plan would have recorded the flashes to detect particles ejected by stellar explosions (supernovas); Soviet scientists had offered the use of 10 000 photodetectors for the experiment.

The first stage of the project would probably use acoustic detection, however, because of its greater sensitivity and economy: sound waves would travel through water far more efficiently than light would, the report said, and the project could use microphones spaced as far apart as 900meters in an array measuring 10km^2, linked by a 56km undersea cable carrying data to a laboratory on Maui and supplying electrical power to the detectors. Recording the precise arrival time of a pulse at each detector would indicate the direction from which the neutrino came. As the neutrinos' paths (unlike those of other particles) would not be bent by magnetic fields in space, the way in which they had come would point to the source, first clue to cosmic ray origins. The high-energy particles and the muons produced when they hit the atmosphere were among the most energetic radiation known. (*NYT,* Feb 15/77, 14)

• Telesat Canada expected to award contracts in Nov. 1977 for three commercial telecommunications satellites, the *Wall Street Journal* reported. The new comsats would use higher frequency bands than the three Telesat Canada satellites now providing television, telephone, and message communications throughout Canada. First of the high-frequency comsats would be delivered early in 1980 and launched later that yr. (*WSJ,* Feb 15/77, 3)

February 16: NASA announced selection of 222 scientists representing the U.S. and 14 other countries to participate in the first Spacelab flight scheduled for 1980. More than 2 000 candidates had responded to invitations to participate in the joint NASA/ESA mission. NASA chose 86 of the scientists, 81 from the U.S. and others from India, Japan, Canada, France, and Belgium. ESA had selected 136 from 10 other ESA-member states plus Austria and Norway.

Prime objective of the first Spacelab flight would be to verify systems and subsystems performance and to measure the environment around the Shuttle; secondary objective would be the scientific, applications, and technology data that would demonstrate Spacelab's ability to perform space research. First-flight emphasis would be on stratosphere and upper-atmosphere research, with other experiments in plasma physics, biology, botany, medicine, astronomy, solar physics, and earth observations, plus areas such as thermodynamics, materials processing, and lubrication.

Manager of NASA's Spacelab 1 payload would be the Solar Terrestrial Division of the Hq Office of Space Science; manager of ESA's payload would be its office of Spacelab Payload Integration and Coordination in

Europe (SPICE) at Porz-Wahn in West Germany. (NASA Release 77-26; MSFC Release 77-27)

• NASA announced it had developed a new system with the U.S. Army Corps of Engineers to monitor the Mississippi River from St. Louis, Mo., to the Gulf of Mexico this spring, providing information to the Corps on the vast water resources of the river valley, including the Atchafalaya River basin in La.

Using off-the-shelf space technology, the system—called Geostationary Operational Environmental Satellite (GOES) Data-Acquisition System—would take vital measurements using data transmitters at strategic locations along the river and its tributaries, and would relay information on river or reservoir water level, rainfall, and water quality through the GOES satellite to a central ground station. There, Corps workers would process, store, and display the data on a board that would give water-control managers real-time data on the river's status. NASA had set up the system at the request of the Lower Miss. Valley Div. of the Corps, to permit quicker response to flood situations and levee-reinforcement calls than in the past. The automated system would start with 80 stations, 20 now operating. (NASA Release 77-25)

• Launch preparations had begun for the 2-spacecraft mission to outer planets Jupiter and Saturn scheduled for late summer 1977, KSC announced. The center described them as "almost a replay" of the Mars Viking launches in 1975.

Launch date for the first Mariner Jupiter/Saturn (MJS) at complex 41 would be Aug. 20, with the second scheduled 10 days later. Preparation for two major launches within a 10-day timeframe had required "the choreography of an intricate ballet," the center said. Checkout of the Titan Centaur rockets 6 and 7 was in progress; operations personnel expected arrival Apr. 11 of a pathfinder spacecraft to be used for pad tests, along with 2 flight craft expected Apr. 25 and May 23. The MJS spacecraft would weigh 3 times more than the *Pioneer 10* and *11* that reached Saturn in 1973 and 1974, respectively, and would carry additional instrumentation such as narrow- and high-resolution wide-angle TV cameras, cosmic-ray detectors, and infrared spectrometers and radiometers to study the outer planets and satellites.

John Noble Wilford reported in the *NYT* that the new vehicles would have a new name—either Voyager or Discovery—fitting the new design for outer planet environments characterized by pale sunlight and great radio distances. The new craft looked like "a giant round ear," Wilford said, consisting of a dish antenna with struts and booms resembling spider legs. The powerful lightweight radio-transmitter tubes that had been the major problem in building the spacecraft had arrived from the contractor a year behind schedule, reducing the time for "life tests" of their durability and performance.

Each of the spacecraft would also carry a 3-computer system to handle failures during flight, in view of the 2.5hr required for round-trip radio signals between earth and Saturn. Trajectories chosen should provide data on particles and magnetic currents (magnetosphere) around Jupiter and Saturn, and if possible Uranus; on the atmospheres of those planets, believed to have gaseous rather than solid surfaces; and on the surfaces of any solid moons of the outer planets. The Uranus encounter would occur in Jan. 1986, nearly 8.5yr after launch from earth. (KSC Release 65-77; *NYT*, Feb 22/77, 15)

February 17: NASA announced that Lewis Research Center had named Kaman Aerospace Corp. of Bloomfield, Conn., to build the world's largest windmill blade: 45.7 meters, more than twice as long as the 19m blades powering the 100kw generator at LeRC's Plum Brook station in Sandusky, O. The $2-million contract would cover design, fabrication, and test and evaluation of the giant turbine-rotor blade, representing the type needed for production windmills generating 1.5 megawatts of electricity, enough for several hundred homes.

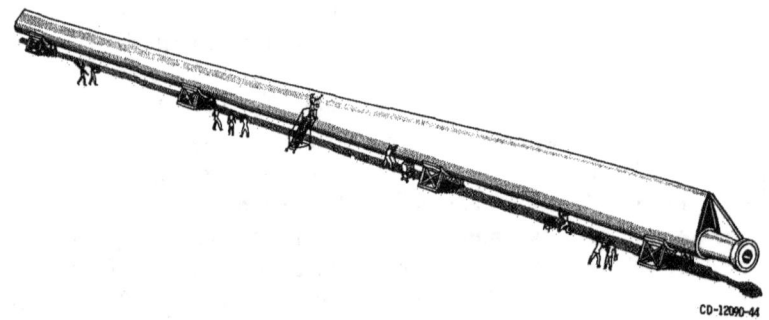

Artist's concept of the 15-story wind-turbine blade, world's largest at 150 ft and weighing about 17 tons, to be built of glass fiber at a cost of $2 million for LeRC by Kaman Aerospace of Conn. (NASA 77-H-156)

The 15-story blade, weighing about 19 400kg and made of glass fiber, would require expensive special tooling and technology advances to permit large-scale production of such blades in future. NASA and ERDA planned detailed design studies on 2-megawatt wind turbines using two of the giant blades to produce power in average winds of 22km per hr. ERDA's decision to develop giant wind turbines had followed studies showing that the big machines could produce electricity at lower cost per kwh. (NASA Release 77-27)

- LaRC announced it had awarded a contract worth about $3 million to Lockheed Aircraft Corp. for development and flight testing of active-control concepts for subsonic commercial-transport airplanes. Scheduled for introduction in the 1980s, the concepts (including maneuver load control, gust alleviation, elastic-mode suppression, augmented stability, and envelope limiting) would help make future aircraft safer, quieter, more comfortable, and more energy-efficient. Work under the contract would be done at Lockheed's Burbank plant over the next 2yr. (LaRC Release 77-4)

- MSFC announced that United Space Boosters, Inc., the firm that would assemble and check out the solid-fuel rocket for the Space Shuttle, had begun a manpower buildup that would eventually include about 150 employees. The present contractor work force of about 60 had been housed in Bldg. 4666 at the center. The firm, a subsidiary of United Technologies Corp. of Sunnyvale, Calif., had received in Dec. 1976 a contract for work at both MSFC and KSC to include stacking, launch operations, and refurbishing the reusable rockets. Responsibilities would include designing tools for applying thermal-protection materials to the boosters, electrical and instrument-verification testing, and installing instrumentation on the booster structure and on the booster aft skirt for tests scheduled at MSFC and at the Thiokol plant in Utah. Employees of USB doing MSFC work at KSC later in 1977 would number from 150 to 200, but the contractor would keep a cadre of about 80 at MSFC through the Shuttle test period. (MSFC Release 77-26)

- LaRC announced that Katherine G. Johnson, aerospace technologist at LaRC's flight dynamics and control division, was among 24 black scientists and inventors honored in Philadelphia who had "advanced technology and helped move the world forward." For Black History Month, Philadelphia's Afro-American Historical and Cultural Museum had unveiled a display including Johnson's portrait among "examples of singular achievement in technology" to "inspire the young, gratify the old, and encourage all who are made aware of these accomplishments." Johnson was cited for her work on interplanetary trajectories, space navigation, and spacecraft orbits. (LaRC Release 77-5)

- President Carter would propose budget changes to reduce emphasis on nuclear power and increase research on energy conservation, predicted Les Gapay in the *Wall Street Journal*. Sources in and out of the Carter administration said that Carter would amend the Ford budget for FY

1978 by cutting money for two major nuclear-power research projects: a multimillion-dollar program to develop a fast-breeder nuclear reactor that would produce more fuel than it consumed, and research on nuclear fusion, a new technology intended to provide commercial electrical power within several decades. Fusion was said to be more environmentally acceptable than other nuclear programs, and the U.S. had been spending several hundred million a yr on this research. Funds cut from the Energy Research and Development Administration would go toward research on energy conservation; ERDA had been studying ways to save fuel in building, transportation, and industrial processes. The Ford budget had asked for a 28% increase in energy research for FY 1978 (from $2.9 billion to $3.7 billion), most of it for ERDA and for nuclear power studies. (*WSJ,* Feb 17/77, 3)

February 18: The $500-million Space Shuttle with no crewmembers aboard took its first ride over the Calif. desert, carried piggyback on its converted Boeing 747 for a 2hr test at nearly 5km altitude. Operating the 747 were veteran NASA test pilot Fitzhugh L. Fulton, Jr.; copilot Thomas C. McMurtry; and flight engineers Victor W. Horton and Lewis E. Guidry, Jr. The Enterprise, prototype of the reusable craft designed to carry cargo and scientists into space in the 1980s, was secured to the 747 by mounts at its nose and tail; the combination craft stood almost 20m from the ground to the top of the orbiter's tail. The explosive bolts provided to separate the two craft in free flight were armed during this first test, in case of emergency.

NASA officials said the weight of the Enterprise (about 65 000kg) did not strain the capacity of the 747, which weighed nearly 200 000kg empty and had carried much greater loads in cargo and passengers. "Most of the time we couldn't even tell the Shuttle was here," Fulton said after the flight, which reached speeds of about 450kph as the crew turned and banked the combined ships, shut down some of the 747 engines, and performed other maneuvers to check stability. The Enterprise would be unmanned during the next 5 combined test flights; in May, a 2-man crew would conduct 6 more tests of the 747/orbiter combination, and in July the orbiter would be released from its carrier for a series of unpowered flights. (*W Post,* Feb 19/77, A3; *W Star,* Feb 19/77, A-8)

• The U.S. Air Force announced plans to launch an Honest John-Hydac rocket from Santa Rosa Is. near Pensacola, Fla., after sunset on Feb. 22 to release a high-altitude multicolor cloud visible throughout the Southeast. Release of a 48kg payload of barium 72km above the Gulf of Mexico would reveal whether or not the resulting cloud would interfere with radio signals between an airplane and a comsat, a USAF spokesman said. A similar test Dec. 1, 1976, created a colored cloud visible as far away as Tennessee. (*W Post*, Feb 18/77, A23)

• NASA announced it had signed an agreement with McDonnell Douglas Corp. allowing the company as a commercial venture to design, manufacture, and test an upper-stage system, including integration services, for Space Shuttle missions having payloads weighing up to 2000kg. Carried into low earth orbit on the Shuttle, the upper stage called SSUS-A (for spinning solid upper stage, Atlas Centaur class) would take payloads into transfer orbit where a kick motor would put it into a circular geosynchronous orbit 35 900km above earth.

A similar agreement last yr with McDonnell Douglas was for an upper stage called SSUS-D (for Delta class) suitable for use with payloads weighing up to 1100kg, like those being orbited on Delta vehicles. NASA had planned to procure SSUS-A systems only from firms with which it had such an agreement; under the agreements, firms would be free to sell stages either to NASA or directly to customers. The agreements would specify performance requirements, define a delivery schedule to meet NASA's Shuttle schedule, and establish unit ceiling prices and a ceiling profit. (NASA Release 77-29)

• Six NASA rockets launched from Australia in Feb. would investigate stars, hot stars, white dwarfs, and exploding galaxies, NASA announced. Goddard Space Flight Center's Richard M. Windsor, working as project manager with the Australian government, the Univ. of Adelaide, and several U.S. universities and scientists, would oversee launch of the 10m-long Aerobee sounding rockets from the Woomera range during the wk of Feb. 16-24. The Aerobee, which could lift a 181kg payload to an altitude of 210km, had carried nearly a thousand scientific payloads since it came into use in 1947.

U.S. investigations would include a Columbia Univ. study of bursting x-ray stars to see if they resembled black holes; a Naval Research Laboratory study to determine whether the Large Magellanic Cloud resembled earth's galaxy, comparing x-ray stars for luminosity, temperature, and fluctuation; another NRL study of x-ray sources in the Coma Cluster; a Johns Hopkins Univ. study of Alpha Centauri A, 4.5 light-years distant, and other stars similar to earth's sun, for evidence of a solar corona in far ultraviolet; another JHU study, this one in UV measurement of early or "hot" stars, to obtain data for use in calibrating satellite-borne UV telescopes; and a GSFC study of the Large Magellanic Cloud using UV filters on a Schwarzchild camera to define distribution of hot stars, galactic dust, high-energy gas, young remnants of supernovas, and other celestial phenomena. (NASA Release 77-28)

February 19: Global communications systems used by the Dept. of Defense were inadequate, inefficient, and incapable of surviving nuclear attack, according to the investigation subcommittee of the House Committee on Armed Services.

A United Press report carried in the *W Star* quoted subcommittee criticism of the Pentagon's top-secret command post which, "while located in the basement of the Pentagon, [showed] little possibility that the National Military Command Center would survive a nuclear attack directed against it." The report cited "dismay over the department's inability to deploy a satellite communications system after 16 years of effort," "security deficiencies" in the computer software of the worldwide Military Command and Control System, weaknesses in the Tri-Service Tactical Communications program, and "urgent need for early deployment of improved voice security equipment throughout the military services." The report commended the use of airborne command posts such as the converted Boeing 747 "doomsday plane" that had taken President Carter to Georgia the previous wk. (*W Star*, Feb 19/77, A-3)

• Canadian scientists recovered a fragment weighing about 2kg from a meteorite seen streaking across the sky over Alberta earlier in Feb., the Associated Press reported. Ian Halliday of the Natl. Research Council noted that the find was only the third time a meteorite had been located after it had been photographed coming down; most meteorites reportedly had burned up or exploded before reaching earth. The AP report noted that four fragments of a meteorite that exploded Jan. 31 over Louisville, Ky., had been recovered, and that all four had struck manmade objects (two rooftops, a house window, and an automobile hood), reportedly a rare occurrence. (*W Post*, Feb 20/77, A19)

• The Univ. of Toyko's Inst. of Space and Aeronautical Science launched a 13kg experimental satellite called *Tansei 3* from the Uchinoura space center into an orbit with 3940km apogee, 826km perigee, using a 3-stage MU-3H rocket developed at the university. The rocket, used in preparation for next yr's launch of Japan's 5th aurora-observation satellite, was 23.8m long and weighed 49 tons. The MST-3 satellite, 7th launched by the university and 9th by Japan, carried control equipment for testing and would sent back scientific data for an estimated 15 days. (FBIS, Tokyo Kyodo in English, Feb 10/77)

February 20: Distant planets Uranus and Neptune had turned out to have spin rates similar to earth's, so that their days correspond to earth days and they had probably come from the same debris that earth did, reported Kitt Peak astronomers Michael J.S. Belton and Bethanne Hayes. Using the observatory's just-over-400cm telescope, world's second largest, the astronomers found that (contrary to previous theory) spin rates of the outermost planets (except for Pluto) resembled not those of Saturn and Jupiter but rather those of earth and Mars. The discovery indicated they were solid ice-like bodies, not big balls of gas, Dr. Belton said.

Using a smaller telescope at Kitt Peak, a separate 4-member team of scientists had reported the first observation of weather patterns in the upper atmosphere of Neptune. Dr. Richard Joyce and 3 scientists from the Univ. of Hawaii had found a marked increase in Neptune's reflectivity between April 1975 and March 1976, suggesting the presence of thin transient clouds high in that planet's atmosphere. Astronomers had been using the planet as a standard measure of brightness, thinking that it experienced no atmospheric changes. (*W Post*, Feb 20/77, F16)

• Meteorites—so rare that museum collections around the world contained no more than 2000, and no curator had reported finding one on his own—had turned up in quantity in Antarctica several years ago when a team of Japanese scientists found an area of the ice-covered continent where hunting meteorites was "as easy as picking mushrooms," the *Chicago Tribune* reported.

Dr. Edward J. Olsen, chief of geology at Chicago's Field Museum and curator of one of the largest meteorite collections, had returned from the first U.S. meteorite expedition to Antarctica reporting that his team had found pieces of 11 different meteorites, weighing about 450kg in all, at two sites more than 200km northwest of the U.S. research station on McMurdo Sound where movement of the ice cover had concentrated meteorites at certain spots.

Scientists valued the meteorites as the oldest objects yet located, dating back as far as 4.5 billion yr, whereas earth's oldest rocks were only 3.7 billion yr old. Meteorites in pristine condition, preserved by the ice, might contain information about organic chemicals in space, Olsen said, and cosmic-ray traces might show whether the sun was running down or getting hotter. (*C Trib*, Feb 20/77, 1-3)

February 21: A lack of sunspots during the past yr and a half might have caused recent unusual cold weather and drought in the U.S., Dr. John A. Eddy of the Natl. Center for Atmospheric Research told the 143rd meeting of the Am. Assn. for the Advancement of Science. Dr. Eddy, an astronomer who had studied the complex relation between solar activity and historic and climatic events, had traced short- and medium-term effects over several thousand yr in history records and as far back as 7000yr by measuring carbon-14 levels in tree rings.

The theory was that sunspots increased the solar wind (solar particles and radiation that fan the earth) which determined the amount of cosmic and galactic radiation striking the earth. Tree-ring studies echoed abnormal conditions of cold weather and drought when the rings exhibited high levels of carbon-14 resulting from reduced solar-wind activity, resulting from fewer sunspots.

". . . The record of climate [has] a one-to-one correspondence that's so good I don't want to believe it," Dr. Eddy told the Denver meeting. History reported the reign of King Louis XIV of France (1643–1715) as a time of high carbon-14 levels; Eddy said the most severe temperature drops on earth in the last thousand yr occurred in that period. Crops failed in the British Isles, a Norse colony in Greenland perished of cold, and Spanish conquistadors had ridden their horses into Mexico across a frozen Rio Grande. The aurora borealis produced by solar particles striking earth's magnetic field did not appear during that period; astronomer Edmund Halley was 60yr old when he first saw northern lights in 1715, when the "little ice age" ended. "The fact is, we should have been into a rise of solar activity in the autumn of 1975," Eddy noted, "and here we are a year and a half overdue into getting that rise." (*W Post*, Feb 22/77, A-3; *NYT*, Feb 22/77, 13)

• Japan's Natl. Space Development Agency announced rescheduling to Feb. 23 of the launch of a 3-stage N rocket carrying the first geostationary meteorological satellite to be orbited by Japan. The launch had been postponed because of trouble with the system controlling the rocket's first stage. (FBIS, Tokyo Kyodo in English, Feb 21/77)

February 22: Pressures from inside and outside the scientific community might result in restricting the freedom of inquiry that had enabled science to flourish since the Renaissance, the annual meeting of the Am. Assn. for the Advancement of Science heard. Walter Sullivan reported in the *NY Times* that a session on the role of science as the key to man's political future had as discussion leaders George W. Ball, senior director of Lehman Bros. investment bankers; Lord Zuckerman, former chief science adviser to the British government; and Charles J. Hitch, president of Resources for the Future, Inc., of Washington, D.C. The panel cited controversies in the Cambridge, Mass., city council over regulating use of genetic material in that city's research laboratories; the Calif. legislature's proposal to regulate research on genetic hybrids; and other attempts to control the directions of research.

The panel expressed the feeling that checks and balances within the scientific community should be able to deal with such problems, and said that misgivings about "too much knowledge" harked back to prehistoric and classical times when mankind feared probing into nature's secrets. Although restraints on export of equipment and knowledge might delay the spread of nuclear weapons, Ball noted that the only real control would be "to create a climate where nations don't find it [nuclear weaponry] useful." (*NYT*, Feb 23/77, B-4)

February 23-27: Japan's Natl. Space Development Agency announced launch Feb. 22 of its engineering test satellite ETS-II from the space center at Tanegashima at 5:50 local time (0850 GMT) on a 3-stage 32.5m N rocket developed with the help of U.S. technology. The rocket weighed about 90 tons and had a thrust of 148 tons at liftoff.

The satellite, a 1.41m-diameter cylinder 93cm long, carried a transmitter/receiver and five small rockets for attitude control, orbital shift, and speed adjustment. At about 6:45 pm local time, the satellite (called *Kiku 2* in orbit) reached an elliptical transfer orbit with 191km perigee and apogee of about 36 000km, the altitude at which it would go into synchronous orbit within a few days. Its synchronous station would be at 130°E over Indonesia, the announcement said. This was the third satellite orbited by the agency and the tenth by Japan. (FBIS, Tokyo Kyodo in English, Feb 23/77; *NYT*, Feb 25/77, A7; *CSM*, Feb 24/77, 2)

NSDA officials announced Feb. 25 that *Kiku 2* would change orbit a day early, on its seventh pass over the Indian Ocean, as data from tracking stations in Japan and on Christmas Is. had confirmed its success in elliptical orbit. The satellite would reach station by March 5 if all went well, the announcement said. (FBIS, Tokyo Kyodo in English, Feb 25/77)

The *NY Times* carried a Feb. 26 item from Associated Press confirming that "Japan's first experimental stationary satellite" had shifted to a circular orbit. (*NYT*, Feb 27/77, 7)

NSDA announced Feb. 27 that *Kiku 2*'s apogee motor ignited at 2:32 pm local time on Feb. 26 to move the satellite from an elliptical to a circular orbit with apogee of 35 756km and perigee of 34 034km, close to preflight predictions and fast enough to permit drift to its permanent station over Indonesia at 35 800km altitude. (FBIS, Tokyo Kyodo in English, Feb 27/77)

February 24: JSC announced plans for NASA Symposium 77 to be held at the center March 1-3 for more than 2000 students from junior and senior high schools in Texas, a first-of-its-kind event for NASA, designed to motivate youth (especially female and minority students) to seek careers in engineering and science. Special guest at the event would be actress Nichelle Nichols, best known as Lt. Uhura, communications officer on the Star Trek television series. JSC and contract employees would conduct workshops and describe their own careers as scientists or engineers, and would take the students on tours of various work areas at the center.

The symposium would include a seminar for college and university administrators from schools having high enrollments of minority and female students, giving information on NASA research, training, and employment opportunities. NASA had conducted programs similar to Symposium 77 at 3 universities, but not at a field center. (JSC Release 77-10)

• The initial phase of Shuttle approach and landing tests (ALT) that began Feb. 18 at DFRC might end ahead of schedule, DFRC announced, if the 4th and 5th flight of the combined 747-and-orbiter vehicles continued to meet program objectives. Donald K. Slayton, ALT program manager, said that if performance measured up to previous successful flights, the 6th test (now scheduled for Mar. 4) would be dropped. Program officials had praised the performance of the 747 and its piggyback passenger, Slayton said; flight data had verified preflight wind-tunnel test results and simulations conducted by Boeing.

Early completion of inert tests with an unmanned orbiter would put the orbiter in its hangar a week ahead of schedule to prepare for its next flight, in the captive-active phase. Five flights scheduled in this phase would carry astronaut crews to perform systems checks and go through crew procedures as the 747 went through maneuvers up to, but just short of, actual release of the orbiter. Tests later in the summer would release the orbiter for an unpowered landing on DFRC's dry lakebed. (NASA Release 77-36; DFRC Release 8-77)

February 25: NASA announced selection of Hughes Aircraft Co.'s Space and Communications Group, El Segundo, Calif., for negotiation of a cost-plus-award-fee contract valued at between $37.5 and $43.5 million, depending on the option selected, for design and fabrication of a thematic mapper to be used on Landsat D, scheduled for launch early in 1981. The thematic mapper would be a remote-sensing instrument offering improved data for land-resources management by making observations with repetitive temporal coverage and delivering to the users information with better spatial and spectral resolution (mapping on a 1:250 000 scale compared to 1:1 000 000 on the present Landsat.)

The basic contract, including assembly, test, and qualification of a prototype flight model and one set of bench-test and calibration equipment, as well as support services, would have two options: either an additional flight unit with an additional set of test equipment and support services, or two additional flight units and support services. Completion and delivery of the prototype would be scheduled for late 1980. GSFC would manage the thematic-mapper program. (NASA Release 77-34)

• NASA announced development of ultrathin large-area silicon solar cells thinner than a sheet of newspaper for increasing the power-to-weight capability of solar arrays, considered a major breakthrough in applying solar energy to missions requiring multikilowatts, even millions of watts, of electricity. The Solarex Corp. of Rockville, Md., had made the thin cells some 40 to 50 microns thick (only one-sixth as thick as cells in present solar arrays) in sizes up to 38cm^2, proving both structurally flexible and less fragile than expected. Energy conversion of the new cells virtually equaled that of conventional cells, 11% compared to 12 or 13% for the thicker cells.

Lightweight cells would mean larger solar arrays for ion-propulsion systems in extended space exploration, and could supply power for platforms and remote-satellite power stations in the future. Solarex had delivered hundreds of the new cells to JPL at Pasadena, Calif., for test, evaluation, and design-application studies. (NASA Release 77-31)

• More than half the 1497 reports received under the NASA aviation safety reporting system (ASRS) in its second quarter of operations included unsolicited recommendations for solutions of a variety of air-traffic problems, NASA announced. The reports, dealing with air operations nationwide, reflected an increase in the quality of reporting for the period ending Oct. 14, 1976, the agency said. NASA's ASRS had been designed as an "early warning system" consisting of voluntary reports submitted by pilots, air traffic controllers, and others in the national aviation system.

During the first and second quarters, an average of 100 reports came in each wk; the number of reports from pilots declined during the second quarter, with reports from controllers now constituting half rather than a third of the total. At least three-fourths of all reports concerned incidents occurring in controlled air space, and some type of flight plan had been filed in 84% of the flights described. NASA had prepared and sent to the Federal Aviation Administration a total of 58 alert bulletins during the report period.

NASA personnel had initially screened all ASRS reports to remove the names of persons reporting; the reports went for analysis to the ASRS office of Battelle's Columbus Division in Mountain View, Calif. Attempts to verify information in 340 of the reports, by telephone contact with the reporter, were successful in 270 instances. Names of persons were removed to protect the right of free comment without fear of reprisal, and 80% of the reports were reviewed within 2 working days of their receipt. (NASA Release 77-30)

• NASA issued requests for proposals from industry to design and manufacture the Infrared Astronomy Satellite (IRAS) recently approved by the U.S. and the Netherlands. Scheduled for launch in March 1981, the observatory would use a telescope furnished by the U.S., a spacecraft built by the Netherlands, and a control facility supplied by the United Kingdom. The IRAS in a 900km polar orbit would survey the entire sky for a yr, at infrared wavelengths not usable by earth-based telescopes because of atmospheric interference, to study stars forming and at the end of their life cycle.

JPL would manage the project and would design and operate a facility to produce an infrared sky map and a catalog containing up to 1 million infrared sources. NASA's Ames Research Center would be responsible for the 60m-aperture infrared telescope. Prof. Gerry Neugebauer of the Calif. Inst. of Technology (operating JPL for NASA) would head the U.S. scientific team, working with as many as 500 U.S., Dutch, and U.K. scientists, engineers, and technicians. Industry proposals were due by Mar. 1, with contractor selection later in 1977. (NASA Release 77-33)

• MSFC reported NASA's Michoud assembly facility had shipped the first of four major pieces of test hardware for the Space Shuttle external tank, an intertank structural-test article scheduled to arrive by barge at MSFC about Mar. 5. It would go from New Orleans to Huntsville by way of Mississippi, Ohio, and Tennessee Rivers, the route NASA had used to ship Saturn stages for testing during the Apollo program. The other test articles (liquid-hydrogen tank, liquid-oxygen tank, and another intertank) would arrive later in 1977.

The external tank, one to be used for each Shuttle mission, would hold containers for liquid oxygen and liquid hydrogen, connected by the intertank; the combined assembly would measure about 47m long. Test data would serve to verify the intertank's structural integrity before static test firing of the Shuttle's main propulsion system at the Natl. Space Technology Laboratories later in 1977. The test firings would use a full-size external tank, three Shuttle engines, and an Orbiter aft-structure simulator, attached to the test stand through the intertank. (MSFC Release 77-30)

• JSC announced it had awarded a cost-plus-award-fee contract worth $12 million to Lockheed Electronics Co. for a third yr of site-support services to JSC's White Sands Test Facility at Las Cruces, N.M. Estimated total value of the 3yr contract was $24 million. (JSC Release 77-11)

• NASA Administrator Dr. James C. Fletcher gave the agency's distinguished public service medal to United Nations Secretary General Kurt Waldheim, at a ceremony at the Natl. Air and Space Museum. Dr. Waldheim received the award for outstanding leadership as chairman of the UN Committee on the Peaceful Uses of Outer Space from 1965 to 1968 and from 1970 to 1971, and for "enduring contributions to the development of a constructive international consensus, during the critical formative years, on the challenging issues presented by man's first uses of outer space." (Text of presentation)

February 28: NASA announced that MSFC had issued requests for proposals from industry to make studies of, and supply data for NASA to use in developing techniques for packaging, transporting, fabricating, erecting, and operating large structures in space, in preparation for a major demonstration in space in 1983 or 1984. An early Shuttle flight would carry an automated beam-fabrication module to make aluminum structural members; later flights would demonstrate use of both aluminum and composite materials. After a few beams were manufactured on early flights, later missions would perform limited assembly, culminating in fabrication and erection of a large structure for continuing use, such as a 100kw solar-power facility to supplement onboard Shuttle power for various experiments.

The capacity demonstrated in these missions would aid in creating a more complex assembly as part of a space-construction base in 1985 or thereafter. The proposal called for completion in 9mo and another 2mo to prepare a report. MSFC would receive proposals until Mar. 18, 1977, and would manage the study project for the Office of Space Flight. (NASA Release 77-35)

• JSC announced it had awarded a contract worth $2 million to Hamilton Standard Division, United Technologies Corp., Windsor Locks, Conn., to supply space suits for Shuttle crews. The system, an extravehicular mobility unit (EMU), would consist of a suit with integrated life-support backpack for use outside the Shuttle orbiter's pressurized cabin. The suit would come in small, medium, and large upper- and lower-torso standardized sections, to replace the individualized suits of earlier manned programs. The contract called for equipment and spares for seven suits and supporting hardware, plus training, manpower, and field support at JSC and other NASA centers, at an estimated ultimate cost of about $18.5 million. (JSC Release 77-12)

• JSC reported it had signed a contract valued at $200K with American Airline's maintenance and engineering center at Tulsa, Okla., for maintenance and operations support of NASA's Boeing 747 Shuttle-carrier aircraft NASA 905. Work would be done primarily at DFRC, site of the first atmospheric and glide-flight landing tests of the Shuttle orbiter. Eventual value of the contract could reach $806K. (JSC Release 77-13)

During February: Japan, one of five countries that previously had agreed to be major investors in a $200 million international maritime satellite organization (Inmarsat), had declined to participate in a preparatory committee set up to act until formal documents were ratified, *AvWk* reported.

Forty countries would be Inmarsat members; besides Japan, the other four major shareholders were to be the U.S., the U.K., the USSR, and Norway. The preparatory committee had met in London in Jan. under sponsorship of the Intergovernmental Maritime Consultative Organization, an arm of the UN, and had set up three panels: one on satellite-system parameters; one on economic, marketing, and financial matters; and one on organization procedures, all of which were to meet in the interval before the next committee meeting set for Oct. 1977. Japan gave no reason for withdrawing. (*AvWk*, Feb 7/77, 23)

March

March 1: NASA announced the first use of a link between the communications technology satellite *Cts* and the portable earth terminal (PET): a teleconference between Housing and Urban Development Secretary Patricia Roberts Harris in Washington, D.C., and a group of California mayors at the Ames Research Center in Mountain View, Calif. Mrs. Harris conferred with the mayors from the PET bus parked outside the HUD building. *Cts*, a joint effort between the Canadian Dept. of Communications and NASA, had been built and integrated by Canada; NASA provided the 200w transmitter and developed the PET at Lewis Research Center. The PET bus, based at LeRC, would travel around the U.S. demonstrating how to save money through teleconferencing. (NASA Release 77-40)

• NASA announced it would join the Interior Dept.'s Bureau of Land Management in using Landsat, the earth resources monitoring satellite, to inventory wild land resources (major streams, ground cover and vegetation types, reservoirs, and fire hazards) in south central Alaska, northwestern Arizona, and southwestern Idaho. The project would examine each of the 3 different regional ecologies for about 1yr. During the 3yr project, the BLM and NASA would set up an automated inventory system to provide accurate and timely data on BLM land as another example of Landsat usefulness in solving public and private resource-oriented problems. (NASA Release 77-37)

• Marshall Space Flight Center announced it had awarded a $634 250 contract to Grumman Aerospace Corp. to design, build, and test a machine that would demonstrate, on earth, the automatic fabrication of beams for space construction. Ground demonstration would be needed to develop a safe and efficient facility for use in space. Beam fabrication, part of the Space Industrialization Program, would be the first step in building large space structures. (MSFC Release 77-33)

• MSFC reported that the contractor for the first high-energy astronomy observatory (HEAO-A), TRW Systems of Redondo Beach, Calif., would ship the spacecraft to Kennedy Space Center on Mar. 7. HEAO-A, a 2660kg unmanned mission scheduled for launch Apr. 15 into low circular earth orbit, would survey celestial x-ray sources and gamma-ray flux. (MSFC Release 77-32)

March 3: President Carter's first budget proposal contained additional funds for NASA, *Nature* magazine noted. The new administration had recommended $15 million for studies of potential missions to Mars after Viking; the previous administration had requested $5 million. Planetary scientists, pleased with the increased budget request as well as with Carter's recent remarks on the Space Shuttle, viewed these actions as presidential support of the space program. (*Nature*, Mar 3/77, 5; *Spaceflight*, Mar 77, 8)

• NASA announced it had used Landsat, the earth resources monitoring satellite, to inventory irrigation systems in Nebraska. The 12 000 central-pivot irrigation systems proved easily recognizable by their circular shape in photos from Landsat's 1480km-altitude orbit. Annual inventory of the irrigation systems had become necessary when the number of central-pivot sites increased by 40 to 60% each yr, affecting property values, land use, and fuel and water allocation. Landsat had proved the cheapest way to acquire an accurate and timely inventory. (NASA Release 77-38)

• Prof. James L. Elliot of Cornell Univ. reported sighting a large number of objects — possible satellites of Uranus, the seventh planet — while he was using the 91cm telescope on NASA's Kuiper Airborne Observatory flying above the Indian Ocean. Another U.S. astronomer, Dr. Robert Millis of the Lowell Observatory in Ariz., reported similar observations while viewing Uranus from Perth in Australia. Each of the scientists claimed that the planet might have as many as 100 satellites ranging from 18 to 24mi in diameter. If confirmed, the new observations would give Uranus more satellites than any other planet in the solar system. (*NYT*, Mar 3/77, 77; *C Trib*, Mar 12/77)

March 4: NASA announced the return to Johnson Space Center of former astronaut Dr. Edward G. Gibson, science pilot on the 84-day *Skylab 4* mission in 1973-74, after a yr with ERNO of Bremen, W. Germany, prime contractor for Spacelab (a future Space Shuttle payload). Gibson, a Ph.D. in physics and engineering, had left NASA in 1974 to become senior staff scientist at Aerospace Corp., El Segundo, Calif. Gibson would be working in JSC's Mission Specialist Office, now headed by astronaut Joseph P. Kerwin, M.D. (NASA Release 77-43)

• NASA's Jet Propulsion Laboratory announced that, as part of the Solar Sail Development Project, it had awarded 6 contracts to build a spaceship powered by solar wind. The craft, using large extremely thin

Artist's concept of solar sail making rendezvous in 1982 with Halley's comet. Launched from an orbiting Space Shuttle, an 800m-sq structure could carry an 800 to 900kg spacecraft on such a mission. (NASA 77-H-128)

metallic sails to capture the energy of solar photons, would tack like a sailboat into the "wind" to head toward the sun or would use photon energy to head for the outer planets. Speed of the ship would vary with distance from the sun, following the inverse-square law that solar intensity would decrease as the reciprocal of the distance from the sun, squared. Cost of the contracts for fabricating sail material and designing sails, booms, and navigation systems would total $800 000. (NASA Release 77-39)

• A 25-article series in *Science News* by Jonathan Eberhart on the Viking mission won the 1976 AAAS-Westinghouse science writing award Feb. 22, *Science* reported. Henry S.F. Cooper's *New Yorker* article on the Space Shuttle received honorable mention. (*Science*, Mar 4/77, 864)

March 7: Dr. James C. Fletcher, NASA administrator, announced that the planetary probes scheduled for launch Aug. 20 and Sept. 1 would be named Voyager 1 and Voyager 2. The two craft would view Jupiter, Saturn, and their respective satellite systems, and might also look at Uranus and Neptune. JPL had designed and built both probes.

In March 1979 Voyager 1 would pass within 357 000km of Jupiter, the solar system's largest planet, and go on to rendezvous with Saturn in Nov. 1980, passing within 209 000km of the ringed planet's cloud tops and within 6430km of Titan, Saturn's largest moon and the only satellite in the solar system known to have an atmosphere. (NASA Release 77-41)

- NASA's upcoming 5yr plan (1978-1982) had projected annual budgets for fiscal yrs 1980-1982 in excess of $4.7 billion, *Aviation Week* reported. The plan, an internal document used to guide the agency, would allow NASA as many as 7 major new mission starts in FY 1979, including a Mars rover and a Halley's Comet rendezvous. (*AvWk*, Mar 7/77, 47)

- MSFC announced it would consider using the Space Shuttle's external fuel tank (ET) as a permanent space platform. James E. Kingsbury, head of MSFC's Science and Engineering Directorate, proposed to replace 57m^3 of an ET's liquid oxygen with provisions for 90 days, then launch another Shuttle to bring a crew, a Skylab-airlock module and multiple-docking adapter, and a solar electric-conversion wing to the orbiting ET to constitute a habitable space station. All the required equipment, now in storage or on display in the Smithsonian's National Air and Space Museum, had been flight-tested. (MSFC Release 77-36; NASA Release 77-42; Htsvl *Times*, Feb 20/77, 8)

- JSC announced it would use CAMAC, a European innovation used in working between computers and hardware to produce standard scientific electronic subassemblies, as a link between Shuttle computers and onboard experiments. CAMAC would reduce costs by permitting actual onboard assembly of major electronic sections from a CAMAC subassembly pool. MSFC had been studying the equipment to ensure reliability during Shuttle operations. (JSC Release 77-14)

- JSC announced award of a $2 752 000 contract to Hamilton Standard Division of United Technologies Corp., Windsor Locks, Conn., for development and production of a Space Shuttle portable oxygen system for use independently or connected to the orbiter's oxygen system. Consisting of a face mask, rebreather loop, heat exchanger, oxygen bottle, and recharge kit, the system would operate during emergencies and for prebreathing before spacewalks to denitrogenize the crew's circulatory systems. (JSC Release 77-18)

- NASA might undertake a Cosmic Background Explorer satellite program, *AvWk* reported. The satellite would examine the character of cosmic background radiation claimed by many scientists to prove the big-bang theory of the universe's formation. (*AvWk*, Mar 7/77, 11)

- A cryogenically cooled infrared telescope carried aboard the Space Shuttle would become the major infrared research instrument in space, *AvWk* reported. NASA studies had indicated such a telescope could gather spectroscopic data on objects 1000 times fainter than now visible. (*AvWk*, Mar 7/77, 11)

- New infrared detectors at Kitt Peak Natl. Observatory might have detected frozen methane on the surface of Pluto, *AvWk* reported. If verified, the observations would indicate a smaller size for Pluto and would change current theories of Pluto's interaction with other planets. (*AvWk*, Mar 7/77, 11)

March 8: MSFC announced a $1 046 100 amendment of a contract with Rocketdyne Division of Rockwell Intl. Corp., Canoga Park, Calif., contractors for the Space Shuttle main engine (SSME). The amended contract, totaling more than $66 million, would include equipment to build SSME. (MSFC Release 77-37)

March 9: NASA announced that Ronald E. Evans, former command-module pilot of *Apollo 17*, would leave Mar. 15 to become executive vice president of Western American Energy Corp. (WAEC) and director of marketing for WES-PAC Energy, coal-producing subsidiary of WAEC. Evans, selected as an astronaut in 1966, flew on *Apollo 17* in 1971 and made a 1hr 6min spacewalk. He had been in charge of plans for the ascent phase of Space Shuttle orbital-test flights scheduled for 1979. (NASA Release 77-44)

- Reduction of data from NASA's 1976 gravity probe (GP-A) experiment had confirmed the equivalency principle, a cornerstone of Einstein's general theory of relativity, MSFC reported. Using data from two identical atomic maser clocks, principal investigator Dr. Robert Vessot of the Smithsonian Astrophysical Observatory (SAO) in Cambridge, Mass., had demonstrated that time actually went faster in a weaker gravitational field, one removed from the earth's mass. The clock that had been carried aboard a rocket to an altitude of 10 000km had run faster. (MSFC Release 77-41)

- MSFC reported plans to request another $18 610 000 in NASA's FY 1978 budget for construction or modification of facilities to be used in building the Space Shuttle external fuel tank (ET) at the Michoud assembly facility (MAF) near New Orleans. With addition of a chemical waste-treatment plant and alterations of the manufacturing, vertical-assembly, final-assembly, and checkout buildings, MAF could produce 28 ETs each yr. (MSFC Release 77-35)

March 10: NASA launched *Palapa 2* (formerly Palapa-B), an Indonesian telecommunications satellite, from the Eastern Test Range at Cape Canaveral on a Delta at 2316 GMT, into a transfer orbit with 36 499km apogee, 231.2km perigee, and inclination of 24.6°. *Palapa 2*, second Indonesian domestic comsat, would use its apogee kick motor Mar. 12 to move into synchronous orbit above the Indian Ocean. Forty ground stations throughout the archipelago would link the islands with the 2 satellites, each capable of handling 4000 phone calls or 12 simultaneous color TV channels. (MORs M-492-208-77-02, [prelaunch] Feb 2/77, [postlaunch] Apr 12/77; NASA Release 77-23; FBIS, Jakarta Domestic Service in Indonesia, Mar 11/77)

• MSFC reported it had established a project office responsible for managing all Space Telescope activities. Manager and deputy would be William C. Keathley and James C. McCulloch. (MSFC Release 77-38)

• NASA planned to send a roving laboratory over the Martian surface during the next Mars landings, *Nature* reported. Dr. Gerald A. Soffen, chief Viking project scientist, said that, although no future Mars missions were officially scheduled, the outlook was favorable given the scientific success of, and public enthusiasm for, the two Viking landers. Speaking at a Paris conference to disseminate Viking information, Soffen invited U.S. and foreign scientists to participate in examining Viking data and in developing experiments for the next generation of Mars missions. (*Nature*, Mar 10/77, 112)

• President Carter had suggested the U.S. and USSR "forgo the opportunity to arm satellite bodies and also forgo the opportunity to destroy observation satellites," the *W Post* reported. Recent Soviet antisatellite experiments had alarmed Pentagon officials, who revealed that the U.S. military space program budget for FY 1978 was $478 million greater than the FY 1977 appropriation. (*W Post*, Mar 10/77, A-11)

March 11: The *W Post* reported that NASA Administrator James C. Fletcher would resign effective May 1. Appointed by President Nixon in 1971, Dr. Fletcher had headed the agency during the end of the Apollo project; the successful Skylab missions; the Apollo-Soyuz test project; and numerous unmanned missions including the launch of the earth resources monitoring system, Landsat, and the Mars probe Viking. Dr. Fletcher planned to return to private life after his resignation. (*W Post*, Mar 11/77, A3; NASA Release 77-48)

• NASA announced it had awarded a cost-plus-fee contract worth about $43.5 million to McDonnell Douglas Technical Services, Inc., Spacelab Integration Division, Huntsville, Ala., for Spacelab integration. The contract, lasting from March 1977 through Dec. 1983, would include design, development, manufacturing, and operational testing of NASA's Spacelab components. NASA planned to provide all operational services including flight operations, crew training, and maintenance, after the European Space Agency delivered the lab. Spacelab would remain in the Space Shuttle cargo bay to perform earth observations and experiments in materials science, physics, life science, communications, navigation, and space systems. (NASA Release 77-47)

• MSFC reported that parachutes to recover the Space Shuttle's reusable solid-fuel rocket boosters (SRB) were undergoing dynamic testing at Martin Marietta Corp., Denver, Colo. Scheduled for completion Mar. 31, the tests simulated in-flight parachute deployment from the SRB. (MSFC Release 77-44)

March 14: NASA announced appointment of Maj. Gen. Kenneth R. Chapman, USAF (Ret.), as assistant administrator for its Hq Office of DOD and Interagency Affairs, effective Apr. 1. He would replace Lt. Gen. William. W. Snavely, USAF (Ret.), who left NASA in Sept. 1976 to accept a position in Saudi Arabia. Chapman, a nuclear chemist, was formerly director of the Office of Nuclear Material Safety and Safeguards at the Nuclear Regulatory Commission (NRC). (NASA Release 77-50)

• NASA announced that LeRC had supplied news media in the Cleveland area with readings of daily solar energy available for use during the winter. Using a pyranometer, a device to record total solar energy reaching a surface, LeRC staff measured the kilowatt hrs of solar energy received on a 150m^2 surface facing south at a 37° angle. Other NASA centers had made similar measurements, all related to solar energy research being pursued by the Energy Research and Development Administration (ERDA). (NASA Release 77-49)

• JSC announced construction of a new lunar curatorial facility, a $2.02 million 4200m^2 addition to the old facility, Bldg. 31. The new facility, being built by Spaw Glass, Inc., of Houston, would provide clean secure storage space as well as processing, experiment, and simulation labs for NASA's 480kg of lunar samples. (JSC Release 77-20)

• NATO had not made a choice between the U.S.-built Boeing AWACS (airborne warning and control system) or the British Hawker Siddeley Nimrod as its airborne-radar plane, the *W Star* reported. Britain and France favored selection of the already operational Nimrod in retaliation for U.S. refusal to grant New York landing rights to the Anglo-French Concorde, and in order to preserve 7000 jobs on the Nimrod production line. Nimrod's performance on North Atlantic patrol might hamper sale of the AWACS to NATO, the *Star* noted. (*W Star*, Mar 14/77, A-11)

- Researchers at Goddard Space Flight Center using data from *Orbiting Solar Observatory 8* (*Oso 8*) determined that iron-line x-rays had come from hot gas between the galaxies in the Milky Way's local cluster, *AvWk* reported, meaning that the space between galaxies in the local cluster was not a perfect vacuum, but the amount of matter between local galaxies was probably about the same as that present within the galaxies. (*AvWk*, Mar 14/77, 11)

- NASA had planned to fly approximately 150 research missions on its Gates Learjet during 1977, *AvWk* reported. Eight different research groups, using the 30cm telescope on board, would conduct infrared observations to obtain upper-atmosphere and magnetodynamic data. (*AvWk*, Mar 14/77, 11)

- A NASA earth resources monitoring satellite, *Landsat 1*, had experienced a malfunction in band 1 of its multispectral scanner on Mar. 3, *AvWk* reported. Although the malfunction affected only band 1 (a green spectral band identifying sunlight reflected from green plants), NASA had shut down the entire scanner pending a safety investigation by controllers. *Landsat 1* had performed satisfactorily since its 1973 launch. (*AvWk*, Mar 14/77, 16)

- The USSR had increased its military use of space, *AvWk* reported. Killer-satellite systems tested during 1976 appeared operational. The Soviets in 1976 had launched 33 reconnaissance missions (the core of their military space program) and apparently had added to their military comsat system as well. (*AvWk*, Mar 14/77, 63)

March 15: DOD would request $2.77 billion for space-related activities in FY 1978, $478.5 million more than its FY 1977 appropriation, *Defense Daily* reported. The increase, resulting from procurement of the Navstar global positioning satellite system and increased DOD activity on the Space Shuttle, would include costs of ground security at JSC and KSC for defense-related Shuttle payloads. DOD had planned to use the Shuttle for critical manned military observations, and had agreed on a formula to reimburse NASA for launching military payloads. (*DD*, Mar 15/77, 57)

March 16: MSFC announced award of a $279 714 contract to Sperry Support Services, a division of Sperry Rand Corp., Huntsville, Ala., for design, construction, and delivery of equipment for the mated vertical ground-vibration test (MVGVT) of the Space Shuttle, scheduled for 1978. The test would confirm the theoretical model used in dynamic analyses of the Space Shuttle, external tank, and solid-fuel rocket boosters. (MSFC Release 77-39)

• Dr. Albert Rango of GSFC and Dr. K.I. Itten of the Univ. of Zurich announced that orbiting earth satellites could monitor winter snowpacks and their runoff as accurately as ground-based or airborne observations, NASA reported. Their findings appeared in the journal *Nordic Hydrology*. Comparing costs and relative accuracy of ground, airborne, and satellite measurements, the two scientists decided that the use of NASA's Landsat earth resources monitoring satellite and the Natl. Oceanic and Atmospheric Administration's environmental satellite *Noaa 2* would provide timely and accurate information at lower cost than the alternatives. (NASA Release 77-52)

March 17: A short-circuit in the power supply of *Viking 1*'s gas-chromatograph mass spectrometer (GSMS) had forced controllers at JPL to turn off the instrument Mar. 12, NASA reported. JPL received confirmation of the command from the Viking system Mar. 14. NASA had shut down the GCMS, which had analyzed the composition of Martian atmosphere and soil early in the mission, to avoid damage to other instruments such as the onboard computer and communications system. (NASA Release 77-51)

March 18: NASA reported that an F-8 aircraft using the digital fly-by-wire control system would simulate the Space Shuttle's flight-control system during a series of 3 flights at NASA's Dryden Flight Research Center. The flights, simulations of orbiter maneuvers after separation from the Boeing 747 Shuttle-carrier aircraft, would determine the thresholds of sensor failure and would test the programming of the Shuttle's flight-computer system. (NASA Release 77-53; DFRC *X-News*, Mar 25/77, 2)

• NASA announced a successful 60-sec firing of the Space Shuttle main engine on Mar. 12 at rated thrust conditions of 2 082 560 newtons. Employees of the Rocketdyne Division, Rockwell Intl. Corp., contractor for the SSME, conducted the test at NASA's Natl. Space Technology Laboratories. (NASA Release 77-54)

• NASA announced it had turned over the Saturn-rocket test stand at NSTL to Rockwell Intl. Space Div. for testing the Shuttle's main propulsion system. Modification of the 90m-high stand to accept Shuttle elements ended in late 1976. Static firing tests would begin in Dec. 1977. (NASA Release 77-55)

March 21: Candidates for the post of NASA administrator, according to the *W Post*, included Christopher C. Kraft, director of JSC; Al Kelly and Rocco Petrone, former NASA officials; and Frank Moss, former chairman of the Senate Space Committee and an early Carter supporter. (*W Post*, Mar 21/77, A3)

• ESA and Canada's Centre for Remote Sensing (CCRS) announced they had signed an agreement to cooperate in studying remote-sensing applications, developing microwave remote sensing, and improving ground stations for processing satellite data. (ESA release, Mar 21/77)

March 22: NASA reported that MSFC had awarded a 1yr $150 000 contract to the Illinois Institute of Technology Research Institute (IITRI) in Chicago to set up a team to find uses for aerospace technologies in manufacturing. The team would work with U.S. industries, industrial associations, and professional groups in areas related to machine tools, electronics assembly, and light fabrication and assembly. The team would be another of the groups created by NASA's technology-utilization office to expand the use of aerospace technology on earth. (NASA Release 77-57)

• TROPEX, an international tropical weather experiment sponsored by the Soviet Union, would use USSR satellites with research vessels from some of the 70 participating countries to gather data on atmospheric conditions, the FBIS reported. (FBIS, Moscow Tass in English, Mar 22/77)

March 23: NASA announced it had invited scientists to submit proposals for use and development of the unfunded Space Telescope [see Feb. 1]. Capable of holding five different instruments at the focal plane of its 2.4m mirror, the Space Telescope would expand the volume of observable space 350 times beyond that of the largest ground-based telescope, offering new information for astronomers and other scientists interested in fundamental questions of cosmology and astronomy. GSFC would manage operations of the telescope's missions and develop its instrumentation; MSFC would be responsible for overall management of the telescope system. (NASA Release 77-58)

March 24: The White House had ordered a review of NASA's 28 advisory committees with the idea of abolishing all those lacking a compelling need for their services, a balanced membership, or open conduct of business within the framework of the law and their mandate, NASA reported. The review should serve to reduce the number of committees except as required by statute and to ensure quality and utility in their recommendations. (NASA Release 77-59)

- ARC announced selection of 10 women volunteers aged 35 to 45 with nontechnical backgrounds as test subjects in an experiment to help set medical standards for future Space Shuttle passengers. The experiment, one of a series using varying ages and sexes, would require 9 days of total bedrest to simulate weightlessness, and occasional spins in a centrifuge to simulate Shuttle reentry. Results would show whether or not age and gender affected the ability to cope with spaceflight. (ARC Release 77-15; NASA Release 77-61)

- The eighth annual session at JSC of the Lunar Science Conference had discussed the possibility of mining asteroids for the raw materials of space industrialization, the *Chicago Tribune* reported. Dr. David Criswell, associate scientist at the Lunar Science Institute, said that two developments would make asteroid mining feasible: an engine capable of towing an asteroid, and a method of locating asteroids rich in needed resources. Materials from asteroids would be 1000 times cheaper than materials launched from earth into orbit. (*C Trib*, Mar 24/77, 4-11)

- The Soviet Union had successfully tested a submarine-based missile with a range twice that of its U.S. counterpart, the *Chicago Tribune* reported. Two SSN-8 missiles fired from a Soviet Delta-class submarine in the Barents Sea had traveled approximately 9120km to targets in the Pacific Ocean. Ships carrying the missiles reportedly could endanger U.S. targets without leaving their home-port areas. (*C Trib*, Mar 24/77, 1-2)

March 25: KSC announced award of a $25 970 contract to the research institute at the Georgia Institute of Technology in Atlanta to develop an automatic manatee-tracking system for the U.S. Fish and Wildlife Service. Attempts to track the manatee, an endangered aquatic mammal, had failed because of poor monitoring systems. Tracking the movements of about 100 manatees in the Banana and Indian Rivers and Mosquito Lagoon near KSC would provide scientists with data on their living habits. (KSC Release 77-77)

March 26: The U.S. should orbit solar-power stations and mine the moon for raw materials, science fiction writer Isaac Asimov told President Carter in an open letter printed in the *Chicago Tribune*. Citing the benefits of orbital stations capable of transmitting solar energy in microwaves to earth stations, Asimov stressed the need for international cooperation in using space and in preventing the extinction of humanity. (*C Trib*, Mar 26/77, 1-7)

March 28: Direct-readout equipment for *GOES*, the geostationary operational environmental satellite, installed in the USAF global weather center at Offutt AFB, Neb., to support Air Force meteorology requirements, would improve the center's forecasting and accuracy, *AvWk* reported. (*AvWk*, Mar 28/77, 11)

• Project Halo (high-altitude large optics), a space-based system proposed by DOD's Advanced Research Projects Agency to detect and track foreign missiles and aircraft, would require 6 Space Shuttle payloads to orbit its structures and equipment, *AvWk* reported. ARPA's Halo research had focused on systems analysis, infrared focal planes, optics, cryogenic cooling, power sources, data utilization, and phenomenology. (*AvWk*, Mar 28/77, 11)

• The USAF Rocket Propulsion Laboratory had successfully test-fired small solid-fuel rocket motors having nozzles molded into the end of the propellant grain instead of externally mounted, *AvWk* reported. Use of integral nozzles would reduce costs of manufacture and would allow more propellant per rocket. (*AvWk*, Mar 28/77, 11)

• Japan's Natl. Space Development Agency planned to launch 2 satellites in 1978 to replace missions that failed during 1976, *AvWk* reported. Scheduled for Feb. 1978 was an ionospheric-research mission to replace an identical craft that malfunctioned in orbit in March 1976. An x-ray research satellite similar to a mission that failed during launch Feb. 4, 1976, might lift off in Aug. or Sept. 1978. (*AvWk*, Mar 28/77, 26)

March 29: NASA announced it had ended an aerial survey of the Pacific Ocean along the west coast of the U.S. that was an operational prelude to the use of SeaSat-A, the oceanographic satellite scheduled for launch in May 1978. The survey had used high-flying NASA aircraft to test four sensors from SeaSat-A: a synthetic aperture imaging radar, a wind-field scatterometer, a scanning multifrequency microwave radiometer, and a radar altimeter. The mission was a rehearsal in research cooperation that scientists would need when SeaSat became operational. (NASA Release 77-62)

• The future of Japan's space program was not a sure thing, the *Wall Street Journal* said. Those who had claimed that a program of launching satellites for resource-rich, technology-poor nations would increase Japan's technological growth and ensure a supply of raw materials had met with complaints that the program wasted tax money and that Japan should have relied on the U.S. and NASA to orbit the payloads. Japan's space agency would have to convince official and public skeptics of the program's value, the paper said. (*WSJ*, Mar 29/77, 44)

• A yr-long (Dec. 1978–Nov. 1979) international program of global atmospheric research using Soviet and U.S. satellites would attempt to construct a model of the atmosphere, Tass reported. (FBIS, Moscow Tass in English, Mar 29/77)

• Soviet satellites might help predict the cotton harvest in South Transcaucasia, Tass reported, through enhanced photographs of the regional cotton crop. (FBIS, Moscow Tass in English, Mar 29/77)

March 30: NASA announced that Project FIRES (firefighters integrated response equipment system), a joint program of NASA's Technology Utilization Office and the Natl. Fire Prevention and Control Administration, had begun with NASA's award of a $478 375 contract to Grumman Aerospace Corp., Bethpage, N.Y., for new standards in firefighting clothing and for development and fabrication of a complete ensemble. The project would use space technologies to achieve greater safety and maneuverability of firefighting equipment. A Users Requirements Committee of firefighters and safety officials would evaluate Grumman's products and fire-test them in 10 cities under actual firefighting conditions before commercial production began. (NASA Release 77-64)

• MSFC announced that NASA had invited researchers to propose materials-processing experiments to be performed on the Space Transportation System (STS), as the Shuttle missions would be called. The experiments, basic or applied research projects in materials science, would exploit the weightless and ultra-high vacuum conditions of space to demonstrate their applications in materials science and technology. (MSFC Release 77-52)

• The Pentagon had been working on an interceptor spacecraft capable of knocking Soviet satellites out of orbit, DOD officials told Congress. The *Washington Post* reported that the program, begun under President Ford, had aimed at producing weapons to protect U.S. space systems from attacks by foreign satellites. (*W Post*, Mar 30/77, A-8)

March 31: Dr. James L. Elliott of Cornell Univ. announced he had found 5 rings around the planet Uranus. Observations above the Indian Ocean with a 91cm telescope on NASA's Kuiper airborne observatory had revealed the rings during Uranus's occultation of a bright star (SAO 158687). Dr. Elliott had concluded earlier [see Mar. 3] that Uranus was surrounded by 100 moons; during the occultation, the same objects appeared as rings, not moons. Uranus and Saturn were the only two planets in the solar system found so far to have ring systems. Dr. Elliott's original mission was to check Uranus for atmosphere and diameter; discovery of the rings was unexpected. (*NYT*, Mar 31/77, 63)

• Developing future telecommunications systems would pose few technical problems, *Nature* reported: the real bottlenecks would be finding the financial resources to build, and estimating the social implications of, such systems. A British Royal Society conference on "Telecommunications in the 1980s and After" had discussed the possible social effects of highly developed telecommunications systems, especially the ability of a system to foster integration in different populations or to increase isolation of individuals within their local groups. (*Nature*, Mar 31/77, 409)

During March: A House subcommittee on aviation and transportation R&D had called on NASA to expand its program on aircraft energy efficiency (fuel conservation), said *Astronautics and Aeronautics* magazine. The subcommittee report with 16 specific recommendations, signed by chairman Rep. Dale Milford (D-Tex), said the U.S. should establish a comprehensive national policy on civil aviation R&D. (*A&A*, Mar 77, 7)

April

April 1: NASA reported it had begun working on its plan to make Ames Research Center the lead center for helicopter research. Approved last summer, the plan resulted from NASA's assessment of its research to see where realignment would improve effectiveness and reduce costs. As lead center, ARC would do research on small- and large-scale helicopter hardware using its aeronautical facilities such as the 40 × 80ft wind tunnel and flight-simulation equipment, and flight-test items like the tilt-rotor and rotor-systems research aircraft.

A helicopter program office at NASA Hq would direct the overall helicopter program and coordinate research between ARC and the Langley and Lewis Research Centers. LaRC, working mainly on helicopter structures, would continue its studies of acoustics, airfoils, aeroelasticity, and avionics. LeRC would continue to emphasize propulsion, its first task being a proposed new program in helicopter transmission technology leading later to work on helicopter engines.

NASA estimated that over the next 3yr the change would result in adding 72 positions to the ARC staff; LaRC helicopter work would phase down during that period but would continue to employ 72. The expected growth in long-haul aircraft technology at LaRC would add up to little long-term impact on that center's manpower or the local economy. (NASA Release 77-8; ARC Release 77-19)

• Dryden Flight Research Center announced it had commenced studies on turbulence caused by insects sticking to aircraft wings. It would use a small jet transport flown at low altitudes to collect insects on the wings, in turbulence tests at higher altitudes and greater speeds. The Johnson and Kennedy Space Centers would assist the studies aimed at developing high-speed fuel-efficient aircraft, by taking advantage of their differing insect populations. (DFRC Release 13-77; KSC Release 93-77)

• KSC reported that the number of visitors in the first 3mo of 1977 was 283 734; the number in March, 114 845, highest so far this yr was 2.8% below the March 1976 level. The 3mo total was 9.3% less than the first 3mo of 1976. The visitor center had displays, exhibits, space science lectures, and space movies, plus a new hall of history opened in 1976. Visitors could see the vehicle assembly building for Space Shuttle preparations and the Apollo Saturn on display nearby, drive around Pad

A where Shuttle operations were in progress, and visit the Cape Canaveral Air Force station. (KSC Release 87-77)

- LaRC announced that Dr. Charles Thiel, director of the Natl. Science Foundation's advanced environmental research division, would speak at the LaRC colloquium Apr. 11 on earthquakes. Dr. Thiel had been in charge of U.S. research on earthquake phenomena for the past 5yr and had traveled the world to view effects of quakes, analyze prediction techniques, and evaluate construction practices in quake-prone regions. His illustrated talk would cover causes and probabilities of quakes, with modern methods of forecasting and monitoring them. (LaRC Release 77-9)

- NASA announced appointment of Dr. John R. Carruthers as first director of a new division in the Hq Office of Applications that would manage materials processing in space. Dr. Carruthers, formerly of Bell Labs, had headed crystal-growth and glass research in developing optical fibers for lightwave communications. (NASA Release 77-67)

- Reform of the aviation industry was in order, Transportation Dept. Secy. Brock Adams told the aviation subcommittee of the Senate Committee on Commerce, Science, and Transportation. Regulatory changes should include freer entry into the system, competition in routes and rates, and provisions for exiting the system. Any changes should protect service to smaller communities, he said, and should address the problems of how market shifts would affect airline users and of financial damage to airport owners and airline investors. Secy. Adams noted that the president should retain his authority over international aviation, since the Constitution required him to conduct foreign policy there as well as in other fields. (DOT Release 44-77)

April 4: The rapid approach of the Martian winter, whose temperatures would fall below $-123°C$, had halted science experiments using the *Viking 2* lander, NASA announced. Problems with the lander's sample arm had forced controllers at the Jet Propulsion Laboratory to cancel a final biology experiment. The meteorology station and the seismic instrument would use power to gather data during the winter; all other power would go toward maintaining survivable heat levels inside the lander. (NASA Release 77-69)

- *Aviation Week* reported that President Carter had nominated Langhorne M. Bond, state secretary of transportation for Illinois, to head the Federal Aviation Administration, and Quentin Taylor, FAA director in New England, as deputy administrator. (*AvWk*, Apr 4/77, 23)

- The Apollo program was a tough act to follow, Space Shuttle pilot Lt. Col. Charles Fullerton said in a *Today* wire service report. Acknowledging the declining public interest in space, Fullerton noted that, after Apollo, "Everything else has to be anticlimactic." (*Today*, Apr 4/77, 1A)

April 5: NASA, which had issued a press kit on plans for an April 15 launch of the high-energy astronomy observatory HEAO-A, said it had postponed the launch because of a malfunction in one of the gyroscopes controlling altitude, essential to the mission. The agency had not set a new date for the launch; it had returned the defective gyroscope to the manufacturer, Bendix Corp., in Teterboro, N.J. (NASA Releases 77-45, 77-71; *Marshall Star*, Apr 6/77, 1)

- NASA announced it would join EPA and the Univ. of Md. in sponsoring a conference on the use of remote sensing to solve Chesapeake Bay ecological problems. The 80 participants would discuss use of satellites (particularly NASA's earth resources monitoring satellite Landsat) to obtain information on land use, resources exploitation, and pollution. (NASA Release 77-70)

- Marshall Space Flight Center announced award of a $918 523 contract to T.H. Taylor Construction Co., Montgomery, Ala., for construction of permanent and semipermanent steel working platforms on the modified dynamic test stand. The stand would hold an entire Space Shuttle assembly during mated vertical ground-vibration tests scheduled for 1978. (MSFC Release 77-54)

- NASA announced that Miles Ross, KSC deputy director since 1971, had resigned to become European regional manager in Brussels for his previous employer, TRW Systems Intl. (NASA Release 77-76; KSC Release 88-77)

April 6: NASA announced it would join the EPA in field-testing a new detector of fecal coliform bacteria at the New York Bight off Craven Point, N.J. The electrochemical monitor, developed by Dr. Judd Wilkins of LaRC, was a buoy with 10 coliform-sampling units and a transmitter to relay data to a monitoring center in a NASA truck on shore about 3mi away. Surveillance of fecal coliform concentrations would measure the environmental impact of public sanitation. (NASA Release 77-73; LaRC Release 77-11)

- NASA reported that Goddard Space Flight Center engineers had tracked a sailboat through the Bermuda Triangle to test the use of satellites for small craft search-and-rescue missions. Signals from a small transmitter on the vessel, relayed by meteorological satellite *Nimbus 6* and applications technology satellite *Ats 1*, allowed GSFC controllers to monitor the ship's position. The experiment also successfully tested a push-button distress-alarm system designed by retired NASA engineer James L. Baker. (NASA Release 77-72)

- MSFC announced it had awarded a $276 597 contract to Bryson Construction Co., Decatur, Ala., for modification of the Redstone Arsenal airfield to accept the Space Shuttle and its Boeing 747 carrier aircraft. The contractor would install a concrete pad 75m long and 55m wide for a derrick to unload the orbiter from the 747. (MSFC Release 77-56)

- LaRC reported that its visitor center was showing a NASA film, "Universe," one of 5 documentaries recently nominated for a motion picture Academy Award. The 28min Graphic Films Corp. movie, showing an imaginary trip through space, offered a new look at the solar system, the release said. (LaRC Release 77-10)

- The Carter administration would not request government funds to develop a supersonic airliner, OMB Director Bert Lance assured Sen. William Proxmire (D-Wis). The *Chicago Tribune* story noted that Proxmire, who had worked to defeat the U.S. supersonic transport program in 1972, was concerned because the House Committee on Science had asked NASA to do a feasibility study on advanced supersonic aircraft. (*C Trib*, Apr 6/77, 1-5)

April 7: MSFC announced it had begun test assembly of large-scale space structures in its neutral buoyancy facility. The NBF, a 12m-deep water tank simulating the weightless environment of space, was in use for the first time since the Skylab program. (MSFC Release 77-57)

- The liquid-oxygen tank of the Space Shuttle's external tank had undergone successful hydrostatic tests at NASA's Michoud Assembly Facility near New Orleans, MSFC reported. Martin Marietta Aerospace, prime ET contractor, had scheduled additional tests to simulate flight pressures. (MSFC Release 77-58)

- NASA officials, assuming a constant purchasing power for the agency over the next 10 to 15yr, had made plans for some 560 Shuttle flights between 1980 and 1991, averaging 60 flights per yr between 1985 and 1991, *Nature* reported. Assuming a maximum payload of about 29 500kg,

Shuttle users (including NASA, DOD, other government agencies, private industry, and foreign organizations) would pay fees from $3000 for a simple automatic payload to $20 million for the entire cargo bay. NASA had claimed a 40% savings in launch costs for the Shuttle compared to expendable rockets; a higher discount for U.S. users would increase its cost-effectiveness. *Nature* said early Shuttle critics, like space scientists who viewed it as a competitor for limited NASA resources, were now supporting the Shuttle as a research tool. (*Nature*, Apr 7/77, 489)

• Spacelab, ESA's manned reusable space laboratory scheduled to fly on the Space Shuttle, would perform experiments in life, physical, and materials science, *Nature* reported. Although Spacelab's multinational management and unique financing scheme (ESA members would pay for the laboratory on a voluntary basis) could limit its scientific potential, participants would receive a proportionate share of construction contracts and experiment facilities. Management problems included the question of apportioning payment of NASA's $10 million launch fee. (*Nature*, April 7/77, 491)

• ComSatCorp, in a notice of its upcoming election of directors, said the nominees for a first term were Howard J. Morgens, director and chairman emeritus of Procter & Gamble; and Charles J. Pilliod, Jr., board chairman and chief executive officer of Goodyear Tire & Rubber Co. The other 10 nominees were current board members. (ComSat Release 77-14)

April 8: President Carter announced that the U.S. would "defer indefinitely" commercial reprocessing and recycling of plutonium from nuclear power plants, and would stop developing a commercial breeder-reactor (a nuclear power facility producing more fuel than it consumed), the *Chicago Tribune* reported. Critics had contended the reactor would increase proliferation of nuclear weapons. The administration would significantly increase funding for solar energy research and synthetic production of oil and gas from coal, the paper said. (*C Trib*, Apr 8/77, 1-2)

• INTELSAT, the Intl. Telecommunications Satellite Organization, reported it had appointed three deputy directors general to serve under Director General Santiago Astrain: Reginald Westlake of the U.K. for finance; Andrea Caruso of Italy for administration; and H. William Wood, formerly with the U.S. Navy and Defense Communications Agency and systems management director for ComSatCorp, to supervise operations and development. Transfer of functions from ComSat, current contractor for management services, to the INTELSAT executive would occur over the next 2yr. (INTELSAT Release 77-5-I)

April 9: Kitt Peak astronomers had suggested designs for a telescope 5 times larger than the 4.5m telescope at Mt. Palomar, Calif., to complement the Very Large Array radiotelescopes in New Mexico and the planned Space Telescope, the *NY Times* reported. An instrument using multiple-mirror technology (several mirrors connected by laser, aligned by computer to within an eighth of the wavelength of the observed light) would have the light-gathering potential of a huge single mirror without the high cost of manufacture or maintenance. The resolution possible with such telescopes would permit observation of weather patterns on other planets. (*NYT,* Apr 9/77, 22)

• Operators of the Anglo-French supersonic airliner Concorde had claimed that new flight procedures would lower Concorde noise levels to those of subsonic aircraft, and should get it approved to land at Kennedy Intl. airport.

A 40-page report to the NY Port Authority, obtained by the *NY Times,* said that Concorde takeoff noise (though it would affect a larger area than noise from other aircraft) was actually comparable to noise from the long-range Boeing 707. The Concorde had also met airport noise standards. Residents of the Five Towns area of Long Island, who had opposed Concorde overflights, would be subject to fewer than originally planned, the report added. Port Authority commissioners, scheduled to meet during the upcoming week, had refused to make a decision without spending more time on the report. (*NYT,* Apr 9/77, 22)

April 11: *AvWk* reported that French space technicians would be allowed on site at the USSR's Kapustin Yar launch facility for the first time when the Soviets launched Signe 3 in June. Swedish technicians had been present there last July 27 when *Intercosmos 16* carried a Swedish solar experiment into space. Third French satellite scheduled for launch by the USSR, Signe 3 was designed by France's Natl. Center for Space Studies (CNES) for astronomical and atmospheric research; it had arrived at the USSR Cosmic Research Institute recently for launch preparation. (*AvWk,* Apr 11/77, 9)

April 12: GSFC controllers and Navy meteorologists at the U.S. Navy Fleet Weather Facility in Suitland, Md., had used since 1971 a microwave radiometer on weather satellite *Nimbus 5* to track an Antarctic iceberg the size of Rhode Island, NASA reported. Satellite tracking had allowed Navy forecasters to minimize ice damage to ships in polar ice fields. (NASA Release 77-68)

April 13: NASA announced that principal investigators for Spacelab experiments would attend briefings at MSFC Apr. 13 to 15 on Spacelab mission management as an Investigative Working Group (IWG). Dr. C.R. Chappell, mission scientist at the MSFC space sciences laboratory, and Dr. Bernt Feuerbacher, ESA project scientist, as IWG chairman and vice chairman, would channel recommendations on selection of payload specialists directly to management. (NASA Release 77-77; MSFC Release 77-61)

• The Dept. of Commerce reported that a Boeing 747 in regular passenger service for Pan American World Airways had been doing double duty as a flying laboratory for 2 government agencies, NASA and NOAA, carrying an electronic package collecting data from the plane's instruments to show that ordinary planes could get weather information in flight and relay it to the ground. This yr, five more packages would be put on planes of international carriers.

The prototype package, weighing about 18kg and the size of an electric typewriter, used about 200w power and transmitted at 401.7MHz. LeRC had built the package under technical management from GSFC to tap into the inertial-navigation systems of widebodied jets like the 747 and DC-10 series, recording data on air temperature, wind direction and speed, and aircraft location and altitude. The package converted the data to a format transmitted hourly to the ground through NOAA's *Goes 1*.

The idea was to enable planes flying over areas where such information was sparse or unobtainable to collect and transmit the data as a help to aircraft operations as well as to weather prediction and analysis. NOAA had begun negotiations with foreign weather services and international airlines to put the packages on their aircraft, especially those flying over equatorial regions. (NOAA Release 77-91)

• The Natl. Air and Space Museum in Washington, D.C., counted its 7 millionth visitor, Mrs. Elizabeth Weber of Queens, N.Y. However, she and many other visitors exemplified the fading public interest in space: none of them recognized the date as the 16th anniversary of Yuri Gagarin's historic feat, man's first spaceflight.

A *NY Times* article contrasted the festive feeling at the museum (the capital's largest tourist draw) and the almost melancholy mood at NASA Hq across the street. NASA employees had cause for despair, said the article, with a space program rarely in the public eye and funding that had delayed long planned projects. Even President Carter, during a cursory 30-minute inspection of the museum, omitted mention of the space program or the space agency. (*NYT,* Apr 13/77, D-10)

April 14: NASA reported that it and the European Space Agency (ESA) had invited scientists to propose experiments for a joint out-of-ecliptic (OOE) mission in 1983 that would study the interstellar/interplanetary medium as a function of solar latitude. Although the mission still lacked Congress's approval, the agency wanted proposals for investigations that would allow definition of the undertaking and give it a prompt start when approved.

As planned, the Shuttle would launch 2 spacecraft (one supplied by NASA, one by ESA) on similar paths from earth to Jupiter, deflected by that planet's gravity into trajectories at right angles to the plane of earth's orbit. One craft would enter over the sun's northern hemisphere, the other over the southern; both would reach solar latitudes near the sun's pole. So far studies of the sun had covered only a narrow region on the ecliptic plane; the OOE, or solar polar, mission would look at the sun over a full range of heliographic latitudes and enable more accurate assessment of the total solar environment. The mission would last about 5yr. (NASA Release 77-79)

• NASA announced award of $350 000 design study contracts both to McDonnell Douglas Corp., St. Louis, Mo., and to a team representing Hughes Aircraft Co., El Segundo, Calif., and General Electric Co. in Philadelphia. The companies would develop for NASA's Jupiter-orbiter mission a probe able to measure the Jovian atmosphere during descent. The mission would reach Jupiter in 1984 if the project were approved for FY 1978. ARC would manage development of the probe under JPL's general direction. (NASA Release 77-78; ARC Release 77-20)

• KSC announced it had awarded a $39 812 contract to Ky. State Univ. for continuing research on effects of prolonged exposure of experimental animals to variations in the atmospheric level of oxygen. Previous studies in the field had been on short-term exposure to 100% oxygen, but little was known about prolonged exposure to oxygen at low to moderate concentrations.

Ky. State's first grant for this work in March 1974 had exposed vinegar flies to higher levels of oxygen than normal and drastically shortened the life span, whereas exposure to lower than normal levels had greatly reduced reproductive capacity. This work had also suggested that certain individuals might prove better adapted genetically for survival in abnormal oxygen environments. The findings would be useful in planning atmospheres for long-term manned space missions aboard the Space Shuttle or on space colonies. (KSC Release 91-77)

• KSC announced award of a $95 025 000 contract to the Space Division, Rockwell Intl. Corp., Downey, Calif., for activation of the Space Shuttle facilities and systems, and for Orbiter support during checkout, launch, and postflight operations, in the first six Space Shuttle missions. Rockwell would also support postflight operations at the DFRC landing site for the first four missions. (KSC Release 92-77)

• MSFC announced it had requested price quotations on designing a space spider (a device capable of building large-scale structures). The spider, containing prefabricated materials, would attach itself to the core of a planned structure and spin a larger structure, of any diameter, around the core. MSFC engineers had sought estimates of costs and data on the best size of spider to fit inside the Space Shuttle. (MSFC Release 77-62)

• Wallops Flight Center announced it would welcome visitors to the rocket launching facilities on Wallops Island Apr. 23-24 as part of a special observance by federal activities in the area. Assateague Natl. Seashore Park, Chincoteague Coast Guard Base, Chincoteague Natl. Wildlife Refuge, the Natl. Oceanic and Atmospheric Administration, and Wallops Flight Center had all planned special exhibits or tours. Visitors to Wallops Island could drive their own cars to see launch pads, rocket storage structures, long-range tracking radars, weather towers, blockhouses, rocket assembly shops, actual launch vehicles, and related range facilities. Cameras would be welcome. The main NASA base would not be included in the visit. (WFC Release 77-2)

• *Nature* magazine reported further on the recognition of ring systems around the planet Uranus announced earlier by Dr. James L. Elliot of Cornell Univ. [see March 31]. The occultation (shutting off from view) by Uranus of a bright object (star SAO 158687 in Libra) had been predicted by the Royal Observatory at Greenwich as visible only from an area on earth extending from equatorial Africa across the Indian Ocean to western Australia. No useful observations of a Uranian occultation had been made before; astronomers considered such observations important because they could furnish data on the size of the planet or stellar object, on limb darkening, and on extent and composition of any atmosphere, as well as an accurate measurement of the object's position, valuable in calculations of celestial mechanics.

Heavy rain at Johannesburg prevented observations there on March 10, but astronomers on the Kuiper observatory plane flying over the southern Indian Ocean east of Kerguelen Is. were able to see an occultation lasting about 25min with secondary occultations lasting 8 to 9min before and after. Observers at Madras and Perth also viewed the event and decided independently that the extended occultation resulted from satellites forming a belt about 48 000km from the center of Uranus. The belt was apparently circular, lying in the plane of the planet's equator, and about 12 000km wide. *Nature* reported that Dr. Elliot was continuing detailed study of the event, and that direct observation of the belts might be possible with a sufficiently large telescope. (*Nature,* April 14/77, 587)

April 15: JSC announced award of a $5 189 500 supplement to an existing contract with Rockwell Intl.'s Space Division, Downey, Calif., covering 8 engineering changes in the Space Shuttle orbiter to accommodate the European-built Spacelab in the Shuttle's cargo bay. (JSC Release 77-26)

• Industrial exploration had made sweeping changes in today's society, ranging from thumbnail-size computers to Space Shuttles operating with the regularity of commercial airlines, Rockwell Intl. chairman W.F. Rockwell, Jr., told the annual Explorers Club dinner in New York City. A microprocessor small enough to balance on the end of a thumb could compute 10 times faster than a 30-ton device introduced in 1946, and would cost $15 to $20, compared to the millions of dollars that the old computer cost, he said. Forecasting the potential of the Shuttle, being developed largely by Rockwell for NASA, he reviewed possible payloads including solar power satellites and navsats; he ended by comparing explorers setting out in search of new discoveries and new knowledge with the explorers in industry searching to improve products and service. (Rockwell Release R-18)

• ERDA announced it had selected Shenandoah, Ga., as the site for building a second solar total-energy experimental plant. The factory, manufacturing knitwear, would be the first in the U.S. to get all its electric power, steam, heating, and cooling from solar energy. A West German firm, Wilhelm Bleyle K.G., would operate the factory and the Georgia Power Co. would operate the solar-energy plant to provide up to 200kw of electricity and 1200kw of heat energy. Construction should begin in 1979 and be completed in 1981.

Robert Fri, acting ERDA administrator, said the knitwear factory would demonstrate the production of hot water and steam as byproducts of solar-power generation, making year-round use of the heat that would otherwise be wasted during the mild weather of spring and fall in that area. ERDA had planned another experimental total solar-energy plant, scheduled to begin construction in 1978 to serve Ft. Hood army base in Killeen, Tex. (ERDA Release 77-70)

April 18: The *Viking 1* lander's gas-exchange experiment had developed a leak during the final analysis of soil from a deep trench at the Chryse Planitia landing site, NASA reported. Failure of the gas-exchange package, designed to detect respiration in soil samples, would not affect the lander's other biology experiments. (NASA Release 77-82)

• MSFC reported it had purchased 5 large solar collectors to heat an exhibit at the Ala. Space and Rocket Center in Huntsville. The collectors would provide 70% of the heat for the Lunar Odyssey, a space voyage simulation in a modified Saturn launch vehicle. (MSFC Release 77-65)

• MSFC announced it would sponsor a 2-day symposium May 11-12 on engineering and productivity advances from space technology. The center's technology utilization office and the Huntsville chapter of the Ala. Society of Professional Engineers would cosponsor discussions by speakers from industry, universities, NASA, and other government agencies on industrial uses of space technology in energy management, materials processing, and electronics. (MSFC Release 77-66)

• A Space Shuttle orbital-maneuvering system engine had completed acoustic environment testing at JSC, and a second engine had fired 1000 times in life-cycle testing at the White Sands, N.M., test facility, *Aviation Week* reported. The latter achievement demonstrated the engine's ability to perform 100 missions, a project requirement. (*AvWk*, Apr 18/77, 13)

• Rockwell Intl.'s Space Division and Autonetics group had won an Advanced Research Projects Agency (ARPA) design competition for a Teal Ruby sensor to detect and track aircraft targets from space, *AvWk* reported. The DOD sensor, using up to a quarter million mosaic infrared detectors with charge-coupled processors, would fly on the Space Shuttle in 1980. A Lockheed-Grumman team had also competed. (*AvWk*, Apr 18/77, 13)

• During confirmation hearings, Dr. Frank Press, President Carter's nominee as director of the Office of Science and Technology Policy, had stated that NASA should broaden the scope of its technology-transfer efforts and its planned use of the Space Shuttle, *AvWk* reported. The administration should continue funding planetary exploration and space science (areas adaptable to the Shuttle's unique capabilities), but should also make NASA's technological knowledge available to industry. Press, former chairman of MIT's Department of Earth and Planetary Sciences, had specialized in earthquake research and detection of underground nuclear tests. (*AvWk*, Apr 18/77, 21)

• Israel Aircraft Industries had modified an Arava twin-turboprop transport plane using winglets designed by Richard T. Whitcomb of

LaRC, *AvWk* reported. Winglets would reduce drag during takeoff and enhance short takeoff and landing capabilities. (*AvWk*, Apr 18/77, 21)

• NASA had decided to plan construction of an orbiting space station from existing hardware because the Carter and Ford administrations had refused funds for major space station development, *AvWk* reported. Four station concepts were under examination: increasing the duration of Space Shuttle missions, using Skylab Apollo telescope-mount components to supply additional electricity for any extended mission; making a Space Shuttle external fuel tank into a Skylab-type station [see Mar. 7]; launching unmanned platforms to be occupied in the future; and modifying the ESA-NASA Spacelab to fly as an independent orbiting laboratory, without Space Shuttle. (*AvWk*, Apr 18/77, 42)

• Comsat General, wholly owned subsidiary of the Communications Satellite Corp., announced that former FAA administrator Dr. John L. McLucas had been elected president and a member of the board of directors. ComSatCorp. president Dr. Joseph V. Charyk said the subsidiary would benefit from Dr. McLucas's "long and distinguished career as administrator, engineer, and scientist." Holding a Ph.D in physics and electrical engineering from Penn. State Univ., Dr. McLucas had been president of the Mitre Corp. before serving as assistant secretary general for scientific affairs of NATO, undersecretary and secretary of the Air Force, and FAA administrator before April 1977.

Programs engaged in by Comsat General included Marisat (maritime satellite communications), Comstar (domestic satcoms), and Satellite Business Systems, in which it shared with IBM and Aetna Casualty plans for a U.S. domestic satcom system. (CGC Release 77-2)

April 19: The gyroscope malfunction [see Apr. 5] that delayed the launch of HEAO-A planned for Apr. 15 would postpone the launch until mid-June or later, NASA reported. HEAO-A, which could observe the electromagnetic spectrum at altitudes inaccessible to ground-based observers, would seek out celestial x-ray sources. (NASA Release 77-83)

• ASSESS 2 (airborne science/Spacelab experiment-system simulation), scheduled on May 16–26 at ARC, would simulate a Spacelab mission using NASA's flying laboratory Galileo 2 (a converted Convair–990 jet transport) to acquire data on earth resources, atmospheric pollution, and infrared astronomy. A mobile van would isolate the crew (mission specialist Karl Henize of JSC and four payload specialists: Robert T. Menzies and David S. Billiu of JPL, and Claude Nicollier and Michael Taylor of ESA) when they were not flying. ASSESS 2 would also serve to rehearse Spacelab's joint NASA-ESA management. (NASA Release 77-80; ARC Release 77-21; MSFC Release 77-67)

April 20: NASA launched *Geos 1,* planned as the first "purely scientific" geosynchronous satellite, for ESA from Eastern Test Range on a Delta rocket at 1015 GMT, into an anomalous transfer orbit (11,710km perigee, 241.5km apogee, 26° inclination) because of premature separation of the Delta second and third stages. Failure to achieve spin stabilization would prevent *Geos 1* from attaining the desired geosynchronous orbit. ESA planners said they would meet April 26 with the mission's prime investigators in an attempt to salvage the $12 million mission. *Geos 1,* chief tool of the Intl. Magnetospheric Project, carried instruments to gather data on effects of the solar wind on earth's magnetic field. (NASA Release 77-66; MOR M-492-302-77-01 [preflight] Apr 13/77, [postflight] Aug 10/77; ESA Release Apr 22/77; *W Post,* Apr 21/77, B10)

• Rockwell Intl. Corp. reported higher earnings and sales for the second quarter of FY 1977 in spite of a 3wk strike in its automotive operations and an energy crisis last winter. Earnings of $36.2 million ($1.05 a share) increased 13% over same quarter earnings in 1976, $32.1 million ($0.95 a share). Robert Anderson, president of Rockwell, said the earnings resulted partly from improvements in the aerospace operations area, which had done more business in military aircraft. (Rockwell Release R-19)

April 21: The USAF reported it had begun joint research with LeRC on use of synthetic fuels in military and commercial jet aircraft, which currently accounted for about 8% of the nation's petroleum consumption. The proposed 10yr study, costing $8 million, would look into the refining required to use as fuel liquids derived from coal or oil shale.

The USAF energy conservation effort arose from increasing fuel costs, which more than tripled for military jets alone in the last 18mo. NASA and the USAF had contracted with Exxon Corp. to define chemical and physical properties of coal and oil shale synthetic crudes compared to fuels derived from petroleum. Usability of the crudes would depend on their adaptability to existing engine and component requirements and on the environmental impact of such fuels. Processing (mostly hydrogenation) of the synfuels would be costly but could be affected by more specific data on engine and fuel system needs. Saving only 1¢ per gallon would save the USAF alone more than $50 million a year, the report said. (USAF Release OIP 070.75)

• Columnist Jack Anderson reported in *Today* that a Senate subcommittee had asked the Pentagon for military equipment, including planes, radar, sensors, and satellites, to be used in the "battle against drug abuse."

Senators Sam Nunn (D-Ga.) and Charles Percy (R-Ill.) had sent a private letter to Defense Secy. Harold Brown saying that drug abuse was costing the U.S. $17 billion a yr and that more than 5000 Americans were dying each yr from drug abuse. DOD alone had the vehicles needed to pursue and overtake the drug traffickers, the senators said, calling the fight against drugs a "war." "When fighting a war," they added, "you use the best resources available and make every effort to win. This is not being done." A spokesman for the Pentagon said an answer was being prepared. (*Today*, Apr 21/77, 16A)

April 22: JSC reported that its engineers had designed, built, and installed an emergency communications console at the medical center in Odessa, Texas, control center for the 17-county Permian Basin emergency medical system. The prototype console would allow physicians and nurses to consult with paramedics in the field, receive incoming electrocardiograms, set up radio-to-telephone patches, page hospital staff members, and automatically dial special care centers and other facilities in the area. After field tests and modification, JSC engineers would make console plans available to interested communities, according to a NASA announcement. (JSC Release 77-27; NASA Release 77-86)

• MSFC announced receipt from Bendix Corp., Teterboro, N.J., of the first two integrated electronic assemblies for the Space Shuttle's solid fuel rocket boosters. Each SRB would contain 2 electronic assemblies to control communications and internal power distribution. (NASA Release 77-68)

April 25: NASA reported it had selected Trudy Tiedemann, former public information specialist at DFRC, as its first female commentator. She would serve as announcer during the Shuttle orbiter flight-test program. (NASA Release 77-84)

• MSFC announced award of a 1-yr $450 540 contract to Sperry Univac, Washington, D.C., for on-site maintenance of the center's Univac 1108 computer system. (MSFC Release 77-69)

April 25-28: Controllers at ESA's European space center at Darmstadt, West Germany, had fired *Geos 1*'s apogee kick motor at 0738 GMT to boost the satellite into a new 12hr elliptic orbit (38 498km apogee, 2131km perigee, 25.85° orbital inclination) that would maximize return of scientific data. Although ESA officials had hoped to delay the orbit change until after a meeting of Geos experimenters scheduled for April 26, the spacecraft's anomalous transfer orbit, caused by a malfunction during launch Apr. 20, had endangered its solar cell power supply. (ESA Release Apr 25/77)

ESA officials and Geos satellite experimenters during a meeting April 26 at the European space operations center in Darmstadt agreed on a sequence for experiment activation and boom deployment. After increasing satellite "visibility" at the Odenwald ground station near ESOC by moving the *Geos 1* apogee from 90°E to 35°E, controllers planned to extend the short radial booms completely and the long booms to 10cm. If this ploy worked, the 20m-long radial booms would extend completely Apr. 30 to help establish the satellite's moment-of-inertia ratio and dynamic stability. (ESA Release Apr 27/77)

Scientists of *Geos 1* believed its new orbit, whose apogee was high as that of the intended orbit, could salvage most of its scientific objectives, *Nature* reported. The new path would put *Geos 1* where ESA had planned to orbit its spare "qualifications" geosynchronous satellite in 1979; launch costs had made use of the spare unlikely, the report said. (*Nature*, Apr 28/77, 767)

April 26: Central Florida officials who had counted on Space Shuttle employment to stimulate the local economy hoped it would differ from the boom-and-bust cycle of the Apollo program, the *NY Times* reported. The Apollo program had brought about 26 000 employees to KSC, but the economy declined with layoffs when the missions ended. NASA had estimated an 11 000-employee strength at KSC for the first Shuttle mission; the Shuttle program should continue for at least 5yr, agency officials said. (*NYT,* Apr 26/77, 13)

April 27: A series of Stanford Univ. experiments had apparently verified the existence of quarks, hypothetical fundamental pieces of matter possessing fractional electric charges, the *NY Times* reported. Dr. William Fairbank, professor of physics at Stanford, described the findings—first evidence of quarks since their theoretical exposition in 1964—at the Am. Physical Society annual meeting in Washington, D.C. If matter consisted of protons, electrons, and neutrons (carrying charges of $+1$, -1, and 0, respectively), quarks—the building blocks of these particles—would have fractional charges of $+1/3$, $-1/3$, $+2/3$, and $-2/3$. Fairbank emphasized that the results were not final, and that the university had scheduled follow-up tests. (*NYT,* Apr 27/77, A18)

April 29: MSFC announced award of a $110 096 contract to General Dynamics Convair Div., San Diego, Calif., to study and define ideas for an automatic structural-beam fabrication machine for use in the Space Shuttle [see Mar. 1]. The beams, made of a composite material, would support large scale space structures and act as electrical and thermal insulators. (MSFC Release 77-72)

• MSFC announced award of a $695 000 contract to Rockwell Intl. Space Div., Downey, Calif., for a study of solar power satellite systems. The study would identify and suggest solutions for technical and social problems that might occur in harnessing solar energy from space. (MSFC Release 77-73)

During April: The Natl. Aeronautic Assn. reported that Kingswood Sprott, Jr., had broken his own world class hot air balloon altitude record March 26 when he reached 41 000ft over Lakeland, Fla. The previous record set March 26, 1977, also by Sprott, had been 38 789ft. (NAA newsletter Apr 77)

• Recent photographs of the inner Martian satellite Phobos, obtained by the *Viking 1* orbiter in Feb., had revealed "the nature of that satellite in unprecedented detail," said *Scientific American*. Earlier pictures by *Mariner 9* or the *Viking 2* orbiter were at a distance of 880km; the new pictures were taken at a distance of 660km to as little as 100km, with resolution that permitted distinguishing objects smaller than 10m across. The surface of Phobos was known to be pitted with craters from the impact of meteorites; the new images showed chains of irregular craters like those seen on the surface of the moon, although Phobos's crater chains paralleled the plane of its orbit around Mars. Pictures from *Viking 2* had shown grooves on Phobos concentrated near its northern pole; the new pictures defined the grooves as crater chains possibly caused by objects hundreds of meters across. Further study of the images should reveal more about the nature of Phobos, the magazine said. (*SciAm,* Apr 77, 57)

• *Nature* reported that the largest radiotelescope in the world—the Soviet RATAN-600—was "at last" officially in operation at Zelenchuk, USSR. A formal message of congratulations from Soviet leader Leonid Brezhnev had been issued March 20. Constructed jointly by the USSR Academy of Sciences, the Ministries of Power and Electrification and of Power Plant Construction, and the Univ. of Moscow, the device had four sections each able to operate independently. Part of the RATAN had been receiving signals since 1974, *Nature* said, and had already undertaken studies of the moons of Jupiter, the galactic nucleus, and the fine radio structure of the sun. (*Nature,* Apr 7/77, 493)

May

May 1: RCA (formerly the Radio Corporation of America) observed the 17th anniversary of the first weather picture received from space through the first Television Infrared Observation Satellite (TIROS), *Today* newspaper reported. The 2601lb TIROS delivered "a pretty shabby product compared to today's images from space," RCA admitted. The company had since built a series of 24 other weather watchers that had taken 3.5 million pictures and logged more than 7 billion miles in orbit. Managed by the National Oceanic and Atmospheric Administration, the spacecraft had gone from pinpointing major storms to predicting runoff from snowmelt, helping ships navigate through icefields, and assisting commercial fishermen to find promising areas for their efforts. RCA was currently developing a fourth-generation TIROS. (*Today*, May 1/77, 18A)

May 2: Aviation Week reported that a controversy within the U.S. intelligence community had kept the U.S. National Security Council and the president from learning that the USSR had developed a directed-energy weapon designed to destroy U.S. intercontinental and submarine-launched ballistic missile nuclear warheads. (Directed-energy weapons was a term coined to include both laser beam weapons and high energy lasers.) Events that convinced some U.S. analysts that USSR weapons were nearing prototype test stage included:
—Detection by a USAF early-warning system of large amounts of gaseous hydrogen in the upper atmosphere with traces of tritium, considered relics of charged particle beam device tests carried out since 1975 at Semipalatinsk.
—USSR ground tests of a small high energy laser destined for spacecraft launch, possibly related to manned space station activities.
—Tests of a new magnetohydrodynamic generator to provide power for a charged particle beam system near the Caspian Sea, monitored by a TRW early-warning satellite stationed over the Indian Ocean.
—Establishment of a new test site at Azghir, the Caspian Sea location, under direct control of the USSR national air defense force.
—Confirmation by a USAF-sponsored team of U.S. physicists and engineers that the USSR had achieved success in 7 areas of high energy physics needed to develop a beam weapon.
—Admission by previously unconvinced U.S. physicists that the USSR

could have developed the technology for a charged particle beam device.
—Hints by a visiting Soviet physicist last summer that the USSR was far ahead of the U.S. in controlling fusion by compression of small pellets of thermal nuclear fuel, and thus in potential weapons based on fusion technology. The information given by Leonid I. Rudakov during his visit to the Lawrence Livermore Laboratory had later been classified top secret by DOD and ERDA.

Maj. Gen. George J. Keegan (USAF, Ret.), former head of Air Force intelligence, had in 1975 reported to CIA head William Colby on Soviet beam technology; the CIA's nuclear intelligence panel had written a report, no copy of which was ever given to USAF intelligence.

AvWk said Colby just before a meeting on strategic arms limitation had notified then Secretary of State Henry Kissinger of a Soviet facility "related to nuclear functions that were unknown"; except for that, no mention of the beam weapon potential had ever been made to the president, the secretary of state, or the National Security Council. *AvWk* also charged that influential U.S. physicists had sought to discredit Gen. Keegan's report, their general attitude being that "if the U.S. could not successfully produce the technology to have a beam weapon, the Russians certainly could not." (*AvWk*, May 2/77, 16)

The next issue of *AvWk* quoted President Jimmy Carter's reply to a query from the *Washington Post*'s Dave Broder on the report of a Soviet breakthrough in high energy weapons: ". . . Is there any such development and does it threaten the U.S. strategic deterrence?" Carter replied: "We have no evidence, Dave, that the Soviets have achieved any major breakthrough in the kind of weapon described . . . the assessment of the report in the aviation magazine has been exaggerated." (*AvWk*, May 9/77, 13)

May 3: Langley Research Center announced award of a 5yr $23 million contract to Wyle Laboratories for electronic instrument support to test and research facilities at the center. The contract, running from Oct. 1977 through Sept. 1979 with options to extend for another 3yr, covered inspection, calibration, maintenance, and satisfaction of measurement requirements on some 30 000 instruments and transducers valued at more than $50 million; also, design, modification, and maintenance of hardware and computer programs associated with digital data systems and special purpose computers used in the research facilities. (LaRc Release 77-15)

• The Senate confirmed the nomination of Dr. Richard C. Atkinson as director of the National Science Foundation. (NASA Legis. Act. Rpt., May 3/77)

May 4: Marshall Space Flight Center announced it had awarded to the Rocketdyne Div., Rockwell Intl., a contract modification for additional work and services to the Space Shuttle main engine program, covering long leadtime hardware and material procurement to provide follow-on engines. Each engine, designed for use up to 50 times before overhaul, would be capable of 1 668 075 newtons (375K lb) of thrust at sea level and 2 090 654 newtons (470K lb) at altitude. (MSFC Release 77-78)

• Ames Research Center reported that on May 10, 10 female volunteers would complete a month-long test [see March 24] to set standards for passengers on future spaceflights. The test included 9 days of total bed rest, to simulate effects of weightlessness, and centrifuge rides to simulate reentry into earth atmosphere. ARC had scheduled a press conference to announce and discuss the results of the test; Dr. Harold Sandler, principal investigator, would attend with members of his staff and the volunteers. (ARC anno May 4/77)

May 5: NASA announced it had awarded McDonnell Douglas Aircraft Co. a 6.5yr, $15.6 million contract to develop and evaluate composite vertical stabilizers for DC-10 transport aircraft. A NASA program called ACEE (aircraft energy efficiency) would pursue the use of composite materials to decrease structural weight of aircraft by about 25% and to reduce fuel consumption by 10 to 15%. LaRC would manage work under the contract at the Douglas plant in Long Beach. (NASA Release 77-89; LaRC Release 77-16)

• NASA announced that Det Norske Meteorological Institute in Oslo, Norway, had asked that the *Nimbus 6* satellite help it track a huge oil slick resulting from an oil well blowout in the North Sea. By May 16 the Norwegians would put 5 random-access measurement-system (RAMS) buoys into the oil slick to send environmental information to the spacecraft. Goddard Space Flight Center engineer William W. Conant, noting that similar buoys earlier this year had helped the U.S. Coast Guard track an oil slick off Nantucket Sound, said NASA would support the Norwegian effort by transmitting locations of all the buoys twice daily for 30 days, adding "We don't normally provide daily teletype information." For more than a yr, the Norwegians had been tracking ice drift in the Spitzbergen region north of Norway, using a RAMS system as an accurate method of tracking in all weather. (NASA Release 77-30)

• Transportation Secretary Brock Adams gave Congress proposals by the Carter administration to help the airlines finance modification or replacement of commercial jet aircraft failing to meet FAA noise-level standards, through revising the present 8% passenger ticket and waybill taxes now earmarked for the Airport and Airway Trust Fund. The Civil Aeronautics Board would let the airlines add a 2% surcharge to be deposited in a special environmental fund to pay for retrofitting existing aircraft or acquiring replacement aircraft that would meet the new noise standards. Airlines whose revenues reached a certain allowable cost level would have the surcharge terminated. (DOT Release 55-77)

May 6: MSFC announced its engineers had found a new way to recondition and lengthen the lifespan of nickel-cadmium batteries used in low earth orbit space missions. Using a device with a converter circuit and load-resistor relays to ensure proper discharge and avoid damage to the cells, the engineers had tested batteries with 2yr normal life in orbit, and had kept them running for 4.5yr, the equivalent of 23 600 orbits. Maintaining stored energy on low-orbit missions had been a NASA concern, especially after a power failure during the *Skylab 1* mission had meant a loss of battery capacity that equipment then available could not correct in orbit. Besides offering longer life and higher reliability, the new device could detect low cell voltage and bypass failed cells. MSFC would build and test a battery of 116 series cells, using the new device for protection and reconditioning. (MSFC Release 77-81)

• MSFC announced it had awarded Martin Marietta Aerospace's Denver division a $928 217 contract for construction work on a test stand for the Space Shuttle program. Modification of the giant hydrodynamic support system last used for Saturn V dynamic tests in the 1960s would allow mating all elements of the Shuttle for vertical ground tests scheduled to begin in 1978. The complete vehicle (orbiter, external tank, and two solid fuel rocket boosters) would occupy the stand together for the first time. (MSFC Release 77-79)

• Earth-orbiting satellites had improved forecasts of water availability and warned of impending low water supply as a valuable tool to water management agencies in Arizona, California, Colorado, and Oregon, NASA announced. Other cooperating agencies included the U.S. Geological Survey, the U.S. Soil Conservation Service, the U.S. Bureau of Reclamation, and the Bonneville Power Administration. The 4 states had begun a combined 4yr operational program using NASA Landsat earth resources satellites and Natl. Oceanic and Atmospheric Administration meteorological satellites to produce imagery of the snow lines along the Sierra Nevada that was faster and less expensive than conventional aerial surveys. Satellites also offered repeated mapping of the same snow cover area with a relatively constant perspective.

A continuing drought in the western U.S. had resulted from below-normal snowfalls in the mountains, source of stream flow water supplies;

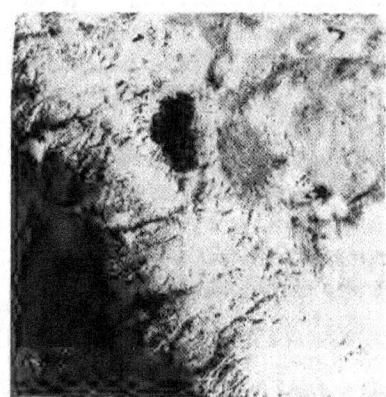 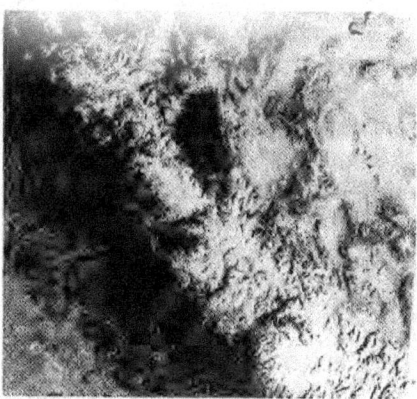

Landsat images of the same area of the Sierra Nevada show the difference in snowcover between Feb. 1975 and Feb. 1977. The 1977 snowline was 2000ft higher than in 1975, a yr of near-normal snowfall. Lake Tahoe is in the center of both pictures. (NASA 77-H-225, -226)

reservoirs on the western slopes of the mountains had already dried up. Some areas had experienced drought for a longer period than others; Calif. had been hardest hit, with minimal snowfall in the Sierra Nevada for the second consecutive yr. (NASA Release 77-91)

• ESA announced it had completed a revised program of switching on and testing the *Geos* satellite launched Apr. 20, since a malfunction of the launch vehicle had prevented the spacecraft from achieving its intended geostationary orbit. The 7 experiments were reported operational, and the two 20m booms had been extended to 2.5m without difficulty. Although the satellite could transmit 8hr of data a day, only half would be outside 5 earth radii (where the experiments had been designed to operate). The booms should deploy fully during the upcoming wk; the journal *Nature* called for prompt launch of the backup Geos. (ESA anno May 6/77; *Nature*, May 5/77, 8; May 12/77, 100)

• *Science* magazine carried an article by Drs. T.B.H. Kuiper of the Jet Propulsion Laboratory and M. Morris of the Calif. Inst. of Technology on the search for extraterrestrial intelligence, based on extrapolations of known physical processes and known behaviors of terrestrial animals and humans. Concluding that interstellar travel would be practicable and that advanced beings might have reasons for making contact with humans, the authors reviewed the options and the interstellar-beacon frequencies proposed for use in such contact. (*Science*, May 6/77, 616)

May 9: Australia's Minister for Science J.J. Webster announced that his department had studied the technical feasibility of installing a temporary

NASA satellite-tracking laser in the western area of the continent for use in geodetic research, mapping, and earthquake prediction. NASA would operate the laser for about a yr, beginning in mid-1978, but neither government had made a final commitment. The proposed site in western Australia would be part of a future network of tracking lasers scattered around the world for use by geologists and geophysicists, in conjunction with NASA's Seasat and existing lasers such as the one near Canberra at the Orroral Valley satellite-tracking station, to gather data for use in research such as refining the measurement of the shape of the earth. (NASA Release 77-92)

• Kennedy Space Center reported it had loaned its hyperbaric chamber to the Univ. of Fla. College of Medicine at Gainesville, for use in research. KSC had installed a hyperbaric (or recompression) chamber in its operations and checkout building to treat astronauts affected by decompression sickness (the "bends") during manned altitude-chamber tests of the Apollo spacecraft. Conditions such as the bends, embolisms, carbon monoxide poisoning, or gas gangrene might require administration of hyperbaric oxygen (under increased pressure in a sealed chamber) for periods up to 24hr.

The end of the Apollo-Soyuz Test Project in 1975 had ended NASA's immediate need for the altitude or the hyperbaric chambers, but KSC had kept them for possible use in Space Shuttle training. (KSC Release 101-77)

May 10: MSFC reported that the NASA barge Poseidon was on its way to Seal Beach, Calif., to pick up a Space Shuttle orbiter simulator and deliver it to the Natl. Space Technology Laboratories at Bay St. Louis, Mo., to stand in for the orbiter during firings next year of the main propulsion test article, consisting of the simulator with an external tank and thrust structure covering 3 main engines. MSFC had used the Poseidon in the 1960s to move stages of Saturn rockets. (MSFC Release 77-84)

• MSFC announced completion of pressure and structural-load tests on the liquid hydrogen tank that would be part of the Shuttle external tank being built at the Michoud assembly facility by Martin Marietta Aerospace. The tests to certify that the tank met design and manufacturing standards included filling the tank with nitrogen at 42psi for 14hr and applying hydraulic loads as high as 600tons where the orbiter and solid-fuel rocket boosters would be attached. Largest of 3 major elements of the external tank, the hydrogen container about 30m (97ft) long and 8.3m (27.6ft) in diameter would now go for cleaning, corrosion protection, and insulating. (MSFC Release 77-83)

May 11: Johnson Space Center announced it had awarded a $399 600 cost-plus-fixed-fee contract to ESL, Inc., of Sunnyvale, Calif., for using Landsat data in a vegetation inventory for eastern Alaska under a NASA project conducted jointly with the Bureau of Land Management. The BLM had begun an automatic inventory system for 474 million acres under its jurisdiction; the Alaska area would represent a tundra ecology. If the joint task were successful, NASA had options for work on grassland and sagebrush ecologies. (JSC Release 77-31)

• NASA announced it had awarded a $2.3 million contract to Douglas Aircraft Co., as part of its aircraft energy efficiency (ACEE) program, for improving wing lift-to-drag ratio to reduce fuel use. Douglas would design and test a supercritical wing like that invented by LaRC's Dr. Richard Whitcomb, shaped to decrease shock waves on wing surfaces, with a high aspect ratio (long span relative to its width). LaRC would manage work to be done under the contract at the Douglas plant in Calif. (NASA Release 77-93; LaRC Release 77-17)

• MSFC announced it had awarded a 45mo contract to the Universities Space Research Association of Houston to provide support for development of the Atmospheric Cloud Physics Laboratory, scheduled to fly as a partial payload twice a year on Spacelab beginning in 1980 or 1981. The Texas-based consortium would review proposals for scientific experiments and provide a liaison between NASA and the cloud-physics community as an aid to the project scientist. (MSFC Release 77-86)

• Strategic arms limitation might suffer another blow, the *Washington Post* said in a column by Victor Zorza, from the USSR's test of a killer satellite viewed by the U.S. as a new weapon intended to deactivate U.S. observation satellites if and when the Soviets decided to launch a first strike. Zorza noted that current agreements did not cover hunter-killer satellites; by the time a new agreement was reached, both powers might have passed another point of no return in the arms race. (*W Post*, May 11/77, A21)

May 12: The U.S. Air Force launched a pair of "jam-proof" Triple-7 defense satellite communications system spacecraft from Cape Canaveral Air Force Station at 10:27am on a Martin Marietta Titan 3C booster, largest developed by the USAF, into an elliptical parking orbit with 35 943km (22 333mi) apogee, 285km (177mi) perigee, and 26.6° inclination. At 5:30pm the USAF confirmed that the two 1300lb spacecraft had separated and had been fired by the Titan third stage into stationary orbit at about 22 300mi where they would remain for 2mo above the Galapagos Islands in the Pacific for system checkout.

The two satellites (fourth attempt in 6yr to launch a Triple-7 package, but only the second successful one), with a DSCS already in orbit and another pair scheduled for launch in November, would form a global communications network that the USAF had been trying to establish for several yr. A Titan 3C malfunction in 1975 had prevented an earlier pair of spacecraft from achieving orbit. (*Today*, May 12/77, 12A; May 13/77, 1A; *C Trib*, May 14/77, 1-9; *NYT*, May 11/77, 19; *AvWk*, May 16/77, 19)

• NASA announced that simulation of a 1980s Spacelab mission using the Galileo II (a Convair 990 jet transport converted to a sophisticated flying laboratory) would begin May 16 with the first of nine daily 6hr flights from ARC carrying four payload specialists and a mission specialist. Known as ASSESS (airborne science/Spacelab equipment-system simulation), the flights would rehearse NASA and ESA personnel in the roles they would perform in actual Spacelab missions, giving the managers practice in coordinating payload management between experiment operators on Spacelab and investigators on the ground [cf. Apr. 19]. The actual missions would be the first chance for scientific personnel without astronaut training to obtain in orbit data on earth resources, infrared astronomy, medicine, and atmospheric pollution.

The Spacelab program that began in 1972 had included 6 simulations to gather data on Shuttle-Spacelab operations; the first set of flights, called ASSESS 1, had consisted of 5 simulations on Galileo over 6 days in June 1975. The new series, ASSESS 2, would fly a different route each of the 9 days, ranging over the Pacific and to the northern and southern borders of the United States, as far east as the Dakotas, and would end May 26. The Galileo II would fly to Paris May 30 for exhibition at the Paris Air Show, then fly to Cologne in West Germany where the ESA experiments would be removed.

Mission specialist on ASSESS 2 would be astronaut Karl Henize of JSC with astronaut Robert Parker, also of JSC, as backup. NASA payload specialists on ASSESS 2 would be David S. Billiu and Robert T. Menzies of JPL, with Leon Weaver of MSFC as backup to both; ESA payload specialist would be Claude Nicollier of Switzerland, with Juergen Fein of West Germany as backup. The other payload specialist position would be shared by Michael Taylor of England (on 6 flights) and Klaus Kramp of West Germany (on 3 flights). The simulation would confine the crew to the aircraft and an adjacent mobile van containing living quarters similar to those available on the Shuttle. (NASA Release 77-94; ESA anno May 10/77)

• The Natl. Space Club announced that Dr. Robert S. Cooper, director of GSFC since July 1976, would speak at its May 26 luncheon. Dr.

Cooper came to NASA from DOD, where he had been assistant director for research and engineering since 1972, heading establishment of the joint services NavStar global positioning system. He had begun at Mass. Inst. of Technology in 1958 as instructor of electrical engineering, was assistant professor there from 1963 to 1966, and had been a manager at MIT's Lincoln Laboratory until joining DOD. (NSC anno May 12/77)

May 13: JSC announced that three life sciences specialists would begin a 7-day simulation of a typical Spacelab mission planned for the Space Shuttle in the 1980s. Living aboard a high-fidelity mockup of the Spacelab and Shuttle-orbiter crew deck would be physician Dr. William E. Thornton, mission specialist, and Dr. Carter Alexander, payload specialist, both of JSC, plus payload specialist Dr. Bill A. Williams of ARC. Dr. Thornton would perform medical monitoring and health services for the simulation crew.

ARC had shipped a rack of 20 life sciences experiments to JSC for the simulation, and JSC had provided 6 other experiments. Life sciences payloads like the simulation experiments would be flown on the Spacelab missions to show the effects of space on living organisms, improve health care for the crews of actual missions, and help in designing life-support systems for space and earthbound clinical research and health care. (JSC Release 77-32)

May 15: NASA announced the expected arrival of a USSR research ship off the Virginia coast near Wallops Flight Center about June 1 to participate in a series of rocket launches to investigate ionization sources in earth's upper atmosphere. The research ship Professor Vize would operate offshore as the Soviet launch platform. The joint American-Soviet particle intercalibration (JASPIC) project would compare techniques used by both nations to measure the intensity of electrons and protons entering the lower ionosphere, and to define the role of these particles in nighttime ionization of the lower ionosphere in the midlatitudes, thought to cause high-latitude auroras.

The researchers would base comparisons on results of 4 sounding-rocket launches from Wallops and 5 Soviet MF-12 rocket launches from the ship. One of the NASA rockets would release a chemical cloud that might be visible over the east coast. A similar joint activity in Aug. 1976 had launched 22 pairs of rockets over a 2wk period. U.S. and Soviet scientists had decided in 1975 that discrepancies shown by the scientific literature in measurements of ionosphere energetic particles might arise from instrumentation differences, and had agreed to joint measurements and subsequent comparisons at the same time and place. (NASA Release 78-71)

• Airline service between the U.S. and the United Kingdom might be disrupted by failure to agree on new rules for sharing the traffic, Transportation Secy. Brock Adams told the *Washington Post.* The Bermuda Agreement that had governed air transportation between the two countries since 1946 would expire in June, and talks had begun in Sept. 1976 on new provisions. The British government was said to want exactly half the passengers traveling between the two nations or their territories to fly on British carriers, although almost two-thirds of the passengers were Americans. U.S. officials were reported in dismay at the British demand for a bigger share of the traffic, although the U.K. had earned $1.4 billion in revenues from aviation and tourism combined, 55% of the total between the two countries.

The U.S. had generally viewed international aviation as part of its overall foreign policy, whereas other nations generally had a single carrier either publicly owned or publicly supported whose officials were interested solely in increasing their carrier's share of passengers; U.S. philosophy had favored more rather than less competition in both domestic and international aviation. U.S. Ambassador Alan S. Boyd, who would head the U.S. delegation to the talks, said he could give "a succinct summary of our differences . . . The United States wants equal opportunity for all carriers; the British want equal benefits." A complication would be the New York port authority's refusal to allow British and French Concordes to land at JFK Airport, declared unlawful by a recent court decision presently under appeal. (*W Post*, May 15/77, K1)

• The *W Post* reported that 9 USSR ships would join U.S. ships during June in "Polimode," a joint study of ocean currents in the so-called Bermuda Triangle where numbers of ships over the years had disappeared without trace, inspiring best-selling books on the subject. A USSR research team had detected by satellite "powerful whirl formations" extending hundreds of miles and resembling cyclones on land, the *Washington Post* said. Tass, the Soviet news agency, said the formations apparently reached from the surface to a depth of several thousand yards. (*W Post, Parade,* May 15/77, 6)

• Threats to the earth's ozone layer remained a matter for concern, Walter Sullivan reported in the *NY Times,* especially fluorocarbons widely used as aerosol propellants and accumulating as chlorine byproducts in the atmosphere where they would break down ozone and expose the earth to more ultraviolet light. Health records from the United Nations had shown that people living in low latitudes where the ozone layer was thinner and sunlight more abundant had a markedly higher rate of skin cancers, especially of the often fatal melanoma.

Although some scientists noted that atmospheric ozone would normally vary with latitude, season, and even with the sunspot cycle, public health officials said any total increase in ultraviolet exposure would probably increase the incidence of cancer.

Other dangers to the ozone layer included supersonic planes and other high-flying aircraft, nuclear explosions, and the heavy use of nitrogen fertilizers. The Natl. Academy of Science had proposed outlawing fluorocarbon gases as propellants except in medical products. (*NYT*, May 15/77, 4-9)

May 16: NASA announced that the 10th annual U.S. industrial film festival had presented the Chairman's Special Award, Best of Festival, to NASA's motion picture "Universe." Finalists among 970 movies from 5 nations were the NASA film, the Smithsonian Institution's "To Fly," and IBM's "Parade of the Tall Ships." Produced for NASA by Graphic Films, "Universe" had also received an Academy Award nomination, a CINE gold-eagle rating, and awards from the San Francisco and Golden Gate Festivals. It would be the U.S. entry in a number of international film festivals during 1977. (NASA Release 77-96)

• The Natl. Aeronautic and Atmospheric Administration announced it would conduct a pilot project this summer with the U.S. Dept. of Agriculture to improve accuracy of crop forecasting, based largely on the amount of solar radiation, available moisture, and air temperatures for a given area. Using devices on NOAA spacecraft to measure cloud cover and relate it to the amount of sunlight reaching ground-based pyranometers, NOAA and USDA investigators would compare hourly readings from the *Goes 1* geostationary satellite with measurements from a network of pyranometers across the Great Plains, to determine the amount of solar energy available to crops in any given period.

The researchers emphasized solar energy as a vital factor in plant photosynthesis and in evapotranspiration (loss of water directly from the soil by evaporation and from plants by transpiration). A Great Plains Agricultural Council representing land grant colleges in Texas, N.M., Okla., Colo., Kans., Nebr., S.D., N.D., Wyo., and Mont., and the USDA agencies serving those states had requested the pilot project as an aid in the production of corn, wheat, and sorghum. (NOAA Release 77-122)

• *Aviation Week* reported that NASA had chosen "key missions" for the six orbital flight tests of the Shuttle and had assigned payloads as well as the weight and volume available for additional experiments. Chester M. Lee, director of Space Transportation System operations, would give Associate Administrator John E. Naugle a list to begin assigning experiments on the basis of the key payloads. For the first mission (tentative schedule, March 1979) the Shuttle would carry only a developmental flight-instrumentation unit that would also ride on other flight tests. On the second mission (July 1979), the first true Shuttle payload would be an instrumented pallet for scientific research amounting to only a small weight in the payload bay but offering the NASA program office a first research opportunity on the Shuttle.

On subsequent missions the Shuttle would carry (Sept. 1979) a remote manipulator built by Canada to maneuver a payload on the end of a jointed arm, plus a scientific payload; (Dec. 1979) a large spin-stabilized upper stage to boost an undetermined spacecraft payload into a higher orbit; (Feb. 1980) an interim upper stage for attachment to Skylab, to boost it into a higher orbit, or an alternate payload such as the long-duration exposure facility; (March 1980) a possible classified payload (such as the USAF Teal Ruby system for detecting and tracking aircraft from space) plus a GSFC-designed multimission spacecraft. The seventh orbital mission set in May 1980 would probably be a NASA-ESA Spacelab launch, unless another flight test was needed. NASA noted that unforeseen circumstances could change the key-payload schedule for the first orbital flight tests. (*AvWk*, May 16/77, 13)

May 17: NASA announced it had investigated a unique propulsion system for sending a spacecraft to rendezvous with Halley's Comet in 1986. The comet had made some 30 visits since 467 B.C., and its next appearance would offer NASA an opportunity to find out more about the origin of the solar system.

A device using solar electric propulsion, suggested by MSFC and tested for several yr at Lewis Research Center, would aim solar energy concentrated by reflectors at conventional solar cells to provide electric power for a system of 8 ion engines capable of taking a spacecraft to an encounter with the Halley comet around Christmas Day 1985 just outside earth's orbit. The spacecraft would reach the comet during its most active state as it passed near the sun, and could send pictures of the comet to earth, possibly sending a probe through the comet's tail toward its nucleus to define its properties. NASA would decide in August 1977 whether to use the solar electric system, or a solar sail being developed by JPL, in its rendezvous with Halley's Comet. (NASA Release 77-97; MSFC Release 77-87)

• NASA announced it would launch on or about May 28 a new environmental monitoring satellite GOES-B, latest in a series to be operated by NOAA, to keep watch over the eastern half of the U.S. and the Atlantic Ocean, just in time for this yr's hurricane season. Built like the first three in the series by Ford Aerospace Corp., GOES-B would replace *Goes 1* as NOAA's east coast satellite at about 75°W; *Goes 1* would move to standby status at about 105°W. (NASA Release 77-88)

• NASA announced it had awarded Computer Sciences Corp. a $41 million contract for support services to communications and instrumentation at KSC, the major site for Space Shuttle launches beginning in 1979.

Services would be in 2 categories: the first, covering 3yr, would include modification, installation, operation, and maintenance of an operational intercommunications system, operational TV system, and checkout-control-monitor subsystem. The second category, 1yr with 2yr option to renew, would cover KSC work on communications, measurements, telemetrics, computer services, data storage and retrieval, program planning, and reliability and quality assurance. (NASA Release 77-104; KSC Release 104-77)

• MSFC reported that the Energy Research and Development Administration had selected 80 new projects for installing solar heating and cooling, on 27 of which MSFC would monitor technical design and construction. MSFC was already monitoring 32 commercial demonstration projects selected by ERDA in April 1976 as the first of a series; ERDA would issue a third solicitation for proposals later this summer, the announcement said.

Dr. Henry H. Marvin, director of ERDA's division of solar energy, said cooperative agreements negotiated by MSFC with the 27 proposers of projects would be signed within approximately 3mo. Government funding on a cost-sharing basis would cover only the solar portions of the projects; more than 90% of the funds would go to projects with substantial small business participation, and more than half the awards would go to teams using solar energy systems supplied by small businesses. (MSFC Release 77-89)

• NASA announced that it and the Soviet Academy of Sciences had agreed to continue joint technical, scientific, and operational activity developed in the Apollo-Soyuz mission July 1975 by starting studies of 2 programs, one for a joint orbital manned flight, the other for an international space station.

The agreement signed May 6 by Acting NASA Administrator Dr. Alan M. Lovelace and May 11 by Anatoly P. Aleksandrov, president of the Soviet Academy of Sciences, envisioned preliminary project documents within 6 to 12mo and definitive documents within a yr, noting that the working groups should proceed on the assumption of a first flight in 1981. Working groups to study experiments, basic and applied, and the operation of a Salyut-Shuttle mission would be named within 3 days after the agreement became effective; a third group, which would consider development of an international space platform, would be appointed within 2mo. NASA stressed that neither side would be committed beyond initial studies, and either could proceed with its own space station interests. (NASA Release 77-98; *NYT*, May 18/77, A15)

May 18: NASA announced it would launch on or about May 26 the third Intelsat IV-A commercial comsat from KSC on an Atlas Centaur for service over the Atlantic Ocean, initially as a backup for *Intelsats IV-A F-1* and *F-2* offering communications services to countries from Central America to Iran. The IV-A spacecraft built by Hughes Aircraft Co. had communications capability almost two-thirds greater than the Intelsat IV series now operating: it had a 7yr design life and capacity to relay more than 6 000 simultaneous telephone calls and two TV programs. (NASA Release 77-99; MOR E-491-633-77-05 [prelaunch] May 20/77; INTELSAT Release 77-10-M)

• NASA reported that its research and technology advisory council (RTAC) panel on general aviation technology had named stalls and spins as the greatest safety problem for designers of light aircraft, accounting for 30% of all fatalities in those planes. LaRC had established a program in response to requests for additional research on the problem, including definition of the aerodynamic characteristics of aircraft at high angles of attack; ways to avoid stalling; ways to recover from spin; and ways to ensure satisfactory spin characteristics. The program would also cover development of test methods such as use of radio-controlled models, and consultation services to industry. LaRC had been a leader in such studies since the 1930s when it began operating a 15ft spin tunnel, enlarged in 1941 to 20ft and used since then in testing more than 400 aircraft designs; this was still the only operating spin tunnel in the U.S. (NASA Release 77-100; LaRC Release 77-21)

• NASA reported that its data-analysis facility at KSC had been using a GE multispectral image analyzer developed for screening Landsat imagery to detect early stages of breast cancer, where diagnosis would normally be difficult and rapid multiplication of cancerous tissue would require massive surgery. The only solution previously available had been repeated diagnoses by x-ray, itself a health hazard.
Radiological examinations of x-rays had been limited by the range of intensity detectable by the human eye, which was unable to distinguish more than 32 shades of gray, meaning that much of the data would be effectively invisible. However, the microdensitometer on the GE analyzer was able to differentiate 256 shades of gray with great accuracy. The

computerized output of the densitometer would enter the analyzer with orders for enhancement, so that data previously invisible to a doctor's eye could be available for earlier diagnosis.

Robert L. Butterfield, electronics engineer at the KSC data-analysis facility, had worked with radiologist Dr. William L. Walls of Titusville, Fla., for 2yr on a process for automatic enhancement of mammograms to detect early stages of breast cancer before radical surgery would become necessary. The aim of the study had been to differentiate women at low risk of breast cancer from those at high risk, based on the rate of change in the gray shades of the enhanced images, to reduce the need for repeated x-ray diagnoses. (NASA Release 77-102; KSC Release 105-77)

- The Intl. Academy of Astronautics announced that the 28th congress of the Intl. Astronautical Federation would take place from Sept. 25 to Oct. 1, 1977, in Prague, in conjunction with the 11th symposium on contributions to the history of rocketry. Theme of the congress would be "Using Space—Today and Tomorrow." Dr. Charles Stark Draper was IAA president. (IAA anno)

May 19: NASA reported that MSFC had received 119 proposals for experiments in materials processing to be performed on Space Shuttle and Spacelab flights. As lead center for the Spacelab program, a cooperative venture of NASA and ESA, MSFC had been given project management responsibility for the first three Spacelab payloads. An MSFC team had evaluated the proposals, and the center would announce selection of between 10 and 20 experiments by the end of July. (NASA Release 77-103; MSFC Release 77-91)

- NASA announced it had issued proposals to buy 6 spinning solid fuel upper stages (SSUS) for boosting spacecraft launched from the Space Shuttle into higher orbits. The decision was the first step following agreement earlier this yr with McDonnell Douglas and Boeing Aerospace, allowing them to design, build, and test the devices and market them independently of NASA. The agency as part of a procurement agreement would normally bear cost of development and production; the current bidding would be limited to the 2 companies that had agreed to develop the stages with private funds.

The purchase would include hardware, analyses, logistics, and services needed to put a spacecraft into higher orbit after launch and orientation by the Space Shuttle. The contractor would build the vehicle, provide services, and operate the systems. First delivery in Dec. 1979 would be a flight demonstrator; 3 others would launch advanced Intelsat V comsats; the other 2 would serve for other missions. (NASA Release 77-101; MSFC Release 77-90)

- Press reports said a 23 900lb solid-fuel rocket had fallen off a Delta launch vehicle standing on its pad at KSC May 18 awaiting the mid-June launch of ESA's high-frequency communications test satellite (CTS). In its 12ft fall, the booster had plunged through a platform and damaged the Delta's liquid-oxygen tank; the entire assembly would have to be removed for repairs. Cause was probably failure of one of the attachments holding the motor in place, a spokesman said; an investigation was under way. The Delta had on its first stage 9 of the solid-fuel boosters, each more than 36ft tall and 4ft in diameter, normally jettisoned about a minute and a half into flight. (*Today,* May 19/77, 8A; photo, 1A; *AvWk,* May 23/77, 27)

- *Nature* magazine reported that the NASA budget had met unexpected opposition in Congress, endangering 2 major scientific missions: the large space telescope, long regarded by space scientists as NASA's top-priority project, and the Jupiter orbiter probe planned as a twin launch in 1982 of craft to orbit the planet and send probes into its atmosphere. Lobbying by NASA and the scientific community might rescue the projects, but loss of either would be a serious blow to the agency's program. *Nature* said this yr's efforts would be crucial because space science money had been squeezed out of NASA's budget for 4yr to pay for the Space Shuttle, and further delay would probably kill future chances for approval of these missions. (*Nature,* May 19/77, 196)

- The *Chicago Tribune* reported that U.S. Secy. of State Cyrus Vance and USSR Foreign Minister Andrei Gromyko had begun high-level arms control talks in Geneva by signing a convention against environmental warfare and extending a previous agreement on space that had governed the 1975 hookup between Apollo and Soyuz spacecraft, calling for cooperation in space science such as exchange of data on Mars, Venus, and the moon. (*C Trib,* May 19/77, 1-1)

May 20: LaRC announced it had developed an ocean-surface wind sensor for SeaSat-A, managed by JPL and scheduled for launch in 1978. SeaSat instruments would record wind and pressure data over the oceans; lack of this information had prevented accurate long-range weather forecasting. LaRC research had shown the sensitivity of microwave scatterometers to sea-surface winds; LaRC's sensor for SeaSat contained a scatterometer able to measure wind speed and direction over a thousand-kilometer area. (LaRC Release 77-20)

- ARC reported successful completion May 19 of a test to qualify the Pioneer/Venus probe for entry into the Venus atmosphere. One of the two spacecraft scheduled for the Pioneer/Venus mission arriving there in

1978 would carry a main probe and three smaller probes to be dropped off at 8000km intervals to measure the Venus atmosphere from top to bottom. At 47km above the surface of Venus, the probe would separate from its heatshield, leaving its parachute behind, to descend as far as possible before the planet's heat destroyed it. Venus surface temperature would exceed 500°C (920°F), hotter than the melting point of zinc.

At White Sands Missile Range, N.M., the U.S. Air Force's Geophysics Laboratory launched a high-altitude plastic balloon measuring 90K m^3 (about 3 million ft^3) to an altitude of about 28km; about 8hr later when wind conditions were right, the balloon released from a height of about 30km an instrumented probe equipped to record system performance while White Sands personnel tracked and recovered the probe pressure-vessel and parachute. At an altitude of 15km above earth, the heat and density of the atmosphere and the probe velocity resembled those expected on Venus at 67km above the surface, the point of actual parachute deployment. (ARC Release 77-30)

• NASA announced it had awarded to Lockheed-California Co. a contract to demonstrate composite-material technology in manufacturing L-1011 vertical stabilizers. NASA would cover $1.57 million of the cost of the 6.5yr contract, Lockheed $1.7 million, like others in the agency's program to increase by 50% the fuel efficiency of civilian transport aircraft. LaRC would manage the contract. (LaRC Release 77-22; NASA Release 77-108)

May 23: President Carter announced the nomination of Dr. Robert A. Frosch, 49, associate director of the Woods Hole Oceanographic Institution in Mass. since 1975, to head the Natl. Aeronautics and Space Administration. Dr. Frosch would succeed Dr. James C. Fletcher, who resigned May 1 after 6yrs as administrator of the agency.

A native New Yorker, Dr. Frosch had earned A.B., A.M., and Ph.D degrees from Columbia Univ. From 1973 to 1975 he was asst. secretary general of the United Nations and assistant head of its environmental program. From 1966 to 1975 he was asst. secretary of the Navy for research and development, working on naval-research projects at the Hudson Laboratories of Columbia Univ. where he had been director from 1956 to 1963. Dr. Frosch had received the Arthur S. Flemming award in 1966 and the Navy's distinguished public service award in 1969. (NASA release 77-107; *NYT*, May 24/77, 19)

• MSFC reported results of a gamma-ray detector launched on a balloon May 11 from the Natl. Scientific Balloon Facility at Palestine, Tex., to study sources of bursting and pulsating x-rays and gamma rays during an

18hr flight. MSFC scientists headed by Dr. Charles Meegan, principal investigator and associate of the Natl. Research Council, had sent a 420 000m³ (15 million ft³) balloon carrying an array of 12 large-area detectors with total payload weight of 590kg (1300lb) to an altitude of 39km (126 000ft) to transmit data back to the launch site for recording. High winds had interrupted the successful mission and brought the balloon down with some payload damage. Investigators would use the data for the scheduled June 30 launch of NASA's high-energy astronomy observatory (HEAO). (MSFC Release 77-93)

• NASA announced that the Arizona Dept. of Transportation and 2 federal agencies had demonstrated solar energy potential by using an array of solar cells to power a dust-storm warning system in central Ariz. The state had installed the system over an 80mi area on Interstate 10 between Phoenix and Tucson, and west on Interstate 8 near Casa Grande, to help motorists avoid accidents in low-visibility conditions resulting from blowing dust. In normal mode, the signs would display directional information; when dust storms struck, the message would change to give motorists the radio frequencies on which storm-alert instructions would be available.

The state agency (working with LeRC, which had managed the applications project for the Energy Research and Development Administration) would replace propane-fueled 60w generators at 40 signs in the system with solar arrays, halving the cost. Each sign would use solar electricity to power the lighting, changeable-message feature, and radio communications activated by a link from Phoenix when dust storms were imminent. (NASA Release 77-106)

• MSFC reported that Teledyne Brown Engineering Co. had built for Shuttle tank testing at the center an "upper reaction structure" resembling a mammoth doughnut, 42ft in diameter, 12ft high, and weighing 124 500kg (137tons), so large that the manufacturer had to remove a wall at its plant to get it out of the building. The structure, riding on a special 32-wheel lowboy trailer with a 16-wheel dolly, would block roads during a 3 to 4hr trip from the builder to the MSFC test stand where Bendix personnel would install it at the 46m (150ft) level, to support the structural test article simulating the external tank and to record test loads and stresses. (MSFC Release 77-94)

May 24: The *Washington Star* reported that the USAF had launched a "spy satellite" from Cape Canaveral at 2:13pm May 23 as part of a series to monitor rocket tests in the USSR and PRC and give early warning of any missile attack. *Defense/Space Daily* said the "experimental payload" launched on an Atlas Agena was either a test of improved early-warning equipment or part of a more advanced early-warning system. Orbit parameters deleted from NORAD and NASA records but reported by the UK Royal Aircraft Establishment were 40 800km apogee, 30 200km perigee, and 9° inclination. The USAF had not announced the launch, visible to residents of the area, before liftoff and issued only a brief statement afterward. (*W Star*, May 24/77, A-3; *D/SD*, May 25/77, 138)

• NASA announced it would launch no earlier than May 25 GOES-B, second in a series funded by NOAA to meet DOC requirements for a geostationary operational environmental satellite system. *Sms 1* and *Sms 2*, prototypes funded by NASA and still in orbit, had been launched in May 1974 and Feb. 1975; *Goes 1*, first operational spacecraft funded by NOAA, was launched Oct. 1975. GOES-B would augment existing coverage of earth weather and would become part of the global atmospheric research program (GARP). (MOR E-608-77-04 [prelaunch] May 24/77)

• ESA announced it had postponed launch of its experimental comsat OTS (orbital test satellite) originally scheduled to go June 16. The spacecraft had left Amsterdam's Schiphol airport May 9 headed for prelaunch checkout at ETR; it had not been mounted on its launch vehicle at the time when one of the solid-fuel booster rockets fell off and damaged the first stage [see May 19]. An inquiry into the April 20 malfunction of a Delta carrying ESA's GEOS had already delayed the OTS launch for at least a wk. (ESA release May 24/77)

• MSFC announced it had completed modifications of the Saturn test stand used in the 1960s for static-test firings of the first stages of the Saturn 1 and 1B and scheduled for structural tests of the Space Shuttle solid-fuel rocket booster. The tests would use a short-stack SRB (lacking two center segments, measuring 18.3cm less than the actual boosters) to simulate prelaunch, flight, and recovery reactions. Each Shuttle launch would need two SRBs each simultaneously generating 11.79-million newtons (2.65 million lb) of thrust at liftoff; separating at burnout from the external tank, the casings would parachute into the ocean for recovery and reuse in as many as 20 flights. (MSFC Release 77-95)

• NASA reported that Dr. Herbert Frey, a NASA-Univ. of Md. scientist working at GSFC, had prepared a paper for presentation June 2 at the meeting of the American Geophysical Union in Washington, D.C., suggesting that asteroid bombardment more than 4 billion yr ago that created earth's ocean basins had resulted in the development of life on this planet.

Building on the work of other scientists, including that of GSFC's Dr.

Paul Lowman on earth's crustal evolution, Dr. Frey used data from Apollo lunar landings and photography from spacecraft studies of the moon, Mercury, and Mars to show that during the reference period all the inner bodies suffered heavy asteroid impacts which must have affected earth as well. The impacts led to geological changes on earth that resulted in plate tectonics and the formation of seabeds. If the presence of life had eventuated from catastrophic bombardment by asteroids, the paper noted, then life elsewhere might also depend upon such events. (NASA Release 77-105)

May 25: Dryden Flight Research Center announced it had developed an autopilot device during flight tests of the YF-12 aircraft that would enable future high-speed planes to fly closer to the preset flight path. Experience gained from aircraft cruising at high speeds and high altitudes had shown deviations up to ±4000ft in altitude and more than 30mph in speed from those planned. An aircraft flying across the country at Mach 3 could encounter normal changes in temperature and pressure as rapid changes that could affect the Mach number; conventional techniques to correct for those changes had caused large altitude deviations and poor-quality passenger comfort. The deviations could degrade aircraft performance or overrun the plane's operating limitations. The new control system combining surface motion and throttle motion, not previously used together, with newly developed data sensors, had kept the YF-12 on a highly precise flight path even at high-speed conditions over extended periods of time. (DFRC Release 19-77)

• KSC announced it had awarded to the Univ. of Fla. a $16 200 supplemental grant to continue a study of the effects of lightning strikes. University personnel had been doing research on lightning under this grant since 1973; working as consultants, they had used advanced instrumentation available at KSC together with engineers from the space vehicle operations and information systems directorates to measure high-frequency radiation from lightning activity within a cloud, intensity of lightning strikes, and the extent of damage to structures or aerial and underground cables. The study would improve handling of Shuttle launch, processing, and landing under marginal weather conditions. (KSC Release 108-77)

• A KSC spokesman said work was proceeding slowly on dismantling the Delta vehicle that had lost one of its solid-fuel rocket boosters May 18 while waiting on the launch pad at Cape Canaveral, *Today* newspaper reported. The second stage of the Delta had already been removed to clear the way for detaching the boosters from the first stage and then removing the first stage, damaged when one of the 9 boosters crashed

12ft through a support platform. The accident, which interrupted an ESA comsat launch scheduled for June, had also delayed launch of a U.S. weather satellite. (*Today*, May 25/77, 12A)

- The Natl. Science Foundation reported that Census Bureau figures for the end of 1975 had shown a 5yr decline in private industry employment of scientists and engineers. Private industry had been by far the largest employer of scientists and engineers, using about 2/3 of all those employed. Engineers constituted 72% (850 000) of all that group employed by private industry, the largest numbers being electrical engineers (292 000) and mechanical and aeronautical engineers (201 000). Computer scientists were the largest group (143 000) among scientists, as well as the largest group (100 000) employed by non-manufacturing industries. Among physical scientists (104 000), the most numerous specialty was chemists (65 000), more than 3/4 of them working in nondurable goods manufacturing industries including chemical industries. Private industry employment of scientists and engineers decreased 5% between 1970 and 1975, compared to a 14% gain in 1965–1970 and a 19% gain in 1960–1965. The drop in employment between 1970 and 1975 was mostly in durable goods manufacturing; nonmanufacturing industries showed an 11% increase in engineers employed during that period. (NSF 77-132)

May 26: NASA launched *Intelsat IVA F-4*, third in the series of Intelsat IVA global comsats, from Cape Kennedy on an Atlas Centaur at 5:47pm EDT into a transfer orbit with 35,904.4km apogee, 548.9km perigee, 21.8° inclination, and 641.87min period. ComSatCorp, U.S. member of the Intl. Telecommunications Satellite Corp. (INTELSAT) and its management services contractor, fired the apogee kick motor at 8:50pm EDT May 27 over the western Pacific, putting the spacecraft into a near-circular synchronous orbit at about 35 784km (22 240mi) altitude to drift eastward to its station over the Atlantic at 34.5°W after a mo of communications tests. INTELSAT, owner of the comsat series, would reimburse NASA for costs of the launch vehicle and launch services.

The Hughes Aircraft-built spacecraft weighing 1515kg (3340lb) at liftoff measured about 7m (23ft) tall and 2.4m (8ft) in diameter and carried a new design of antenna that would provide coverage from Central America to Iran by using shaped beams. The 320MHz bandwidths pointing east and west were sufficiently isolated to allow the frequency spectrum to be used twice, once eastward and once westward, doubling the use of that spectrum. The spacecraft with a 7yr lifetime would offer about 6250 two-way voice circuits plus 2 TV channels, two-thirds more capacity than that of the Intelsat IV series. It would initially serve as backup to the two Atlantic area comsats, *Intelsat IVA F-1* at 24.5°W and

Intelsat IVA F-2 at 29.5°W. (NASA Release 77-99; KSC Release 103-77; MOR E-491-633-77-03 [prelaunch] May 20/77, [postlaunch] Aug 30/77; INTELSAT Release 77-11-M)

• DFRC announced that the first manned test flight of the Space Shuttle orbiter, previously scheduled for May 26, had been postponed to a date no earlier than June 9, depending on successful completion of orbiter ground tests. The manned test would begin the second phase of Shuttle approach and landing trials to verify aerodynamic and flight-control characteristics of the orbiter. Cause of the postponement was delay in orbiter systems checkout and verification testing after completion in March of the first flights of the orbiter linked to its 747 carrier. (DFRC Release 18-77)

• KSC announced it had awarded to Planning Research Corporation, McLean, Va., a $25 720 364 contract extension for engineering support services to the Space Shuttle program and other activities where the design-engineering directorate at KSC had design responsibilities. The 12mo cost-plus-award-fee extension through May 1978 had brought the value of the contract originally awarded in May 1974 to $72 616 061. The company employed more than 1100 workers at KSC and Cape Canaveral Air Force Station in Fla., and at DFRC and KSC's Western Launch Operations Division in Calif. (KSC Release 111-77)

May 27: MSFC announced it had issued requests for proposals to build a tethered satellite system for launch on an orbiting Space Shuttle. The idea had been to provide long-duration measurement, observation, or performance time in earth's atmosphere between 80 and 120km altitude, where density would be insufficient to support the flight of aircraft or balloons, but too thick for free-flying satellites. Exploration here had been limited to short-term flight of sounding rockets or low-level satellites.

One concept had been to use a 100km cable to "troll" a tethered satellite from an orbiting Shuttle; the pull of gravity, with constant tension on the cable, would keep the captive craft at the end of its string, oriented toward earth. Typical missions for a tethered satellite would be global mapping of earth's magnetic or gravity fields; cargo transfer between space vehicles; retrieval of satellites or space debris without Shuttle maneuvering; transfer of large amounts of energy to a remote experiment, or from a remote (possibly hazardous) power source to a space station. MSFC would manage tethered satellite activity for NASA's Office of Space Flight. (MSFC Release 77-82)

• A *NY Times* editorial said that criticism of the Apollo-Soyuz mission on the basis that the U.S. learned too little and the USSR too much was misguided: the U.S. space program from its beginning was open "to any Russian who reads American publications." It was the Soviet program that was secret, and only because of Apollo-Soyuz had American observers been able to visit key Soviet space installations. Proposed cooperation in space would cost comparatively little and bring considerable benefit from sharing costs and improving relations. (*NYT*, May 27/77, A-24)

• *Science* magazine said the stir caused by *Aviation Week's* claim [see May 2] that the USSR had tested a charged-particle beam capable of neutralizing U.S. strategic missiles had been "the sensation of the week" in Washington, giving rise to secret briefings for congressmen, an "infrequent" statement from the CIA, and even a presidential assurance that the U.S. was not in jeopardy. *AvWk* responded that the president had been "screened from vital technical developments" by the CIA and the Defense Intelligence Agency. *Science* said *AvWk* has made use of technical and intelligence information "too esoteric for easy evaluation" that *Science* had been unable to verify independently.

An article in *Nature* magazine on the same subject said *AvWk* had based its charges of Soviet weapons testing on "hydrogen with traces of tritium in the upper atmosphere—which could mean almost anything." (*Science*, May 27/77, 957; *Nature*, May 26/77, 304)

May 31: JPL announced that its controllers had turned off the biology instrument on Viking lander 2 May 28 and that on lander 1 on May 30. Lander 1 had started biology investigations July 28, 1976, and lander 2 on Sept. 11, both operating continuously since that time, depleting their high-pressure helium, nutrients, and other consumables. They had carried out all the tests originally scheduled and "several that hadn't been planned," the announcement said. Both of the landers would continue to operate except for the biology and organic analysis instruments, and both of the orbiters had continued to operate normally. Biologists in laboratories across the U.S. were trying various combinations of soil and chemicals to duplicate results from the biology instruments on the Mars landers. (JPL anno May 31/77)

During May: Natl. Research Council's *News Report* said a committee chaired by MIT's Wilbur B. Davenport, Jr., had recommended that NASA resume support of comsat services research and development but stay out of the business of operating such services. The committee said that agency support of an operational program of public service satellite communications would "obscure the real costs of the public service";

agencies that needed to use or provide such services "would not be faced with the hard reality of determining whether a program is worthy of execution when compared with its actual cost," the report said. (NRC *News Report,* May 77, 1)

• The Natl. Aeronautic Assn. newsletter reported that the USAF industry team that produced and demonstrated the B-1 aircraft system was 1976 recipient of the Collier trophy, rather than the NASA Viking project with its "spectacular success" in landing spacecraft and instrumentation on Mars. The *AFSC Newsreview* said that the B-1 team represented the Air Force, Rockwell Intl. Corp., General Electric, Boeing Co., Cutler-Hammer's airborne instruments laboratory division, and more than 3000 other subcontractors and suppliers.

The NAA noted that 50yr earlier the solo flight of Lindbergh across the Atlantic, considered a shoo-in, had failed to receive the trophy given for "the greatest achievement in aviation in America, the value of which has been demonstrated by actual use during the preceding year"; that trophy committee had unanimously awarded it to Charles L. Lawrence, designer of the Wright Whirlwind aircooled radial engine that had powered not only Lindbergh's Spirit of St. Louis, but also a number of other notable flights in 1927. (NAA newsletter, May 77, 1; *AFSC Newsreview,* May 77, 4)

June

June 1: NASA announced it had appointed Gerald D. Griffin, deputy director of Dryden Flight Research Center since 1976, to be deputy director of Kennedy Space Center as of July 1. He would replace Miles Ross, who resigned from NASA in May. Griffin had worked at NASA Hq for 3yr, as asst. administrator for legislative affairs, then as deputy associate administrator for operations in the Office of Space Flight. At Johnson Space Center, Griffin had been a flight controller during the Gemini program, a flight director on all 11 Apollo missions, and lead flight director on *Apollo 12, 15,* and *17.* Among his awards were NASA exceptional service medals for his work on *Apollo 12* and *15,* the presidential Medal of Freedom group achievement award for *Apollo 13*, and the Hq creative management award. (NASA Release 77-109; DFRC Release 20-77)

• DFRC announced it would drop-test at the Natl. Parachute Test Range, El Centro, Calif., the parachutes for recovering the reusable solid-fuel rocket boosters on the Shuttle. The tests would be a series of air drops to evaluate design, deployment, performance, and structural integrity of full-scale parachutes loaded with two configurations of drop-test vehicles offering various limit and overload conditions for the drogue and main parachutes at reefed and full open-canopy shapes. Engineers from Marshall Space Flight Center, which had responsibility for developing the Shuttle boosters, and from Martin Marietta, developer of the parachute system, would use the results to evaluate the adequacy of the system. (DFRC Release 21-77; MSFC Release 77-98)

• The *Washington Post* reported that biology instruments on both Viking landers on the surface of Mars had been shut off, according to a spokesman at the Jet Propulsion Laboratory, who added that they had detected no signs of life, although combinations of soil and chemicals like those on the landers were under test in laboratories across the U.S. trying to duplicate the unexpected results of the instruments on Mars. Those results were still being studied, and one test "leaves the question very much open." Don Bane of JPL said the two landers were continuing other operations such as meteorology and telephoto transmission. (*W Post*, June 1/77, A-10; NASA Release 77-111)

• The *W. Post* reported that the latest test of a satellite-destruction weapon conducted by the USSR despite an appeal by President Carter had apparently ended in failure. In a test May 23, first by the USSR in about 5mo, the Soviet interceptor spacecraft had missed the target satellite by about 50mi, according to U.S. intelligence sources. The officials said that, without knowing the objectives of a test, it would be hard to assess results, but the consensus was that the operation had failed. Five USSR tests in the past 18mo had failed to destroy a target satellite; the attempts had not been aimed at any U.S. satellites but were confined to their own. Both nations had used observation satellites to monitor each other's military activities, including strategic-weapons testing, missile-base construction, and similar indicators. Still in the research stage was a U.S. system to knock out Soviet satellites, with first flight tests at least 3yr away and first deployment estimated in about 5yr. (*W Post*, June 1/77, A-21)

June 2: NASA announced it had selected Boeing Services Intl., Seattle, Wash., for final negotiations leading to award of a cost-plus-award-fee contract to provide ground-systems support for launch operations managed by KSC and for USAF operations at the Eastern Test Range and Cape Canaveral, to include operation and maintenance of launch systems and facilities; estimated cost of the 3yr contract would be $80 500 000. (KSC Release 77-113; NASA Release 77-113)

• U.S. interplanetary spacecraft had confirmed that the planet Jupiter was emitting high-energy electrons over a vast region of space, Walter Sullivan reported in the *NY Times*. The spring meeting of the Am. Geophysical Union in Washington, D.C. heard Dr. John A. Simpson of the Univ. of Chicago and others describe the Jupiter emissions as similar to pulsars—electromagnetic radiation appearing as pulses sweeping through space like airport beacons—but carrying the "signature" of the planet's spin rate, picked up every 10hr as a rotation was completed. Dr. Simpson said Jupiter was not a direct analogy of pulsars but would offer common features that might provide clues to an understanding of both.

Observations of the Jupiter emissions had come from *Pioneer 10*, which had passed Jupiter in Dec. 1973 and was now beyond the orbit of Saturn, and from *Pioneer 11*, which had first passed Jupiter in Dec. 1974 and was scheduled to approach Saturn in Sept. 1979. The volume of Jupiter's magnetosphere had been estimated as 5 times that of the sun's; Io, innermost moon of Jupiter, had an orbit within the magnetosphere and its radiation belt (a greatly intensified version of earth's Van Allen belt), and the unusual brightness of Io on emerging from Jupiter's shadow had been attributed to particle activity. The 2 Voyager spacecraft to be launched this yr toward Jupiter and Saturn would be programmed to investigate further when they arrived nearby in March and July of 1979. (*NYT*, June 2/77, B4)

June 3: NASA reported that *Pioneer 11* would cross the orbit of Jupiter on June 10 for the second time on its trip to the outer reaches of the solar system. It had first flown past Jupiter in Dec. 1974 when the planet's gravity had kicked it into a path that would go past Saturn in Sept. 1979. This was the first time a spacecraft had used the gravity of an outer planet to accelerate it toward a different planet; the Voyager spacecraft to be launched this summer would also use this maneuver.

Pioneer 11 had covered two-thirds of its 2.3-billion-km (1.5-billion-mi) journey from Jupiter to Saturn; passing Saturn, it would head out of the solar system in the direction that the solar system takes through the galaxy, nearly opposite the path of *Pioneer 10*, now between the orbits of Saturn and Uranus, also on its way out of the solar system. Ames Research Center had managed the Pioneer project for NASA's Office of Space Science; TRW Systems, Inc., had built the spacecraft. (NASA Release 77-112; ARC Release 77-33)

• DFRC announced that a malfunction during a June 2 test of the Space Shuttle orbiter's auxiliary power system would delay the orbiter's first manned captive flight from June 9 for 1 or 2 weeks, depending on successful completion of ground tests. (DFRC Release 22-77)

• United Nations Secretary-General Kurt Waldheim and delegates from 14 nations had recorded messages to be carried with music and natural sounds on a copper phonograph record by the Voyagers scheduled for launch from Cape Canaveral Aug. 20 and Sept. 1, *Today* newspaper reported. Messages were from Australia, Canada, the U.S., Chile, France, Belgium, Austria, Sweden, Egypt, Nigeria, Sierra Leone, Iran, Pakistan, and Indonesia. They included poems in French and Swedish, a passage from the Koran in Arabic, and various texts in Esperanto, English, Flemish, German, Spanish, Persian, Urdu, Indonesian, the Creole language of Sierra Leone, and the Efik language of Nigeria.

In his message, Waldheim said: "As the Secretary General of the United Nations, an organization of 147 member states who represent almost all of the human inhabitants of the planet Earth, I send greetings on behalf of the people of our planet. We step out of our solar system into the universe seeking only peace and friendship, to teach if we are called upon, to be taught if we are fortunate." NASA spokesman Timothy Ferris said the messages, aimed at extraterrestrial intelligences, would be "like a note in a bottle for the extraterrestrials, if they exist." Voyager 1, which would reach Jupiter in 1979 and Saturn in 1980, would pass out of the solar system several yr later. (*Today*, June 3/77, 10A; *NYT*, June 3/77, A4)

• Scientists at the McDonald Observatory in Los Alamos, N.M., had discovered that the red-giant star Betelgeuse, familiar feature of the constellation Orion, was 3 250 000 times larger than earth's sun. Pictures taken at the observatory, using a new process with a special one-of-a-kind television camera tube produced by RCA, had shown the star expanding and throwing off matter into space at a prodigious rate; the photographs for the first time had viewed and measured the shell of matter around a red giant.

Betelgeuse and its shell had proved to be 400 times larger than the orbit of Pluto; it had measured nearly 3 trillion miles across (2800 billion miles). Earth's sun in comparison was 864 000 miles in diameter at its equator, and earth itself was only about 8000 miles in diameter at the equator. The pictures had verified a theory that red-giant stars were a primary source of raw material for making new stars, according to astrophysicist Maxwell Sandford of McDonald laboratory. The red giants would routinely lose matter to relieve internal pressures built up by consumption of nuclear fuel, and the pictures had shown the loss to be much greater than previously thought. (*Today*, June 3/77, 11A)

• NASA belt-tightening measures that might cut pay for about 60 high-ranking JSC employees might have the same effect at KSC, *Today* newspaper reported. The *Houston Post* reported earlier that the JSC employees had heard their Civil Service job grades would be reduced, with possible loss of $5000 to $10 000 in annual pay to follow 2yr later. A JSC spokesman said the grade reductions had resulted from a slowdown in space programs requiring reorganization of the center. Similar grade reductions were a possibility for KSC, which had been told by NASA Hq to cut its operating budget from $119 million to $109 million. KSC Director Lee Scherer had ordered a freeze on new hiring to allow a complete review of operations and consideration of measures to save money. KSC's director of public affairs, Charles Hollinshead, said activities subject to possible funding cuts after Oct. 1 would be "things we can cut that don't impact the launch schedule [for expendable rockets] or Space Shuttle preparations." (*Today*, June 3/77, 16A)

June 4: The *NY Times* reported that the General Accounting Office, Congress's investigative arm, had told the lawmakers to consider delaying completion of the Space Shuttle fleet, originally planned to include five ships. Two had been built so far. GAO said Congress should go ahead will building a third Shuttle but should put off construction of the other two until "there is sufficient confidence in the shuttle development program and more information is available on the space transportation system operations cost and plans for future space activity."

Today noted that this report was "the most sweepingly and harshly critical" of a number of negative assessments of the Shuttle issued by GAO. Among its charges were that the Shuttle would cause sonic booms over Calif. and Fla. upon landing; that its tires would blow out and cause a crash when it tried to land; that it could make too few orbits; that there was no proof it would reduce the costs of carrying cargoes into space; and that hardly anyone would want to use its services except DOD and NASA itself. GAO pointed out that other nations, led by Japan, had been developing their own launch vehicles and were likely to use them instead of paying to use the Shuttle; it also said NASA had erred in setting its fees low to encourage users to change over from expendable launchers. NASA's plan to recover expenses later, when costs had fallen with increased use, might never be possible, the GAO said.

Former NASA administrator James C. Fletcher said he disagreed with most of GAO's comments: NASA would solve the technical problems as it had solved others, he said, and delaying acquisition of the fourth and fifth Shuttles would be wasteful. "If they are procured while tooling and assembly lines are in place, they unquestionably can be procured at far less cost than would be incurred by delay." A 3-yr delay would run up the program cost by from $3 billion to $4.5 billion, he said. (*NYT*, June 5/77, 43; *Today*, June 4/77, 8A)

• The *NY Times* reported that a new USAF A-10 antitank attack jet crashed June 3 at the Paris Air and Space Show, killing Howard W. Nelson, chief test pilot for Fairchild Industries. Nelson, who had flown 105 missions during the Korean War and had over 10 000hr total flight time, had flown 500hr in A-10s; he was over the runway and hit the plane's tail on the ground in view of his wife and thousands of spectators. No one else was injured. This was the 6th crash in 12yr at these shows, held yearly to exhibit the latest aircraft to the world market; at the 1973 show, a Soviet Tupolev SST jetliner had crashed into a group of houses near the field killing 5 crewmembers and 8 residents.

A Fairchild spokesman said the USAF would set up a board to investigate the crash. The USAF had bought 30 A-10s at $4 million each, and a squadron of these was scheduled to go into operation June 10 at Myrtle Beach AFB, S.C. This was the first major accident since the A-10 was first flown in 1972. (*NYT*, June 4/77, 3; *Today*, June 4/77, 8A)

June 6: MSFC announced that SPAR IV, fourth of the space-processing applications rockets, would be launched June 21 carrying four experiments including a new acoustic-levitation device for containerless processing and a device to magnify in-flight photographs of the materials being processed. Principal investigators would be Dr. Taylor Wang of JPL, using acoustic pressure to handle a drop of water without touching it; Dr. Donald Uhlman and Dr. Bennett Joiner of MIT, studying solidification of 7 different materials in low gravity; Dr. John Papazian of Grumman Aerospace, studying castings produced in low gravity; and Dr. Arthur Lord of Drexel Univ. with Jerry Wouch of General Electric, using the electromagnetic levitator to make an alloy in low-g and study its properties. (MSFC Release 77-80)

• INTELSAT announced it had awarded to Miteq Inc., Happauge, NY, a $34 800 contract for up-converter and down-converter units to be used with time-division multiple-access test-bed equipment. (INTELSAT Release 77-13-M)

• The Natl. Space Development Agency of Japan had decided to build a third version of its N launch vehicle before the mid-1980s, *AvWk* reported. The N-3 would be able to put into synchronous orbit a payload weighing 500kg (more than 1000lb). N-1, first of the series, had used a liquid-propellant first stage like that in the McDonnell Douglas Delta, with three Nissan-designed strap-on motors instead of the 9 on the Delta. (*AvWk,* June 6/77, 203)

June 7: The *NY Times* reported that Dr. Robert A. Frosch, testifying at the Senate hearing on confirmation of his nomination as NASA administrator, had told the committee that no U.S. manned landing on Mars could occur before 1990. If NASA had an operating Space Shuttle by 1980 that could be used to assemble a vehicle in space and to store the fuel needed for a trip to Mars, work on a manned Mars ship could begin in 1984 or 1985 at the earliest, Frosch said, with completion a year or two later. (*NYT,* June 7/77, 13)

• The *Chicago Tribune* reported that A. Thomas Young, NASA's head of lunar and planetary programs, had presented to the Washington, D.C. meeting of the Am. Geophysical Union a planetary-exploration plan to launch two roving vehicles to Mars in 1984, a possible robot-vehicle landing on Mercury, and flights to Venus, Jupiter, Saturn, Halley's Comet, and four asteroids. The plan also included launch of automated ships to Mars in 1990 to collect samples of rock and soil and return them to earth. (*C Trib,* June 7/77, 4-16)

June 8: NASA announced it had published a book, *Space Settlements: A Design Study* (NASA SP-413), describing construction and operation of permanent settlements in space where up to "10,000 people work, raise families, and live out their lives." The 185p volume contained results of a

10-wk study in 1975 sponsored by NASA and the Am. Soc. of Engineering Education at ARC and Stanford Univ. Chapter topics included physical properties of space, human needs in space, habitat design, space manufacturing, agriculture, and settlement locations. Concluding that permanent communities could be built and inhabited away from earth, the report said the obstacles to further expansion of human frontiers were "principally philosophical, political, and social rather than technical." (NASA Release 77-118)

• The *Wall St Journal* reported that the government of France had announced plans for a major overhaul of its "financially troubled aerospace industry," that would include partial nationalization of Avions Marcel Dassault-Breguet, maker of the Mirage jet fighter. A sharp drop in orders for civilian and military aircraft had forced companies to dismiss workers and cut down operations; the government hoped to introduce new products, step up sales, and increase cooperation with other European aircraft firms. (*WSJ,* June 9/77, 13)

June 10: ARC announced that a team of scientists using the 91cm (36in) infrared telescope on the Kuiper Airborne Observatory and the ground-based 230cm (90in) infrared telescope at Steward Observatory, Univ. of Arizona, had made the first discovery of planet formation in process and the first identification of a flat disc-shaped luminous stellar object. The object, MWC 349 in the constellation Cygnus, exhibited a surrounding mass of intensely glowing gas with a diameter 20 times that of the central body, emitting 10 times as much light; the luminous area would disappear in about a century as disc material spun into the central body. The find had been significant because it might show how planets had been formed in earth's solar system.

Members of the team were Drs. Rodger Thompson and Peter Strittmatter of the Steward Observatory, and Drs. Edwin Erickson, Fred Witteborn, and D.W. Strecker, all of ARC. Infrared instruments could view the object through dust shrouding the disc: the Kuiper telescope had detected infrared spectra that could not penetrate earth's atmosphere, and the Steward telescope could offer higher resolution. (ARC Release 77-33; NASA Release 77-119; *NYT,* June 16/77, A21; *W Post,* June 16/77, B-11)

• NASA announced that the first manned test flights of the Shuttle orbiter Enterprise would begin at DFRC about June 16. In these captive flights (phase 2 of Shuttle approach and landing tests) all orbiter systems would be ON to verify crew procedures and determine the best separation profile for upcoming manned free flights. In the initial phase that ended Mar. 2, the orbiter fastened to its Boeing 747 carrier had made five captive inert (unmanned) flights at DFRC to check out its systems; in phase 2, the orbiter would make four flights still attached to the 747 but with crew aboard to check flutter, steering, and other performance items. (NASA Release 77-117)

• Two main items in NASA's $4 billion budget request for FY 1978—the Jupiter-orbiter mission and the Space Telescope—had met with strong opposition in Congress, John Noble Wilford reported in the *NY Times*. The items were two of three major new starts sought by the agency; the third, a new earth resources monitoring spacecraft (Landsat-C), had not been opposed, nor had additional funding for Space Shuttle development.

NASA officials had not expected opposition to the Jupiter project, which had the endorsement of numerous scientists and the White House and had been authorized for eventual development by both houses of Congress. However, Rep. Edward P. Boland (D-Mass) had persuaded the House appropriations subcommittee on independent agencies, which he chaired, to block funds for the Jupiter project in view of charges by many astronomers that NASA had spent "a disproportionate level" of its funds on planetary science rather than on deep-space astronomy, which would be the aim of the Space Telescope. The House Appropriations Committee had followed suit and the full House was expected to vote accordingly.

NASA had concentrated its Jupiter appeal on the Senate appropriations subcommittee headed by Sen. William Proxmire (D-Wis), longtime critic of NASA spending, who had nevertheless been impressed by JPL officials who argued that the Jupiter project should receive immediate funding because of a favorable launch situation due in 1982 when Jupiter would be relatively close to earth, which would not occur again until 1987. Without the new start, JPL would have to lay off 300 to 400 scientists and engineers because it would have no missions beyond the launch of two Voyagers this summer. If the Senate should approve the Jupiter project and reject the telescope, a conference committee would have to settle the matter; NASA might have to choose between the two projects, or to start both with greatly reduced first-year funding, Wilford noted. (*NYT*, June 10/77, A-17; *Nature*, June 23/77, 659)

June 11: A new estimate of the cost of building three more Shuttle orbiters might run as high as $600 million apiece, Thomas O'Toole reported in the *W Post*. NASA had told Congress 3mo ago that the remaining Shuttles would cost $550 million apiece; the increase had resulted from the rising price of aluminum and titanium, the two key metals in Shuttle construction. NASA had bought the materials for two vehicles 3yr ago when prices were lower but had held off buying for the last three.

Only one of the five Shuttles presently proposed had been completed, at a cost of about $500 million. A second being assembled at the Rockwell Intl. plant in Palmdale, Calif., would cost just over $500 million. Christopher C. Kraft, Jr., director of JSC (the Shuttle-management center), said that NASA expected "by the end of this month a new proposal from Rockwell for the last three spacecraft. . . . I don't know how much Rockwell's going to suggest but they're for sure going to cost more than the last estimate we got."

NASA had hoped Congress would approve starting construction of the third Shuttle before the end of 1977. This vehicle was to be sent to Vandenberg AFB in March 1982 for Pentagon use. The fate of the last two Shuttles was not certain, since GAO had told Congress to delay further action on them. NASA and the USAF had expressed concern that delay would mean not only the shutdown around the country of contractor and subcontractor work forces that would be both difficult and costly to reestablish, but also further and more rapid escalation of prices: O'Toole quoted a NASA estimate that a 2yr delay would boost to about $1 billion the price of each Shuttle affected. (*W Post*, June 11/77, A-2)

• The *NY Times* reported that ERDA's Sandia Laboratories in N.M. had produced the first fusion neutrons in the U.S. by firing a high-energy electron beam at pellets of heavy hydrogen. Use of electron beams was a late development in the search for economical methods of fusion-energy release. Existing atomic-energy plants had operated by fission, or the splitting of large atoms; fusion would squeeze small atoms to make heavier ones. Each reaction would convert a small amount of matter into large amounts of energy. Whereas fission generally used uranium, a relatively scarce material, fusion could use the heavy form of hydrogen called deuterium abundant in sea water. The process used at Sandia had differed from that used by researchers at Moscow's Kurchatov Inst. who last year reported success in releasing fusion energy. (*NYT*, June 11/77, 22; ERDA Release 77-99)

June 12: A new $1.6-million infrared telescope with a 92-in-dia. mirror (the world's largest) to be installed soon in an observatory atop 9656-ft Jelm Mt. in Wyo. would be sensitive enough to measure the heat of a warm iron on the moon, according to *Today* newspaper. Professors Robert Gehrz and John Hackwell of the Univ. of Wyo. had supervised design of the telescope, which would incorporate a sensor also of their design: a computerized device sensing heat energy collected by the large mirror and converting it to an infrared contour chart depicting heat intensity as a topographical contour map would depict terrain. Jelm Mt.

was considered the best site for the observatory because of the low water vapor in the area, a requirement for good IR reception. The two scientists had persuaded the state legislature of Wyo. in 1975 to appropriate $975 000 for the project; remaining funds had come from the Natl. Science Foundation. (*Today,* June 12/77, 12A)

June 13: NASA announced it had awarded Boeing Commercial Airplane Co., Seattle, Wash., a $1.7-million contract for work on control technology for the Boeing 747 civil-transport aircraft, part of a program to increase fuel efficiency of future aircraft by up to 50%. Work would include design, wind-tunnel tests, and evaluation of modifications such as winglets, wingtip extensions, and active controls for load maneuvers, gust alleviation, and suppression of structural vibrations. LaRC would manage the contract for work to be done within 2yr at the Boeing plant in Seattle. (NASA Release 77-120; LaRC Release 77-25)

• INTELSAT announced that the Republic of Chad had become the 96th member of the organization, and the 23rd from Africa. The INTELSAT operating agreement had been signed June 9 by the Societe de Telecommunications Internationales du Chad. (INTELSAT Release 77-14-I)

• NOAA reported it had used two of its operational environmental satellites with NASA's *Landsat 2* to obtain imagery that could give early warning of river ice melts causing ice jams and flooding. David F. McGinnis, Jr., and Stanley R. Schneider of NOAA's Natl. Environmental Satellite Service (NESS) had studied the Ottawa River separating Quebec and Ontario provinces and joining the St. Lawrence River below Ottawa and Montreal over 12 days in April 1976 to find which sections were covered with ice. By the end of the period, all but three of the 14 sections studied had been free of obstructing ice. Disappearance of ice monitored by the team in the Chaudiere River of SW Quebec had resulted from hydraulic transport rather than melting, which had occurred in the Ottawa River because of the blocking effect of dams, islands, and sharp bends in the river. (NOAA Release 77-145)

• The Energy Research and Development Administration reported that its scientists had developed a new way to reclaim used automotive oils, removing solid and liquid impurities in a solvent-and-distillation treatment. After heating the used oil to drive off volatile hydrocarbons and water, the process would add a solvent to make sludge of the contaminants; the remaining oil would go through normal processes of distillation, improvement of color and odor, and reformulation with additives. Iowa State Univ. had run a vehicle-fleet test for more than 10mo using the re-refined oils with no abnormal wear or performance; the product had not generated polluting byproducts. Most commercial re-refining had used an acid-based technology producing wastes more polluting than the used oil itself. (ERDA Release 77-14)

• ERDA announced that the world's largest aircraft would airlift one of the heaviest loads ever transported by air June 19 when a USAF C-5 Galaxy carried a U.S.-built 40-ton magnet from Chicago to Moscow for tests of an experimental magnetohydrodynamic (MHD) process for generating electricity. The USSR's U-25 facility near Moscow, operated by the Soviet Institute for High Temperatures, was the largest MHD test facility in the world. Dr. William D. Jackson, head of ERDA's MHD program, said that using the Soviet plant would speed commercialization of the process in the U.S. at reduced cost to the American public. The magnet would aid in tests conducted jointly for about 2yr under the U.S.-USSR Energy Agreement of 1974. (ERDA Release 77-100)

• The new administration of President Carter would not initiate a working Landsat earth resources monitoring system until it found out how much money Landsat data users would invest in such a system, *AvWk* reported. The president's science adviser Frank Press, director of the White House office of science and technology policy, would set up a committee to study the user-cost question. Press had told the Senate subcommittee on science, technology, and space that it would be "premature to commit the federal government" to support of an operational system because of many uncertainties about remote sensing; he said that development of a sector to be served by Landsat data had not evolved "as rapidly as some anticipated several years ago," and that "no comprehensive assessment" had been made "of the overall market structure for Landsat data applications. . . ."

The subcommittee was considering a bill cosponsored by Sen. Wendell H. Ford (D-Ky) and former astronaut Sen. Harrison M. Schmitt (R-NM) that would establish an operational Landsat system, the space segment to be run by NASA and the ground data distribution by the Dept. of the Interior. *AvWk* noted, however, that the Interior testimony had been altered at the last minute to deny support of the bill. (*AvWk*, June 13/77, 91)

June 14: NASA announced that, because of recent flight and checkout problems, it had revised the launch schedule of its Delta rockets: GOES-B, June 16; GMS (Japan's meteorological satellite), July 14; Sirio (Italian comsat), Aug. 11; OTS/ESA (comsat), Sept. 8; ISEE-A/B, Oct. 13; Meteosat (ESA metesat), Nov. 3; and CS (Japanese comsat), Dec. 12. (NASA Release 77-123)

• NASA announced that LaRC had undertaken a program called CLASS (cargo logistics airlift-systems study) to see whether tailoring an advanced air freighter design to an integrated transportation system could mean air-cargo growth and higher carrier profits. NASA had negotiated twin study contracts with Lockheed-Georgia Co. and with Douglas Aircraft Co. to survey air-cargo operations at major airports; to estimate present and future air-cargo demand, based on shipper response to reduced cost and improved service; and to study the need for improved efficiency in air-cargo operations, and for research and technology to support the design of future cargo aircraft. The studies had begun June 1 and would continue for 11mo. LaRC would exchange information in the study with government and industry organizations including the Dept. of Transportation, Federal Aviation Administration, Dept. of Defense, and private companies such as airframe manufacturers, air shippers, commercial airlines, and freight forwarders. (NASA Release 77-122; LaRC Release 77-26)

June 16: Rockwell Intl.'s B-1 division announced it had made significant savings in awarding a contract to Honeywell Inc.'s avionics division for equipment to test electronic systems on the USAF B-1 strategic bomber. B-1 division president Bastian Hello said the new equipment would use large amounts of off-the-shelf items and would be able to test a wide range of systems supplied by Rockwell, Boeing, and the AIL division of Cutler Hammer. The contract would direct Boeing and AIL to get test equipment from Honeywell under separate purchasing agreements. Rockwell should receive first deliveries of the equipment in 1978. The USAF would use the test equipment to maintain B-1 operations; Boeing (offensive-avionics integrator) and AIL (defensive-avionics contractor) would use the equipment in supplying electronic instruments to the program. (Rockwell Release LA-6)

June 16—Aug. 18: NASA launched *Goes 2,* second in a series of geostationary operational environmental satellites, from the Eastern Test Range at 6:51am EDT June 16 on a Delta vehicle into a transfer orbit. On second apogee the boost motor fired at 11:26pm EDT to put the spacecraft in a nominal orbit where it would move to a position specified by NOAA at 75°W, and be turned over to NOAA for operations in about 30 days. Orbital elements were 36 859km apogee, 188.6km perigee, 23.7° inclination, and 651.8-min period.

NASA announced on Aug. 18 that it had turned *Goes 2* over to NOAA for operational use on July 29 at 1600Z. NASA engineers had completed a planned orbital checkout to ensure proper operation, having moved the satellite to 66.5°W over the equator where most of the tests occurred. Of the original 80lb of hydrazine carried for stationkeeping, more than 67lb remained unused. The engineers had found minor problems in the data-handling equipment used to relay signals from earth-based data-collection platforms, and in the x-ray section of the solar-environment monitor; the data-collection problem had resulted from interference by VHF equipment that could be turned off for most of the mission, but the x-ray problem arose from faulty cabling that might require more calibrations. *Goes 2* was officially certified as successful on Aug. 5–15, having achieved its mission objectives. (MOR E-608-77-05 [postlaunch] June 20/77, [postlaunch] Aug 18/77)

June 17: NASA announced that preparations at KSC for launching the Space Shuttle had required only two new major constructions: an orbiter-landing facility, one of the largest runways in the world (roughly twice as long and twice as wide as an average commercial landing strip), with a microwave scanning-beam landing system to assist the orbiter to an automatic landing; and a large hangar called the "orbiter processing facility" with two high bays for checking out and servicing the orbiters immediately after landing.

Existing facilities modified for Shuttle use were the Vehicle Assembly Building, 2 of its 4 high bays being altered for Shuttle assembly, and the launch-control center being equipped with an automated launch-processing system developed for Shuttle checkout and launch, using a tenth of the manpower for Shuttle that had been used for Apollo: 45 persons, compared to more than 450. Final countdown for Shuttle launches would take only 2.5hr compared to the 28hr needed in countdown for an Apollo/Saturn V. Much of the work had been completed, NASA reported, and the eventual cost of the modifications would be about $240 million, less than a fourth of the 1960s cost of building the Spaceport for Apollo. (NASA Release 77-124)

• MSFC announced it had awarded to Contractors Cargo Co., South Gate, Calif., a $227 120 fixed-price contract for moving the Space Shuttle orbiter Enterprise eight separate times during its testing at MSFC in 1978. Now in the midst of approach and landing tests (ALT) at DFRC, Enterprise would fly on its Boeing 747 carrier to MSFC on March 17, 1978, to be mated with an external tank and solid-fuel rocket boosters and erected on a dynamic test stand for vertical ground-vibration testing. The contract called for design, testing, and delivery to MSFC of an orbiter ground-transporter to travel about 6 km between the airfield and the pretest-preparation building, a trip to be completed within 2hr on March 18, 1978.

The "prime mover," a 480-hp tractor with 10 wheels, would be attached to 2 rear dollies each having 32 wheels and a forward dolly with 16 wheels. The segments would be joined by a "strongback" construction, and the entire rig would weigh about 24 000kg (52 900lb). Length of the transporter would be nearly 50m (about 160ft), the width of about 24m would accommodate the orbiter's wingspan, and the height from ground to top of a vertical stabilizer would be about 16m (about 54ft). For an operating demonstration before the arrival of the orbiter at MSFC, the transporter would carry a weightload of nearly 100 000kg (220 000lb), half again the weight of the orbiter. Contractors Cargo was the same firm that had moved Enterprise from the Rockwell plant at Palmdale, Calif., to DFRC in Jan. 1977. (MSFC Release 77-106)

• Dr. Robert A. Frosch, awaiting confirmation as NASA's fifth administrator, probably had "more experience in research administration" than any of his predecessors in that position, *Science* magazine said. The article pointed out that Dr. Frosch—former assistant secretary of the Navy for research and development, former assistant executive director of the UN environmental program, former associate director of the Woods Hole Facility, and director at age 28 of Columbia Univ.'s Hudson Laboratories—had been stressing science in his initial public statements. "I'd like to be remembered as the guy who was able to help NASA imagine new uses for space and aeronautics and who helped the agency do good science," Frosch told *Science* in an interview.

The magazine noted, however, that the new administrator would take office at a time when Congress was cutting NASA projects and taking the agency "completely by surprise." NASA had enjoyed good relations with the chairman and ranking minority member of the Senate committee authorizing NASA programs, who were also ex officio members of the appropriations committee; but the old authorizing committee had been dismantled, and NASA had come under the jurisdiction of a subcommittee of the Senate commerce committee headed by Sen. Adlai E. Stevenson (D-Ill). Also, aside from his introduction to the president ("He sure gives a tough job interview," Frosch said), the new administrator had not been dealing with the president or his immediate staff. Overall administration attitude toward the space agency was a big unknown, the magazine stated, and rumors had been circulating that NASA would be taken over by the Commerce Dept. or the Dept. of Defense, although a spokesman for Carter's government reorganization chief had denied the rumors. Although Dr. Frosch might preside over a stronger science program for NASA, the article concluded, he might also preside over a change in the agency to "some new, perhaps unrecognizable form." (*Science,* June 17/77, 1301)

• Newspapers and news services reported the death June 16 of Dr. Wernher von Braun, 65, former director of NASA's Marshall Space Flight Center, whose name had become synonymous with "America in space" when the team he headed used a Jupiter-C rocket developed by him for the U.S. Army to launch *Explorer 1,* the western world's first earth satellite, on Jan. 31, 1958. Dr. von Braun had undergone surgery in 1975; he had spent most of his time since Oct. 1976 in the hospital.

Dr. von Braun, an enthusiast of space travel and astronomy from his childhood in East Prussia (now part of Poland), in his early teens came across a picture of a rocket traveling to the moon that illustrated an article by Hermann Oberth, pioneer rocket theorist who later was part of the von Braun team at Huntsville, Ala. When von Braun obtained Oberth's book on rocketry, he realized he would need mathematics to progress in his studies, and went on to obtain a doctorate in physics from the Univ. of Berlin in 1934 at the age of 22 with a thesis on rocket engines.

He had continued his interest in the amateur Society for Space Travel (Verein fur Raumschiffahrt, VfR) whose rocket experiments impressed the German army seeking weapons not forbidden by the Treaty of Versailles that ended World War I. Employed by the ordnance department after 1934, von Braun had continued his work with rockets, up to the success of the V-2 used against Britain, until 1945. The von Braun team at Peenemunde then decided to go south and surrender to U.S. forces rather than be captured by the Soviet army. About 120 of von Braun's associates were taken with him to the U.S. in Operation Paperclip to demonstrate their achievements with captured V-2 rockets. The group went to the U.S. Army's Redstone Arsenal in 1950 to work on a ballistic rocket called Redstone. When *Sputnik 1* went into orbit in 1957 and the Navy's Vanguard rocket blew up on its pad, a version of von Braun's Redstone called Jupiter-C (Juno 1) put *Explorer 1* into orbit for the U.S. in 1958, and another version carried Alan B. Shepard, Jr., on the first U.S. suborbital flight in 1961. Weeks later, when President Kennedy called for a moon landing within the decade, von Braun got the task of creating the rocket; his Saturn V won the race to put a man on the moon's surface in 1969.

When NASA was established in 1958, Dr. von Braun and his team had transferred to that agency from the army and he had become director of the Marshall Space Flight Center at Huntsville, Ala. After the moon landings, Dr. von Braun went to NASA Hq in 1970 as deputy associate administrator to promote post-Apollo space activities for the U.S. But public interest and support had declined, and Dr. von Braun resigned in May 1972 to become vice president for engineering and development at

Dr. Wernher von Braun, director of MSFC, photographed in 1969 near the actual Saturn V rocket AS-506 that would launch the first U.S. astronauts to the moon. (MSFC 9-00841)

Fairchild Industries, Inc. In 1975 he founded and became first president of the Natl. Space Institute, a private group to increase public understanding and support of space activities; in 1976, when illness overtook him, he remained chairman and was active in formulating policy. He had resigned from Fairchild late in 1976, effective in Jan. 1977.

At a memorial service June 22 in Washington Cathedral, tributes came from former astronaut Michael Collins; Dr. Ernst Stuhlinger, longtime associate of von Braun in development of rockets; and Dr. James C. Fletcher, former NASA administrator. An editorial in the *W Post* that described Dr. von Braun as "an American national hero in the 1960s after being an American national enemy in the 1940s" said that his life should be judged "as one of the new breed of international scientists . . . You can think of him as a hired gun if you like. But you can also think of him as he apparently thought of himself—as a man indentured only to a dream . . . And, unlike most of us, he saw a large part of it come true." (MSFC Release 77-111; AP wire service obit, June 17/77, nos. 17 & 20; *W Star*, June 17/77, A-1; *NYT*, June 18/77, 1; *W Post*, June 18/77, A7; June 23/77, C11, A24 (ed); *C Trib*, June 18/77, 1; *Today*, June 18/77, 1A; June 23/77, 8A)

• FBIS carried an announcement by Tass that the USSR had launched "this morning" the French satellite *Signe 3*, product of a 2yr collaboration between the Toulouse Space Research Center and the USSR's Space Research Inst. of the Academy of Sciences. The spacecraft had carried French-made instruments to observe ultraviolet emission from the sun; objective of the mission was to continue study of "already discovered sources of gamma radiation in galaxies and also to replenish the catalog of stars." The announcement said that originally the French were to design the spacecraft and provide the equipment, while the USSR ensured launch; however, "the Soviet colleagues suggested so many design novelties . . . that it is actually more correct to regard the *Signe 3* as a Soviet-French satellite," according to French project director Antoine Miosi. This launch had begun a second decade of USSR-French cooperation in space exploration: Soviet spacecraft had carried French-made instruments, and the French Araks experiment had orbited a Soviet electron accelerator to study polar lights. (FBIS, Tass in English, June 17/77)

June 19: The moon landing effort was possible only because the U.S. had laid the basic theoretical groundwork in the 1940s and 1950s, the *W Post* reported, quoting Richard C. Atkinson, new head of the Natl. Science Foundation. The man who "probably controls more basic research spending than anyone else in the world" said that Americans should not hope science would come up with an overnight solution to the energy crisis that would preserve their current lifestyle—"Science just doesn't operate with those turnaround times," he said. The fact that many people expected such solutions indicated "a basic incomprehension of what science is all about," an ironic byproduct of the mind-boggling scientific achievement in the past 20yr. "After Sputnik . . . there was a big rush to catch up here . . . They put in a lot of high-powered curricula . . . Now we have a small core of students really well educated in science and a huge group that just [drops out] . . . Those people never have any exposure again. Even college graduates now are really poorly informed on science matters."

Liberal education, Atkinson said, should include solid grounding in basic sciences so that citizens could understand issues like the fate of nuclear reactors and the gene-manufacturing possibilities of recombinant DNA. "I worry," he said, "that the nation may turn to applied research so strongly that it neglects basic research, and basic research remains our fundamental commitment." (*W Post,* June 19/77, C-4)

June 19-20: Newspapers carried reports of the first manned flight of the Space Shuttle orbiter Enterprise June 18, riding piggyback on its Boeing 747 transport during the 54-min trip over DFRC, and carried to a landing. The orbiter, designed for reuse as many as 100 times, remained attached to its carrier by 3 struts, two aft and one on the nose, for its investigation of structural flutter on the craft's brakes and flaps. Flying under the control of JSC in Houston, the two astronauts (Fred W. Haise, the civilian commander, and rookie USAF Lt. Col. Charles Gordon Fullerton) rode the piggyback craft to 15 000ft at 209mph for most of the flight.

Earlier unmanned tests of the piggyback orbiter in Feb. and March had gone smoothly; since then, engineers had installed control systems, guidance instruments, and power units whose repeated malfunctions had delayed the manned tests originally scheduled to begin May 26. The June 18 test had been delayed for 24hr by problems with the entry hatch and by the failure of guidance-system computers. The astronauts reported trouble with a cabin-pressure valve and with radio communications, and said they had experienced more airframe surface noise than expected. NASA had scheduled a second "manned inert" test for June 28, with the first free flight set for late July or early Aug. (*W Star,* June 19/77, A-6; *C Trib,* June 19/77, 1-3; *NYT,* June 19/77, 19; *W Post,* June 19/77, A-18; June 20/77, A-6)

June 20: JSC announced it had signed a $40 million letter contract with Rockwell Intl. Corp. Space Division for modifications to the first Space Shuttle orbiter Enterprise and for starting design of a third orbiter. Funded in part by the Economic Stimulus Bill, the contract had been directed at generating more employment opportunities. (JSC Release 77-37)

•JSC announced it had signed a supplemental agreement with Rockwell Intl. Corp. Space Division deleting two items from the Space Shuttle orbiter contract: a neutral-buoyance trainer, and vibroacoustic tests on the aft fuselage of the orbiter. This work had been scheduled for Rockwell's plant in Downey, Calif., with support from JSC and KSC field offices. The decrease of $25 334 750 had lowered the value of the cost-plus-award-fee contract with Rockwell to $3 013 971 603. (JSC Release 77-38)

• *AvWk* reported that the USSR and France had agreed on a joint mission to Venus in 1983 for which France would design a balloon 26 to 30ft in diameter to float in the planet's atmosphere for up to 100hr collecting data, and the USSR would design an orbiter weighing more than 600kg (1322lb). Jacques Blamont, chief scientist for the French space agency CNES (Centre National d'Etudes Spatiales), said a Soviet mission planned for Venus next year would also carry French multiwavelength ultraviolet spectrometers on at least two spacecraft scheduled for flybys. (*AvWk*, June 20/77, 27)

June 21: JPL announced that NASA had followed its recommendation and selected a 12-bladed spinning solar sail spacecraft as a candidate for the automated interplanetary shuttle to be used within earth's solar system in the 1980s and beyond. Also known as a heliogyro, the set of extremely long blades or sails made of reflective aluminized plastic would be deployed in two tiers of six each by centrifugal force after its spacecraft had been launched from the Space Shuttle late in 1981. The spacecraft and its scientific payload, mounted at the center of the heliogyro, would be propelled by photon radiation from the sun to rotate once every 3min. The spinning-sail concept had been chosen by program engineers and designers for development as being more feasible than a square-sail concept that would have used an 850sq.m (half-mile square) sail configuration. NASA would choose in mid-Aug. between the heliogyro and a proposed ion-drive (solar-electric propulsion) spacecraft for a Halley's Comet mission in 1986. JPL had been working with other NASA centers and a dozen industrial and research groups on parallel efforts to develop low-thrust long-life spacecraft for the 1980s. (JPL Release 826)

June 22: MSFC announced it had received 26 proposals from researchers in the U.S., the U.K., and Italy for cloud-physics experiments to be performed on Spacelab. Cloud-physics research would try to explain cloud formation and release of precipitation. Preventing or limiting the effects of adverse weather such as severe storms or floods would enable man to avoid yearly losses of lives, property, and agricultural crops. MSFC teams would evaluate the proposals from engineering, management, cost, and scientific peer group standpoints; in July NASA would select experiments that would fly in 1980 or 1981. (MSFC Release 77-114)

• MSFC announced it had successfully tested the recovery system for the Shuttle solid-fuel rocket boosters June 15 at the Natl. Parachute Test Range. Engineers had used an SRB simulator, a drop-test vehicle weighing more than 21 773kg (48 000lb) released from beneath the wing of a B-52 aircraft at a speed of about 85 meters per sec (190mph) and an altitude of about 5500m (18 000ft). Primary purpose of this test was to measure the maximum load on the drogue chute that would open first to stabilize the vehicle and pull out the other chutes. The measured load, about 114 000kg (a quarter of a million lb), was about that expected. The next test, scheduled for late July or early Aug., would measure loads on the other parachutes. Each Shuttle launch would use two SRBs, jettisoned at about 43km (27mi) altitude to descend by parachute into the ocean and be towed to shore for reuse. (MSFC Release 77-115)

June 23: The USAF launched the Navy's navigation technology satellite *Nts 2* from Vandenberg AFB on an Atlas F booster into a 12hr circular orbit of 63° inclination at about 11 600mi altitude. First of a system to be called NavStar, the *Nts 2* carried two cesium-beam time standards (atomic clocks) to produce precise time signals and 14 solar-cell experiments for its builder, the Naval Research Laboratory. The joint-service NavStar system upon completion would consist of 24 satellites (8 in each of 3 circular-orbit planes) to provide users with longitude, latitude, and altitude information around the clock in any weather. It would give a customer his location within 33ft, his speed within about 4in/sec, and the correct time. Uses of NavStar would include tactical-missile navigation and precision delivery of weapons; aid to space, air, land, and sea travel; grid mapping and geodetic surveys; aerial rendezvous and refueling; and search-and-rescue operations. The USAF had announced plans to launch five of its own satellites later in 1977 and 1978 for NavStar testing. (*D/SBD,* June 20/77, 279; June 27/77, 320; *AvWk,* Oct 11/76, 47; NRL Release 20-7-77C; AFSC *Newsreview,* Aug 77, 3)

June 24: ComSatCorp (Communications Satellite Corporation) announced that the U.S. Navy's navigation technology satellite launched June 23 from Vandenberg AFB had been the first satellite to use a nickel-hydrogen battery for energy storage. *Nts 2* had carried two 7-cell storage batteries said to be rechargeable, highly reliable, and longer-lived than the nickel-cadmium batteries used by satellites, having three times the amount of energy stored per unit of weight and five times the cyclic lifetime of the nickel-cadmium type. The Naval Research Laboratory had sponsored development of the new batteries by ComSatCorp Laboratories in Clarksburg, Md. (ComSat Release 77-19)

• MSFC announced arrival June 24 at the Natl. Space Technology Laboratories of a huge steel structure to simulate the Space Shuttle orbiter in main propulsion-test article firings later in 1977. The article would include the simulator with an aft fuselage, an external tank, and three Shuttle main engines; first test would take place in Dec. The simulator, about 29m (95ft) long and weighing about 110 000kg

(243 000lb), had moved from the Michoud assembly facility near New Orleans to NSTL in Miss. on the barge Pearl River, having traveled from Calif. to Michoud via the Panama Canal on the NASA barge Poseidon [see May 10]. (MSFC Release 77-118; NASA Release 77-132)

• NASA announced that *Skylab 4* commander Gerald P. Carr would leave JSC June 25 to join a Houston firm, Bovay Engineers, Inc. Carr, one of 19 astronauts selected in Apr. 1966, had shared a spaceflight-duration record of 2017hr (84 days) with Dr. Edward G. Gibson and William R. Pogue, and had retired from the U.S. Marine Corps in Sept. 1975 to become a NASA civilian employee. His departure left 27 astronauts on the active list at JSC. (NASA Release 77-129)

June 25: MSFC officials had said the Space Shuttle main engine might cost nearly twice the original estimate, *Today* reported. In 1972 when Rockwell's Rocketdyne Division got the contract to build the engine and deliver 24 units, the estimated cost was $500 million; latest estimate was $450 million more. Project director James R. Thompson and his deputy John Harlow said the increases had resulted not only from design problems and inflation but also from delays and changes ordered by NASA and Congress. (*Today,* June 25/77, 6A)

June 27: NASA announced a new way to control the flight of its sounding rockets and point their instruments more accurately in astronomical and other scientific studies: a small TV camera in the nose cone would allow an operator on the ground to remotely aim the onboard attitude-control system precisely at the desired stellar object. In the first use of "manned" rocket observation Apr. 15 at the White Sands Missile Range, N.M., scientists from Johns Hopkins Univ. had made the first sighting of a quasar in the ultraviolet spectrum. The quasar (quasi-stellar object, a celestial source resembling a star but probably a galaxy) was 3 billion light-yr from earth; it was the faintest and most remote object observed from above earth's atmosphere. Dr. Arthur Davidsen of JHU's department of physics had operated the rocket controls while watching the TV screen; his colleagues were Professors William G. Fastie and George Hartig. Observations in the ultraviolet range would let scientists compare properties of relatively near and more distant quasars. (NASA Release 77-131)

• The U.S. superconducting magnet flown to a USSR magnetohydrodynamic facility near Moscow [see June 13] had been bolted into place June 20, ERDA reported. The successful installation, said Dr. William D. Jackson, head of ERDA's MHD program, was comparable to the 1975 linking of the Apollo and Soyuz spacecrafts. Final leg of the magnet's journey had been a 10-mile trip on a U.S. truck flown in with the magnet on a USAF C-5 Galaxy aircraft, escorted to the facility by Soviet police. Operation of the system would begin in Oct., and about 20 U.S. scientists and engineers would take part in the 2yr experiment. (ERDA update, June 27/77)

• Interest expressed by the USSR in joint development with the U.S. of spaceborne systems for navigation, air surveillance, and air-to-ground communications had come at an awkward time for the U.S., *AvWk* reported. Congress earlier in 1977 had killed the Aerosat program to launch two air-to-ground comsats sponsored jointly by the FAA and ESA with Canadian participation; ESA might seek its own joint effort with the USSR. Also, the Dept. of Defense's NavStar system, designed mainly for military navigation users through a secure signal system, had offered a "clear" system for civilian use that the DOD had been trying to sell to airlines and other potential civilian markets. Soviet officials (who had told a visiting delegation from the U.S. radio-technical commission for aeronautics that they had been keeping tabs on NavStar development) were apparently more enthusiastic prospects for NavStar than U.S. users. (*AvWk*, June 27/77, 17)

June 28: NASA announced that the second manned inert flight of the Shuttle orbiter Enterprise atop its Boeing 747 carrier, with astronauts Joe Engle and Richard Truly at the controls, was so successful that the agency had canceled a fourth test and would limit the captive flights to three. The third flight would be put off until July 28 to allow replacement of main landing-gear actuators and a leaky power unit, and installation of 100gal hydraulic reservoir tanks. The orbiter would remain mated to the 747 during the work.

The faulty auxiliary power unit had been activated on schedule about 16min into the flight; a small leak noticed before the captive-active tests had not been thought serious enough to require replacement. After the 747 had taxied to a ramp area for shutdown, the leak appeared to have increased.

The June 28 test had included tasks originally scheduled for the second and third manned tests: low- and high-speed flutter and speed-brake evaluations; a separation-maneuver test; and a fly-through of the microwave scanning-beam landing system that would guide the orbiter to earth in its first free flight. The flutter test had begun about 3min after takeoff, moving the control surfaces first of the orbiter, then of the carrier aircraft. The pilots deployed the orbiter's speed brakes to 60, 80, and 100% open positions, pausing between each setting for rudder-deflection test and flight assessment. The high-speed flutter tests at an altitude of about 20 000ft and a speed of 270 knots, followed by speed-brake tests between 20 000 and 16 000ft, were in the same sequence as the earlier tests except that the brake settings were in 10% increments.

Upon completion of these tests (34min into the flight) the carrier climbed to 20 600ft for a separation data run. "Pushover" occurred at 43min into the flight; the data run terminated with an "abort separation" order at 14 000ft. The carrier returned to 17 500ft for the automatic fly-through, followed by landing after a total flight time of 63min.

Early data from this test indicated that separation conditions would be satisfactory for free flight; that the mated configuration was flutter-free for the flight envelope; and that operating the speed brake would produce no significant buffeting. (Postflight rept, orbiter ALT #7, captive-active flt #2, June 28/77; *W Post*, June 29/77, A-3; *B Sun*, June 29/77, 3; *LA Times*, June 29/77, 2; *W Star*, June 29/77, 2; *Today*, June 29/77, 10A; *AvWk*, July 4/77, 18)

June 29: NASA announced the E.S. (Todd) Groo, associate administrator for center operations since 1974, would leave the agency July 23. Ray Kline would succeed him, remaining assistant administrator for institutional management. Deputy NASA Administrator Alan W. Lovelace said Groo had benefited the agency by clarifying the missions of the field centers and instituting training for future managers. (NASA Release 77-134; NASA anno June 29/77)

June 30: The *NY Times* reported that a team of two Soviet crewmen, accompanied for a time by a woman plant expert, had completed a 4mo stay in an isolation chamber to test an artificial environment designed for long interplanetary space missions. The team had breathed oxygen and consumed water and food produced by miniature wheatfields and vegetable gardens occupying about half the space of their 1260ft^2 test chamber called Bios 3. The plants, grown by hydroponics (mineralized solutions rather than soils), had exuded enough moisture for drinking and household needs and had provided a third of the proteins and fats and half the carbohydrates in the crew diet, supplemented with dehydrated rations of animal fats and proteins stored for the mission.

The *Times* quoted an *Izvestia* story describing the experiment and reporting "no significant changes in the physiologic, biochemical and psychological functions of the subjects. Their weight has also remained virtually the same." Dr. G. Lisovsky, head of the Controlled Biosynthesis Laboratory of the Krasnoyarsk Physics Inst. in Siberia, said that earlier experiments in the series begun shortly after the Gagarin flight of 1961 had tried chlorellas (green single-cell algae) as a source of protein and B-complex vitamins for long space trips. The algae had yielded sufficient oxygen and water but did not provide a balanced diet; researchers had then used early-maturing wheat and vegetables. Agronomist Mariya

Shulenko had done the cooking while checking out the hydroponics; after she left, Gennady Asinyarov and Nikolay Bugreyev had shared the job. (*NYT,* June 30/77, 28)

During June: John F. Yardley, NASA's associate administrator for space flight, was one of 92 persons honored by election to membership in the Natl. Academy of Engineering. His citation was for "contributions to engineering theory and practice and leadership of organizations that pioneered major space programs." (NAS News Report, June/77, 7)

• *Nature* reported that a May 13 meeting of the Royal Astronomical Society in London had discussed Mars's moons Phobos and Deimos, the 100th anniversary of whose discovery would occur in Aug. Observations, both ground-based and from satellites, had revealed oddities: Phobos, the only satellite in earth's solar system that circled its central body faster than that body could spin on its axis, would appear to an observer on the planet's surface to rise in the west and set in the east, whereas its fellow Deimos would appear to rise in the east and set in the west like earth's moon. Also, the orbit of Phobos was coming nearer to the surface of Mars (a body approaching within 8700km would be pulled apart by differential tidal forces, the article said). A Soviet astrophysicist, I.S. Shklovsky, had claimed in the 1950s that the low density of Phobos indicated it was a hollow artificial satellite produced by an advanced civilization on Mars. Viking spacecraft images had shown that the moons were blackish gray rather than the red of Mars, indicating different composition; astronomer Joe Veverka of Cornell Univ. concluded that the moons had originated in the asteroid belt. Viking had found that the 10km-diameter Stickney crater on Phobos covered 40% of the maximum diameter of Phobos itself. The two Mars moons had almost identical shapes (triaxial ellipsoids) and both rotated synchronously with their orbital periods and their long axes pointing directly at Mars (*Nature,* June 30/77, 758)

• Reductions made by the House subcommittee on defense appropriations in funding for the defense satellite communications system (DSCS) and the USAF/Navy Fleet Satellite Communications (FltSatCom) program would force the DOD to lease its communications channels from commercial carriers, *AvWk* charged. The subcommittee had eliminated the DOD's entire $60-million request for R&D funds to proceed with a contract awarded in Feb. to General Electric for a qualification model and two flight-demonstration DSCS-3 satellites. It also had denied money for a fourth and fifth FltSatCom satellite to be built under a contract awarded to TRW in 1972; the Navy had been leasing capacity since March 1976 on the Marisat system managed by Comsat General. The subcommittee, headed by Rep. George Mahon (D-Tex), had advocated continued leasing despite DOD testimony that leasing would be more expensive, and had refused the funds requested for two other DOD satellite systems: it had rejected outright DOD's international global-positioning system to be built by Rockwell, and had reduced the $83.2 million requested by the USAF for a Hughes-built satellite data system by more than half, postponing procurement of a fifth spacecraft. (*AvWk*, June 6/77, 60)

- JPL officials were considering an 8mo extension of the Viking mission through early 1979, instead of terminating it May 31, 1978, *AvWk* reported. Viking mission director G. Calvin Broome had estimated he would need a team of about 180 for a continuation mission; he currently had about 300 people working on it, and had had as many as 700 at the height of the mission.

In the extended mission, the two landers would continue photographing the sites at Chryse Planitia and Utopia Planitia and transmitting data to earth; their biology instruments had been shut down earlier [see June 1]. The orbiters would also continue to photograph the planet, as the flight team had conserved sufficient gas supplies for propulsion and attitude control. (*AvWk*, June 20/77, 79)

July

July 1: NASA announced it would conduct sounding-rocket tests with the USSR Hydromet Service for 2wk in Aug. to compare U.S. and Soviet measurements of upper-air temperatures and detect discrepancies in their meteorological instruments. NASA would launch a series of rockets from Wallops Flight Center and the Hydromet representatives would launch rockets from the research ship Akademik Korolev off the U.S. east coast. Under a 1971 agreement on space sciences and applications, the two nations in 1972 had begun exchange of data that had shown the temperature field in the eastern hemisphere to be consistently colder than that of the western hemisphere. Both sides had evaluated their rocket systems for sources of error; direct comparisons made in 1973 at Kourou, French Guiana, had again shown discrepancies in temperature and wind data. Improvements in instruments and techniques since 1973 had made further comparisons necessary, NASA said. (NASA Release 77-133)

• Kennedy Space Center announced it would host for the second consecutive yr a gathering of more than 85 meteorologists and atmospheric physicists for TRIP 77 (Thunderstorm Research Intl. Program 1977), including 21 principal investigators and their associates, continuing their studies of lightning and thunderstorms and the hazards resulting from them. The 1976 program had yielded much data on electrical charges within and between clouds and from clouds to earth's surface. (KSC Release 127-77)

• Marshall Space Flight Center announced it had resumed the atmospheric variability experiments (AVE) begun in 1964 to study processes associated with the onset and development of severe storms. Originally conducted to aid in scheduling Saturn rocket firings at MSFC, the experiments were designed to detect unusual weather conditions common to different localities before severe storms occurred, to improve short-term predictions based on data from weather balloons, weather satellites, and ground-observation posts. Sixth of the AVE series recently conducted over 24hr in an area including Texas, Alabama, Illinois, and North Dakota had monitored severe storms with hail and tornadoes in Missouri and western Tennessee, and a frontal system with low-pressure disturbances in Alabama. (MSFC Release 77-122)

- KSC announced it had awarded a $480 193 fixed-price contract to Reynolds, Smith and Hills of Jacksonville, Fla., for modifications to Complex 39B to adapt it for Shuttle launch operations as had been done to Complex 39A. KSC had used 39B to launch *Apollo 10; Skylabs 2, 3,* and *4;* and the U.S. spacecraft of the Apollo-Soyuz mission. (KSC Release 125-77)

July 5: NASA announced that its aviation safety reporting system begun in 1976 for the Federal Aviation Administration had been a success, as evaluated by a NASA Advisory Council subcommittee established in Feb. 1977. During the system's first year, pilots, air traffic controllers, and others working in national aviation had voluntarily submitted about 5500 safety-problem reports that had been fed into a computer to note trends and monitor corrections. Next evaluation would be in June 1979. (NASA Release 77-137)

July 6: NASA announced publication of the first book of a series on the Skylab mission, *Skylab, Our First Space Station* (NASA SP-400), a general introduction to and overview of the mission's problems and successes. The 10 chapters tell how Skylab grew from existing technology; how it nearly failed when its micrometeorite shield and one of its solar panels were lost, and how it was saved; how 3 manned visits occurred, one of them for a record 84-day occupation; and how it proved the possibility of living and working for extended periods in a space environment. (NASA Release 77-135)

July 7: NASA announced it had awarded Boeing Commercial Aircraft Co. of Seattle a contract to design and ground-test composite structures for the Boeing 727 aircraft. NASA's aircraft energy-efficiency program to decrease transport-aircraft fuel consumption by 50% had included contracts with several manufacturers to reduce aircraft weights by using composites (high-strength filaments in a polymer matrix, lighter than metal structures). Boeing would build five shipsets of composite elevators (five right- and five left-handed) using manufacturing techniques that might become standard procedure. Costs would be shared, $8 100 000 contributed by the government and $890 000 by the contractor. Langley Research Center would manage contractor effort over 3.5yr. (NASA Release 77-139; LaRC Release 77-27)

- MSFC reported that employees Whitt Brantley, Jr., and Robert W. Rood had invented a solar energy device capable of 300 000 Btu per hr, enough to heat or air condition several buildings. It would offer a 50 to 70% efficient solar energy option for uses requiring temperatures above

200°F, costing as little as $5 per ft² for units 25 to 50ft in diameter. (MSFC Release 77-123)

• NASA announced that a team of astronomers at the Harvard-Smithsonian Center for Astrophysics using data from NASA's *Uhuru* satellite had discovered what seemed to be "superclusters" of galaxies more than 150 million light-years in diameter, bounded by areas of extremely hot gas acting as x-ray sources. The mass of gas required to produce x-rays would be 5 to 10 times greater than all material seen at other wavelengths, and would suffice to bind the galaxies into superclusters. Primarily hydrogen and helium, the gas was thought to be "primordial material" from an explosion that created the universe.

In the debate over origins, "open universe" theorists supported the idea of a big bang 20 billion yr ago expanding outward forever; "closed universe" advocates said gravitational collapse would halt expansion and the material would fall back on itself, perhaps to repeat the cycle. Traditional optical and radio techniques had not found enough material in space to supply the gravity needed to "close" the universe; the new data indicated the missing mass might exist as vast amounts of hot gas between galaxies.

Uhuru, first of a series of small astronomical satellites (SAS) for studying cosmic x-rays, had been launched Dec. 12, 1970, as *Explorer 42*. The astronomers had noted the x-ray emissions while compiling the fourth *Uhuru* catalog from observations in the early 1970s. The project was managed for NASA by Goddard Space Flight Center. (NASA Release 77-138)

July 8: NASA announced it and ERDA had contracted with Boeing Engineering and Construction Co. to design and build on a utility company site chosen by ERDA the largest windmill in history, a 2500kw generator with 300ft blades, to provide data on such systems coupled with conventional power plants. Lewis Research Center would manage the project for ERDA. (NASA Release 77-141)

July 10: Television programs beamed by satellite to the whole world were unwelcome to some delegates at the Vienna meeting of the UN Committee on Peaceful Uses of Outer Space, the *W. Post* reported. Its reprint of a *London Observer* article said that the major technical problem remaining for global TV transmission was overcrowded frequencies, but political problems remained.

A majority of UN members including the Soviet bloc had pressed for regulation of TV transmission, claiming the right to intervene between their citizens and the broadcasts they could receive. The Soviet group objected to "propaganda" especially on its Eastern European borders where "Western TV programs, soap operas and all, are often more welcome than the home product." Developing countries unable to afford their own TV broadcasts claimed that unrestricted foreign TV from satellites would result in "cultural imperialism."

The delegates seemed to want data from satellites scanning the earth to be generally available, but differed over what should go to whom. A U.S. satellite detecting a new Soviet oil field might have to ask permission before reporting the find; discovery by any satellite-operating nation of resources (uranium, for instance) in a developing country might start a "gold rush" by major powers wanting to "help" that country exploit those resources. The committee agreed to formulate guidelines for UN General Assembly approval, but western European and North American nations wanted to prevent anything resembling censorship. Because of a Belgrade conference in session "only a few hundred miles further east," the Soviet bloc also was sensitive to any measures that could be interpreted as infringing human rights or the free exchange of information. (*W Post,* July 10/77, A-14)

July 11: MSFC announced it had selected General Electric Co. of Valley Forge, Pa., to negotiate a cost-plus-award-fee contract for a prototype Atmospheric Cloud Physics Laboratory (ACPL) to be flown on a number of Spacelab missions carried by the Space Shuttle [see June 22]. The prototype with software and spares was due for delivery to MSFC by Oct. 1, 1979, and to KSC for Shuttle launch by Jan. 1, 1980. (MSFC Release 77-124)

July 12: KSC reported that an experimental homemade airplane built by John Murphy of the center's Technology Utilization Ofc. had used a NASA-developed winglet set to increase its efficiency by about 8%. Murphy had begun work on the plane in Dec. 1976 from plans by a California designer and had completed it on June 12. Except for the engine and engine mount, the plane consisted entirely of fiberglass and styrofoam and weighed only 630lb. Its 100hp Continental aircraft engine could carry two passengers at 200mph, requiring 1000ft for takeoff and about 2000ft for landing. Murphy, a graduate of Georgia Tech, had been flying for 30yr and owned another aircraft that he had built himself. He had completed 24 of the 50hr flying time needed for Federal Aviation Administration approval, so that he could fly it to the annual meeting of the Experimental Aircraft Assn. in Wisconsin later this summer and show it off to the designer. (KSC Release 128-77)

• LaRC announced it would close its Technology Utilization House to the public Aug. 1 to prepare for the arrival of the family that would live in the house for a year to demonstrate the innovations built into it, many of them deriving from NASA space-program technology. The family consisted of a professor from Florida State Univ., his wife (a registered nurse), and their teenage daughter and 12yr-old son. The new equipment included solar energy collectors, special insulation, a water-recycling system, automated heating and cooling, emergency lighting, an improved fireplace, and a special security system. Materials in the contemporary style 3-bedroom house, chosen to save energy and water and thus reduce costs, could save more than $20 000 over a 20yr period. (LaRC Release 77-28)

• Political interference had delayed development of science and technology in China, according to Chien Hsueh-shen, director of the mechanics institute of the PRC Academy of Sciences. The *W. Post* quoted his statement in the journal *Red Flag* that China would "catch up with and surpass the world advanced levels." Chien, now China's "top missile scientist" according to the *Post,* had been a colonel in the U.S. Army Air Corps before returning to China in 1955. Meanwhile, provincial radio reports monitored in Hong Kong said that "political infighting" at a PRC base for nuclear weapons and missiles had "crippled defense construction" for months in 1976. (*W Post*, July 12/77, A-15)

July 14: NASA launched a 280kg (620lb) geostationary meteorological satellite (GMS) for Japan from Cape Canaveral at 10:56 EDT on a Delta vehicle into an elliptical orbit, to be boosted at third apogee into a geosynchronous orbit at about 35 500km altitude over the equator south of Tokyo. *Gms 1,* known as *Himawari* in Japan, was that country's contribution to GARP, the global atmospheric research project sponsored by the Intl. Council of Scientific Unions and the World Meteorological Organization. The cylindrical craft had carried a visible infrared spin-scan radiometer (VISSR) and a space environment monitor to identify and photograph weather patterns from Hawaii to Pakistan and detect solar energy activity that might disrupt communications on earth. More than half the world's tropical storms had begun in the western Pacific and eastern Indian Ocean, said Kazuo Watanabe, project manager for Japan's national space development agency; *Gms 1* could provide detailed weather photos and other data every 30min to give instant warning of typhoons in the area. Watanabe noted that Japan had been depending on data from a U.S. satellite in polar orbit that furnished images only twice a day. (NASA Release 77-110; MOR M-492-101-77-01 [prelaunch] July 12/77; Hughes Aircraft Co. release July 12/77; *NYT,* July 15/77, A-8; *W Post,* July 15/77, A-22)

• NASA Administrator Dr. Robert A. Frosch announced that President

Carter had declared July 16–24 as a U.S. Space Observance, the dates chosen by Congress to coincide with the anniversary of the *Apollo 11* moon landing as well as Apollo-Soyuz and the *Viking 1* landing on Mars. Dr. Frosch said that the president's recognition of NASA achievements "should serve as inspiration and encouragement to all of us." (NASA anno July 14/77)

July 15: NASA announced that more than 8000 persons had applied for the 30 to 40 openings as Space Shuttle astronauts after a yr-long recruiting effort. Candidates selected would report in 1978 for a 2yr training period at Johnson Space Center. Of the 8037 applicants, 6735 were for mission specialist and 1302 were for pilot; 1142 of the applicants were women. NASA would announce selections in Dec. 1977. (NASA Release 77-145; JSC Release 77-39)

• LaRC announced it would dedicate on July 19 the site of a new wind tunnel, the National Transonic Facility, scheduled for completion in 1980 at a cost of about $85 million. The only facility of its kind in the world, the tunnel would use cryogenic nitrogen to test aircraft models in a realistic flight environment; cooling the medium to $-300°F$ would permit accurate simulation of Mach numbers and atmospheric-density ratios. Although NASA and DOD would be principal users of the tunnel, the schedule would be shared by other federal agencies, private industry, and the scientific community. Speakers at the dedication would be Dr. John J. Martin, assistant secretary of the Air Force for research and development; Donald F. Hearth, LaRC director; and Oran W. Nicks, LaRC deputy director. A display of aircraft developed from LaRC transonic research would include a B-747 freighter; a wide-bodied DC-10 passenger plane; USAF fighters F-14, F-15, and F-106; a Learjet business aircraft; LaRC's B-737 flying laboratory; a T-38 jet trainer; and a UH-1H helicopter. (NASA Release 77-144; LaRC Release 77-29)

July 16: About 300 people had assembled at Cape Canaveral, the *W. Post* reported, to celebrate the anniversary of the launch of *Apollo 11*, the mission that first landed humans on the moon. On July 16, 1969, a Saturn V rocket launched from Kennedy Space Center had carried 3 astronauts, none of whom attended the observance. (*W Post*, July 17/77, A-18)

July 18: MSFC announced that one of its orbiting satellites, *Pegasus 1*, silent since Jan. 13, 1968, had suddenly come to life; GSFC's network operations group had picked up signals from its beacon June 27 as interference with data transmission by another satellite. *Pegaus 1* had been launched Feb. 16, 1965, to acquire meteoroid data on its 2300ft^2 of instrumented surface and transmit its findings to earth stations. After a silence of nearly 10yr, its revival was a puzzle to NASA officials who were checking circuit drawings still available. After notifying MSFC of the occurrence, GSFC network operations had sent a command from Australia's Orroral station to turn off the satellite's AM beacon, only to find that the FM beacon was also working. The transmitters, operating when the vehicle was in full sunlight, were sending strong signals; the network group had recorded signal strengths of both transmitters before commanding them off. (MSFC Release 77-130)

• MSFC reported it had fired for 2min the largest solid-fuel rocket motor ever developed, at Thiokol Corp.'s installation 24mi from Brigham City, Utah, and 2mi from the nearest occupied building. The Thiokol motor, the propulsive part of the Space Shuttle booster, contained 502 454kg (more than 1.1 million lb) of propellant, half again as much as the next largest motor ever fired at the site. A composite fuel, mixed and poured into the 4 segments of the SRM, had been cured for 4 days before the segments were joined horizontally at the test site by special handling devices. The segments would go by rail to KSC for vertical assembly. Each Shuttle launch would use two of the boosters, designed for reuse up to 20 times. (MSFC Releases 77-125, 77-133; NASA Release 77-142)

• NASA announced it had awarded a 12th grader from Calif. a prize in addition to the grand prize for 1977 in a nationwide student science competition. Jonathan Holman, who had worked with control mechanisms in blood clotting, received his award from his older brother Jeffrey, top winner in the 1976 competition and now a student at the Mass. Inst. of Technology. Jonathan had chosen his experiment because his family had a history of hemophilia, in which the blood fails to clot properly. The competition was sponsored by the Natl. Consortium for Black Professional Development. NASA coordinator for the program was Everett P. Colman, EEO officer at LeRC, who said it was a part of NASA's effort to increase the number of people with scientific skills the agency would need in future. (NASA Release 77-140)

• MSFC reported that NASA had issued a request for proposals to integrate mission payloads, in preparation for the first three Spacelab missions and the sixth Space Shuttle orbiter flight test. Integration would include combining a number of experiments into a compatible payload; controlling the configuration during instrument development, testing, and flight operations; and disassembly of the payload after the mission. Proposals, due Aug. 15, would be reviewed by a source evaluation board. (MSFC Release 77-131)

July 19: The Natl. Science Foundation reported that research and development spending in the U.S. would reach an estimated $40.8 billion in 1977, 9% more than the 1976 level of $37.3 billion. The federal government expected to increase its R&D outlay from $19.8 billion in 1976 to $21.8 billion in 1977, emphasizing defense, space, and energy. (NSF Release PR77-73)

• The Natl. Oceanic and Atmospheric Administration (NOAA) announced it had completed the first phase of a project funded by the Natl. Science Foundation to learn more about earth's magnetic field and near-space environment. The task consisted of placing special magnetometers, devices to measure earth's magnetic-field densities, at more than 25 strategic and remote sites in Alaska, Canada, Brazil, and several Pacific islands. Two NOAA geostationary spacecraft of the GOES series (one over the Pacific and the other over Colombia, South America) would transmit data from the devices to Boulder, Colo., for immediate distribution to earth, atmospheric, and space scientists.

Using the network of magnetometers with instant relay of the data to earth would permit continuous and precise monitoring of magnetic-field changes over the western hemisphere. The information would aid research on the possible connection between magnetic-field variations and climate changes; on reducing magnetic interference with communications and power systems; and on the use of magnetic fluctuations as a tool in locating geothermal, mineral, and other natural resources. (NOAA Release 77-195)

July 20: WFC announced it would receive this week the first of two rotor-systems research aircraft (RSRA) for contractor flight tests. Sikorsky Aircraft, which would conduct the tests, had designed the RSRA (a joint project of NASA and the U.S. Army) to fly as a regular helicopter, as a compound helicopter (both rotors and wings), or as a conventional fixed-wing aircraft. As the RSRA's engines or wings were changeable, NASA planned to use it as a "flying wind tunnel" for trying out various rotor systems: articulated, hingeless, or gimbaled, or with variable geometry, variable diameter, or controllable twist. (WFC Release 77-7; NASA Release 77-149)

• NASA announced that LeRC, under the agency's general aviation turbine engine (GATE) program, had awarded study contracts totaling $300 000 to 4 different companies for using in small lightweight planes with less than 1000hp the advanced turbine technology used in large commercial aircraft. The companies were Teledyne Corp.; Garrett AiResearch; Detroit Diesel Allison; and Williams Research Corp.

LeRC spokesman Bill Strack said most engines in the light-aircraft class (reciprocating engines of less than 400hp) dated from 10yr ago and lacked the advantages of turbines: reduced vibration and noise, less maintenance, lighter weight, and greater safety and reliability. The large production base existing for piston engines might make it hard to produce small turbine engines cost-effectively, he said. (NASA Release 77-147)

• NASA announced it would conduct the third "manned inert" flight of the Space Shuttle orbiter Enterprise on or about July 26 at Dryden Flight Research Center. At the controls of the orbiter atop its Boeing 747 carrier would be astronauts Fred W. Haise and C. Gordon Fullerton, making their second piggyback run in this test series. They had made the first captive flight June 18; the second, on June 28, had Joe Engle and Richard Truly as pilots.

In the first flight, lasting 56min, Haise and Fullerton had checked out the orbiter flight controls at low altitude and low speed. For the third test, the 747 would carry the orbiter to an altitude of 7895 meters (25 905ft) where it would "push over" in a practice separation; the orbiter and 747 crews would go through all the motions except actual separation. The flight would be a dress rehearsal for the first free flight of the orbiter, scheduled on or about Aug. 12. (NASA Release 77-146; JSC Release 77-41)

July 21: INTELSAT announced that Paraguay had become the 97th member of the organization on July 18, and the 22nd nation in the western hemisphere to join. Paraguay's ANTELCO (Administracion Nacional de Telecomunicaciones) had signed the operating agreement. Created in 1964, INTELSAT claimed to provide more than half the world's transoceanic telecommunications and domestic communications for 9 countries through its global satellite system. (INTELSAT Release 77-19-I)

• *Today* reported that Congress had approved June 20 the largest NASA budget in nearly 10yr, including new starts on three major programs. In conference committee, the House (which had cut $20.7 million requested by NASA to begin work on an unmanned 1982 Jupiter probe) had agreed with the Senate on $17.7 million for the project, which NASA said was enough to start with. The budget also contained $36 million for the large orbiting telescope, a frequently delayed major scientific project, and $26.6 million for Landsat-D, fourth of a series of earth monitoring satellites. The newspaper said the only real disagreement was over the Jupiter mission, which Rep. Edward P. Boland (D-Mass) wanted to kill [see June 10] because of its eventual cost, estimated to reach $435 million by the time it was scheduled for launch in 1983. The House conferees had finally accepted the Senate's figure. Planned to take advantage of a favorable alignment of the earth with Jupiter in 1982, the mission would not only be the first to orbit Jupiter but would also drop a probe into the planet's atmosphere. During its 20mo in orbit, the mission would also view Jupiter's moons at close range. (*Today,* July 21/77, 20A)

July 25: NASA announced it had on July 21 and 24 adjudged the *Explorer 52* and *55* (Atmosphere Explorer) missions successful. Launched Oct. 6, 1975 (as *AE-D*) and Nov. 20, 1975 (as *AE-E*), the two spacecraft had carried instruments to study energy transfer and chemical processes in the atmosphere. *AE-D* had "performed properly" until Jan. 29, 1976, when a power system failure ended control of the spacecraft, which reentered earth's atmosphere March 16. Its perigee had moved from equatorial at launch, northward to the winter pole and southward to the summer pole, producing "excellent measurements" at all latitudes. *AE-E*, still in operation, "continues to acquire and contribute worthwhile data" on the Atmosphere Explorers' findings, which had resulted in 250 formal presentations at scientific gatherings and 130 publications in technical journals. (NASA MOR S-852-75-04/05)

• NASA announced it had selected Ball Bros. Research Corp., Boulder, Colo., to negotiate a $13.8 million contract for the telescope system on the infrared astronomy satellite (IRAS), a joint U.S.-Netherlands mission to survey the celestial sphere in the infrared spectrum. The U.S. would provide a large cryogenically cooled instrument and the Netherlands would provide the spacecraft and would supervise integration. NASA would launch IRAS from WTR early in 1981 on a Delta for its yr-long scientific mission. The Ball Bros. contract would cover design, fabrication, test, and launch support for the telescope, plus spares, an engineering model, and ground-support equipment. The Jet Propulsion Laboratory would manage the project, and Ames Research Center would be responsible for the telescope system (NASA Release 77-154)

• NASA announced it had selected Lockheed Missiles and Space Co., Sunnyvale, Calif., and Perkin-Elmer Corp., Danbury, Conn., to negotiate contracts totaling more than $131 million for two major elements of the Space Telescope. Lockheed would design, fabricate, and integrate the telescope support-systems module, provide systems engineering and analysis, and support NASA in ground and flight operations. Perkin-Elmer would design, manufacture, and deliver the optical assembly and equipment, including systems engineering, support of the launch, verification of the orbit, and plans for mission operations. The 8ft-diameter telescope, scheduled for launch on the Space Shuttle late in

1983, would offer an astronomical capability beyond anything possible from the earth's surface or any telescopes launched so far because of its improvements in resolution, power, light sensitivity, and wavelength coverage. It would be refurbished periodically in orbit or could be returned to earth on the Shuttle for repair or upgrading. The support module would provide electric power, communications, data processing and storage, attitude sensing and control, and environmental control to the telescope assembly and up to 5 additional scientific instruments accommodated by the telescope. MSFC, which had overall project management responsibility, would manage the contracts for the support module and the optical assembly; GSFC would manage scientific instrument development. (NASA Release 77-153)

• ARC announced that its 1977 summer study under the leadership of Dr. Gerard P. O'Neill on space settlements and industrialization using nonterrestrial materials would offer a briefing Aug. 2 on the technical areas covered: life-support systems, habitat design, materials transport and processing, and space manufacturing. (ARC anno July 25/77)

July 26: NASA announced that the launch of HEAO-A (the high-energy astronomy observatory) originally scheduled for April 1977 had been set for Aug. 12. Problems with the spacecraft's gyroscopes, required for attitude control, had caused the delay. Analysis and testing had produced design changes to improve gyroscope reliability for the expected 6mo mission lifetime. HEAO-A would be first of three astrophysical observatories orbited over the next 3yr to study radiation from stellar sources throughout the universe. (NASA Release 77-156)

• NASA announced it would hold a briefing at Hq Aug. 4 on the scientific studies to be conducted by the Voyager spacecraft scheduled for launch later this year toward the giant planets Jupiter, Saturn, and possibly Uranus. Participants in the briefing would be Dr. Noel W. Hinners, NASA's associate administrator for space science; Dr. Edward Stone, Calif. Inst. of Technology, project scientist; John Casani of JPL, project manager; Dr. Bradford A. Smith, Univ. of Ariz., leader of the imaging science team for the Voyager project; and Dr. James W. Warwick, Univ. of Colo., principal investigator for planetary radioastronomy. (NASA Release 77-155)

July 27: NASA announced that the Soviet ship expected to participate in comparisons of atmosphere-measuring instruments [see July 1] would arrive off the Virginia coast about Aug. 7 to serve as the platform for USSR rocket launches. The tests would consist of paired launches of about 22 U.S. Super Lokis and about 33 USSR M-100Bs, plus 6 Super

Loki sphere systems to be launched by Wallops; statistical evaluation would require from 14 to 22 successful paired launches. (WFC Release 77-8; NASA Release 77-151)

• NASA announced that a disaster simulation in Texas using the communications technology satellite—world's most powerful comsat—had turned into an actual emergency operation providing communications between Red Cross headquarters in Washington, D.C., and flood-stricken Johnstown, Pa. ComSatCorp and the Red Cross had set up an experiment to demonstrate emergency use of satellites in natural disasters; the Texas results had led Red Cross officials to ask for assistance in the actual emergency at Johnstown.

The *Cts* equipment transferred from Houston to Johnston July 23 and set up at a vocational high school near the Univ. of Pittsburgh was a portable station with a 4ft antenna to send and receive signals to and from the synchronous satellite orbiting over the equator just west of South America. During the emergency it had sent messages through the *Cts* to ComSat offices in Clarksburg, Md., connected by land line to Washington. The disaster-related communications included requests for medical and physician services, supplies and administrative services, and transmission of situation maps. Johnstown's residents also could use the *Cts* link to communicate with relatives around the U.S. when the Red Cross was not using the lines. The emergency had preempted Canadian and U.S. experiments scheduled for July 25–26. (NASA Release 77-157)

• WFC announced it had completed tests using rockets, balloons, and supporting ground observations to obtain data on the natural state of earth's ozone layer and on stratosphere pollution. As part of the joint WFC-GSFC applications sounding-rocket program (ASRP), the tests July19–22 had measured daylong variations in the vertical distribution of ozone to establish a model for studying the effects of pollutants. (WFC Release 77-11)

• LaRC and WFC announced arrival of an LaRC airplane at Wallops for tests of spin characteristics, using four different tail designs developed at Langley as part of a larger program begun in 1972 to improve safety in light airplanes by investigating spin and stall [see May 18]. About 30% of fatalities in light planes resulted from spins or stalls. The low-wing, single-engine test plane had been modified to make use of the various tails and of a spin recovery parachute to counter the effects of some of the configurations. Two high-wing lightweight planes would share the tests. Besides full scale planes, the program would also use spin-tunnel models and powered radio-controlled free-flight models instrumented to give detailed information on each flight in real time. (WRC Release 77-10; LaRC Release 77-33)

July 28: NASA announced plans to launch the first Italian experimental comsat Sirio on or about Aug. 17 from ETR on a Delta vehicle. A large segment of Italian industry interested in space technology had worked on the spacecraft and would use the experience for future international space programs. In synchronous orbit over the equator off the west coast of Africa, the comsat would transmit voice, data, and TV in the super-high frequency (SHF) range from Europe to North America; its experimenters would study propagation of SHF during adverse weather conditions including rain, snow, and fog. Use of SHF had become necessary because conventional frequencies were overcrowded.

Italy's research agency, Consiglio Nazionale della Richerche (CNR), had contracted with NASA for launch and support services until the craft reached its station 30 to 45 days after launch. The name Sirio was an acronym for Satellite Italiano Ricerca Industriale Orientata (industrial research-oriented Italian satellite). (NASA Release 77-152)

• KSC announced it had awarded a $7 325 000 fixed-price contract to Algernon Blair Industrial Contractors, Inc., of Norcross, Ga., for converting a Saturn/Apollo mobile launcher to a mobile-launcher platform for the Shuttle program. Work would include removal of a 400ft umbilical tower and crane and replacement of the single exhaust opening in the original platform with three openings required by the Shuttle main engines. One of KSC's three mobile launchers had almost completed conversion and the new contract would cover work on a second. The reusable Shuttle had been scheduled for its first manned orbital mission in the spring of 1979. (KSC Release 137-77)

• Comsat General Corp. announced that the U.S. Navy had agreed to extend its use of three Marisat satellites for an additional 2.5yr, ending in 1981, at a cost of about $138 million [see During June]. The Navy had begun using the Atlantic Marisat in March 1976, the Pacific satellite in June 1976, and the Indian Ocean satellite in Jan. 1977. Each of the three Marisats had a design life of 5yr and could operate in three frequencies: UHF (ultrahigh frequency) for Navy service, and the L-band and C-band for commercial maritime users. The Navy had leased all UHF capacity in all three Marisats for communications between its own fixed and mobile terminals. The Marisat system was owned and operated by four companies under a joint-venture arrangement approved by the Federal Communications Commission: Comsat General, which acted as system manager, with nearly 87% ownership; RCA Global Communications;

Western Union Intl.; and ITT World Communications. (Comsat General Release CG 77-3; *WSJ,* July 29/77, 6)

July 29: NASA declared the launch of Japan's geostationary meteorological satellite *Gms 1* on July 14 successful, as of July 26-27. The satellite was being maneuvered over the South Pacific to a station about 140°E above the equator, due south of Tokyo. (MOR M-492-101-77-01 [postlaunch] July 29/77)

• NASA announced that the first 20 of about 200 Shuttle astronaut applicants would report to JSC Aug. 2 for a week of individual interviews and physical exams. Ten of the group were from the USAF, 7 U.S. Navy, 2 Marines, and 1 civilian. Although all the initial group had been pilot applicants, NASA expected the number interviewed at JSC would divide evenly between pilot and mission specialist applicants. NASA would complete processing by mid-November and hoped to select as many as 20 astronaut candidates in each of the two categories in December. Candidates would report to JSC in 1978 for 2yr of training and evaluation; final selection as astronaut would depend on satisfactory completion of training. (JSC Release 77-42)

• ERDA announced that the longest space mission ever planned—a 10yr trip by two Voyager spacecraft to the outer planets—would get power from RTGs (radioisotope thermoelectric generators) developed by ERDA as successors to nuclear systems that had powered earlier missions to Mars and Jupiter.

ERDA's predecessor, the Atomic Energy Commission, had begun the research on nuclear power for space missions. ERDA plans included developing high-efficiency thermoelectric materials and dynamic converters for NASA missions such as a Jupiter orbiter in 1982 or a Mars rover in 1984, and DOD satellites in the 1980s. (ERDA Release NF-77-20)

• ERDA reported that the U.S. atomic energy detection system had recorded seismic signals, presumably from a Soviet underground nuclear explosion, at 1pm EDT on July 26. The signals had originated in central Siberia, north of the Arctic circle. (ERDA Release 77-129)

During July: NASA reported that researchers at LaRC had devised a nuclear pumped laser output 100 times more powerful than that available from noble or inert gases, 3.5w in a volume of only 1 cubic centimeter. A major step in laser technology, the experiment had converted kinetic energy from the splitting of atoms directly into photons (light energy, in the form of laser beams) without the energy-wasting intermediate step of converting heat to electricity.

The aim of the research had been to produce and transmit large quantities of power in space, using high-power laser beams as highly efficient energy conductors with transmitters and receivers measured in meters rather than in kilometers for similar microwave equipment. Laser beams could connect central power plants in space with settlements, materials processing centers, or scientific operations in earth orbit; laser powered vehicles could transport payloads to geosynchronous orbit from lower altitudes. (NASA Release 77-143)

• An idea tried 61yr ago had come of age, NASA reported, with the piggyback flights of the orbiter Enterprise on its Boeing 747 carrier. In 1916, to put a Bristol Scout airplane at an altitude where it could attack Zeppelin bombers over England, the British had loaded it on a three-engine Felixstowe flying boat for a single flight May 17 of that year. Both aircraft had landed safely, but the experiment had not been repeated. (NASA Release 77-148)

• NASA noted that *Landsat 1,* first satellite to focus specifically on earth and its natural resources, had completed its 5th yr in orbit. Launched July 23, 1972, with a life expectancy of only 1yr, Landsat had monitored air and water pollution, geological resources, agricultural production, and potential water supply. Dr. Wernher von Braun had predicted that this program alone could give the U.S. "a return exceeding its total space program investment." (NASA Release 77-150)

• In a *W. Post* column, science writer Daniel S. Greenberg commented on the "leaky embargo system that is supposed to deny the Soviets western technology of strategic value." The principal feature of a new DOD plan drawn up for State and Commerce Dept. approval before presentation next spring to the Coordinating Committee of Nations ("NATO minus Iceland, plus Japan") on restricting strategic export to Communist bloc nations was that "it isn't confined to hardware," Greenberg noted. The new embargo scheme, he said, was aimed at technological know-how; its enforcement would admittedly encompass "the entire span of advanced American technology. . . . Hardware can be stopped at the loading dock, but know-how, being intangible or confined to paper, calls for different barricades."

The truth about the embargo, Greenberg charged, was that "virtually no one is interested in it except certain military and industrial elements in the U.S. Furthermore, most of what we won't sell is easily available from sales hungry producers elsewhere. And, finally, there is ample evidence that . . . the Soviets, despite their many technological shortcomings, can home produce whatever they deem necessary." The proposed embargo, Greenberg concluded, might retard the Soviets a bit, but the presence of DOD "gumshoes in the American scientific and technological enterprise is a stiff price to pay" for that delay. (*W Post,* July 12/77, A19)

• JPL reported that Donald Lynn and Jean Lorre of its image processing laboratory had shared in a scientific investigation of the Shroud of Turin, preserved in Italy since 1578. At the request of groups in N.Y. and N.M., Lynn and Lorre had used advanced NASA techniques (mathematical and contrast enhancement) to obtain a noticeably clearer image from negatives and color slides taken at a 1973 display of the cloth and the image appearing on it. (JPL *Universe,* July 1/77, 1; *Natl Inq,* July 19/77, 4)

August

August 1: NASA announced it had scheduled the first free flight of the Shuttle orbiter for 8am PDT Aug. 12 at Dryden Flight Research Center. Briefing the press Aug. 11 would be Deke Slayton, ALT program manager at Johnson Space Center, and John Young, chief of the JSC astronaut office; Chet Lee, Hq director of Space Transportation Systems operations, speaking on payloads, pricing policies, and users; and Aaron Cohen, manager of the orbiter project at JSC, on orbiter systems. (NASA Release 77-158; DFRC Release 26-77)

• NASA announced it would put on each of the Voyagers to be launched in Aug. a phonograph record, "Sounds of Earth," containing natural noises of surf, wind, and thunder, and of birds, whales, and other animals. The 12in disk would include greetings from earth in a number of languages [see June 3] and samples of music from different cultures and eras. The disk contained electronic data convertible by an advanced technological civilization into diagrams, pictures, and printed instructions designed to give beings that intercepted the craft an idea of 20th century earth and its inhabitants. "The spacecraft will be encountered and the record played only if there are advanced spacefaring civilizations in interstellar space," said Dr. Carl Sagan of Cornell Univ., originator of the message sent into space on *Pioneers 10* and *11*. The copper disk in its aluminum jacket could survive as long as a billion yr; the Voyagers might take 40 000yr to approach another star, and other predictable approaches would occur in 147 000 and 525 000yr. (NASA Release 77-159)

• Langley Research Center announced that 21 area high school students had completed a 4wk hands-on career exploration program at the center, applying classroom concepts to actual situations. Each student had been assigned to a sponsor in science, computers, engineering, or mathematics to supervise the unpaid 8hr workdays. (LaRC Release 77-34)

• Despite U.S. revulsion at Soviet domestic oppression of scientists and the resulting condemnation of such practices, the U.S. and the USSR had found "enough common ground" in the sciences to renew cooperation for another 5yr, the *Christian Science Monitor* noted. Renewal of the 1972 accord signed by President Nixon and Leonid Brezhnev was

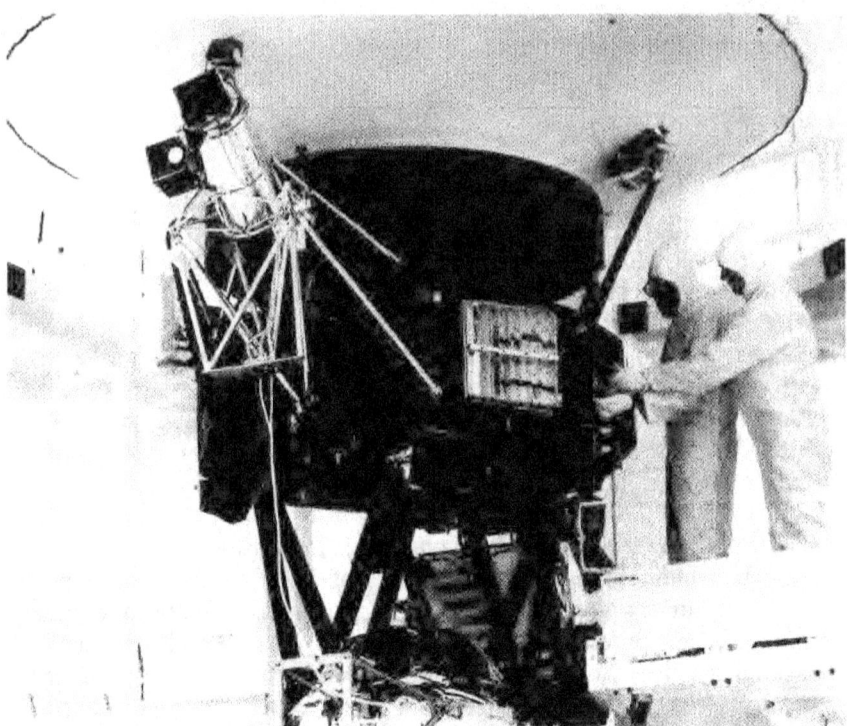

Technicians attach "Sounds of Earth" recording to Voyager 2 *spacecraft before encapsulating it for launch, scheduled later in August. (NASA 77-H-509)*

"neither casual nor automatic," *CSM* said. In view of charges that the agreement mainly benefited the USSR, a committee of the U.S. Natl. Academy of Sciences had reviewed scientific détente and had told presidential science adviser Frank Press that "positive benefits" for the U.S. made continued cooperation valuable. Subsequent negotiations, *CSM* noted, went on without mention of human rights. ". . . There is only so much one nation can do in holding up a moral standard for another," the article concluded, adding that ". . .Two powerful nations holding seemingly irreconcilable political views should . . . continue to seek ways to live peacefully together on the same small planet." (*CSM*, Aug 1/77, 11)

August 2: JSC reported the first 20 Shuttle astronaut applicants were due for screening. Center Director Christopher C. Kraft, Jr., said that the quality of the applicants would make it "difficult to narrow the field . . . paring that number will be a real challenge." (NASA Release 77-161)

• NASA announced it had appointed Isaac T. Gillam, IV, deputy director of DFRC, succeeding Gerald D. Griffin, recently appointed deputy director of Kennedy Space Center in Fla. Gillam, who had directed Shuttle operations at Dryden in the current approach and landing tests, had joined NASA in 1963 after serving in the USAF as a pilot, missile launch crew commander, and ROTC instructor. After receiving a B.A. from Howard Univ., he had done graduate work at Tenn. State Univ. and had been assistant professor of air science there. NASA had appointed him assistant manager of the Delta program in 1966 and program manager in 1968. In 1973, he became program manager for small launch vehicles, and had gone to DFRC in 1976. He had received NASA's distinguished service medal for his work with launch vehicles. (DFRC Release 27-77; NASA Release 77-164; NASA anno Aug 4/77)

August 4: NASA announced it had scheduled launch of two Voyager spacecraft toward the outer planets, one on Aug. 20 and the other about 12 days later. The Voyager 2 would go first and the Voyager 1, flying a faster trajectory, would go later; it would reach Jupiter 4mo ahead of Voyager 2, probably in March 1979. By the time it reached Saturn, Voyager 1 would be 9mo ahead of the other craft. NASA planned to use the gravitational field of Jupiter to send the Voyagers on to Saturn by means of a "slingshot" technique; Voyager 1 would reach Saturn in 1980, 3.2yr after launch, on a trip that would take 6.1yr without the gravity assist. Estimated cost of the Voyager project (not counting launch vehicles, tracking and data acquisition, or flight support) would be $320 million. (NASA Release 77-136)

• NASA announced that IBM Corp., under contract to LaRC, had developed the most efficient solar cell for space use reported to date. The new solar cells, derived from compound semiconductors gallium aluminum arsenide and gallium arsenide, had an efficiency of 18% instead of 13 to 14% for commercially available silicon-derived cells. High-efficiency cells could fill a power requirement with fewer numbers, meaning less weight to be launched and maneuvered as well as reduction in mission cost. The new cells had a "window layer" of gallium aluminum arsenide on the surface of the gallium arsenide cell. Absorption of sunlight in an ultra thin layer just below the surface of a solar cell had meant loss of some of the energy at the surface, accounting for the low efficiencies of previous cells; presence of the window layer had reduced this loss. (NASA Release 77-163)

• JSC announced that a summer workshop at the Univ. of Calif., San Diego, would discuss scientific and technological exploitation of lunar and asteroid materials. Participants selected for the workshop were geoscientists and geoengineers from government, private industry, and the university community. Univ. of Calif. professor and lunar researcher Dr. James Arnold, chairman of the workshop committee, would present to NASA's Office of Space Science the group's recommendations on research to be attempted and on the exploitation possible at present levels of understanding. (JSC Release 77-43)

August 5: NASA announced that Bradford Johnston, associate administrator for the Office of Applications since June 1976, would resign effective Sept. 30. Dr. Robert Frosch, NASA administrator, said that Johnston's contributions to useful applications of space techniques had made the program more responsive to national needs. (NASA anno Aug 5/77; NASA Release 77-166)

• The *W. Post* reported that London's Intl. Assn. Against Painful Experiments on Animals had protested Soviet government use of white rats in space research, following USSR launch of a spacecraft carrying several white rats for an international study on weightlessness. (*W Post,* Aug 5/77, A-13)

August 8: NASA reported a first-time continuous operation by NASA-funded Univ. of Fla. researchers of a nuclear pumped gas laser reported earlier by LaRC scientists [see During July]. Confirming the results of the Georgia Tech experiment, the Fla. group had transmitted laser energy for 15min and reported that longer transmission was possible. Earlier nuclear pumped lasers had produced only short bursts of energy. (NASA Release 77-165)

• NASA announced it had patented a way to control cell division that might mean a new treatment for cancer or replacement of lost cells in the brain or other parts of the human nervous system. Studying at LaRC the effects of space radiation on living cells, Dr. Clarence D. Cone, Jr., former head of LaRC's molecular biophysics lab, now chief of the nearby VA molecular biology laboratory, had changed electrical potential across hamster cell membranes. Studies had shown cell division could be either blocked or stimulated by various agents, most of which could not produce both effects. Cone's method of surrounding the cells with various concentrations of sodium, potassium, and chloride ions, patterned after ionic mechanisms used by the body to control cell division, could either turn on or turn off the process.

Cone's successful attempts to stimulate cells to divide or proliferate, or prevent dividing and proliferating, had not damaged the cells, and showed that the processes were reversible. Manipulating the ionic levels in human body cells might permit reversal or inhibition of uncontrolled proliferation, such as that of cancer cells. (NASA Release 77-162)

August 9: JSC announced it had awarded Lockheed Electronics Co. of Houston a 2yr cost-plus-award-fee contract estimated at $4 004 715 for technical support services at the Slidell Computer Complex's earth resources facility in La. Lockheed would furnish about 80 persons to operate and maintain the data acquisition laboratory, data systems laboratory, data processing laboratory, and data preparation laboratory. (JSC Release 77-45)

• NASA announced that Dr. Hans Mark, former director of Ames Research Center, had been sworn in at the Pentagon, Washington, D.C., Aug. 9, as Undersecretary of the Air Force by Defense Secretary Harold Brown. Dr. Mark, ARC director from Feb. 1969 to July 1977, had taught nuclear engineering at the Univ. of Calif. at Berkeley and engineering at Stanford Univ.; he had also taught at Boston Univ. and the Mass. Inst. of Technology.

Born in Germany in 1929, he had become a U.S. citizen in 1945 and had been a consultant to the Institute for Defense Analysis (1958-1961), the USAF Scientific Advisory Board (1969-1976), and other defense-related and scientific organizations. C.A. Syvertson would be acting director of ARC until a successor was named. (ARC Release 77-38; NASA anno Aug 15/77)

• NASA announced that Dr. Bruce T. Lundin would retire as director of Lewis Research Center Aug. 26 after 34yr with NASA and its predecessor, NACA. LeRC deputy director Dr. Bernard Lubarsky would be acting director until a successor was named. In 1943, Dr. Lundin had begun working at the NACA Lewis Research Center on improvement of World War II aircraft engines.

When NASA was established in 1958, Dr. Lundin became assistant director of LeRC; in 1961 he became associate director for development, working on space propulsion and power generation. In May 1968 he was deputy associate administrator at NASA Hq, then acting associate administrator for advanced research and technology. He became director of LeRC as of Nov. 1, 1969, and had led the center's research on solving the nation's energy problems. He had not announced his future plans. (NASA Release 77-167; NASA anno Aug 9/77)

• LaRC announced that the NASA/ASEE engineering systems design team would give a final presentation Aug. 12 on the role of aerospace technology in agriculture. The team of faculty from local colleges and universities would discuss agricultural aircraft, remote sensing, and impact on the social and physical environment. The Old Dominion Univ. would distribute the team's written report later in 1977. (LaRC Release 77-36)

August 11: NASA announced it had selected Hughes Aircraft Co.'s Space and Communications Group for negotiation of a $33 million contract to build, test, and deliver three geostationary operational environmental satellites, GOES-D, -E, and -F, as follow-ons in the synchronous meteorological satellite (SMS) series providing near continuous high quality observation of the earth and its environment. The GOES spacecraft would carry a visible infrared spin-scan radiometer atmospheric sounder (VAS) to obtain day or night cloud cover photos of the portion of the earth viewed from geostationary orbit, with temperature readings at various levels of the atmosphere; the $33 million contract did not include the three VAS instruments being produced under contract by Santa Barbara Research Center. They would also carry a space environment monitoring system of three separate sensors to monitor solar emissions.

Goddard Space Flight Center would be project manager for the GOES systems developed and procured by NASA on a reimbursable basis for the National Oceanic and Atmospheric Administration, which would operate the spacecraft after launch and checkout by NASA. (NASA Release 77-170)

• JSC announced that the second group of 20 Shuttle astronaut applicants would report Aug. 15 for individual interviews and physical examinations. Like the first group, all those in the second group had applied for pilot positions. The second group included 11 USAF applicants, 7 Navy, 1 Marine, and 1 civilian. (NASA Release 77-171; JSC Release 77-44)

• LaRC announced that the Charles W. Swain family of Tallahassee, Fla., would move into the Tech House Aug. 15 for a yr-long demonstration of NASA designed systems in and around the house to save energy and water and make the home more secure. They would pay a prorated rent and all utilities during their stay. Dr. Swain, a humanities professor at Fla. State Univ., would work with LaRC personnel in career counseling and employee relations. Mrs. Swain, a registered nurse, hoped to continue in her profession while living at LaRC. Their two children would attend local high schools. LaRC had chosen the Swains from a list provided by the American Council on Education, a nonprofit organization sponsoring university professors for 1yr assignments to the federal government. (LaRC Release 77-37)

• Calif. Gov. Edmund G. Brown, Jr., hailed the advent of the space age at a Space Day gala in Los Angeles sponsored by 11 aerospace companies and the United Bank of Calif. More than a thousand persons attended the exhibit, prelude to the Shuttle orbiter free flight Aug. 12 over the Mojave desert. The program included two space movies, as well as speeches by Carl Sagan on planetary exploration, Gerard K. O'Neill on space colonies, and Jacques Yves Cousteau on spaceflight and the oceans of the world. (*W Post,* Aug 12/77, A-9)

August 12: NASA launched the first high-energy astronomy observatory, HEAO-A, from ETR on an Atlas Centaur at 2:29am EDT into a near circular orbit with 456.3km apogee, 431.4km perigee, 93.5min period, and 22.75° inclination.

Heao 1, nearly 19ft long and nearly 9ft in diameter, weighting 5626lb at liftoff (including 2960lb of experiments and an 831lb separation system), had a 6mo design lifetime but would carry enough consumables and orbit at an altitude to allow extension of its mission up to 1yr. It would operate in 2 modes, pointing and scanning: its batteries would point continuously at the sun as the vehicle moved 1°/day in the ecliptic longitude, and its detectors would scan in wide circles perpendicular to the plane of the ecliptic. TRW had built the spacecraft equipment module (SEM) containing all the subsystems used by all the experiments for operation and control, and a mission unique experiment module (EM) filling the requirements of each payload segment. Dr. Fred Speer, HEAO manager at Marshall Space Flight Center, said the spacecraft, rotating end over end, would survey the entire sky within its 6mo lifetime.

This launch was the first in a three mission program to study x-rays, gamma rays, and cosmic rays emitted by stellar sources, the other launches to be in 1978 and 1979. The HEAOs would carry large and heavy scientific instruments needed for observations at the high-energy and low-flux levels of these phenomena. Estimated cost of the 3 observatories was $237 million.

The original plan in 1968 had been to launch a 40ft vehicle weighing 22 000lb; budget problems had reduced the size of the undertaking. However, the x-ray detector built by NRL as 1 of 4 experiments carried by *Heao 1* was the largest instrument ever flown on an unmanned satellite, weighing almost 1000lb. The NRL package, called the large area x-ray survey, was a 7-module array 20 times as large as any of its predecessors that would map the sky by recording incident x-ray fluxes and would determine the spectrum, time, and intensity variations of newfound x-ray sources. NRL's Dr. Herbert Friedman noted that known x-ray sources now numbered about 200; *Heao 1* might increase the list to several thousand. Celestial sources of x-ray emission included white dwarfs, neutron stars, and black holes, products of stellar catastrophe, as well as quasars and radio galaxies; also, a class of ministars almost too small for detection by the biggest optical telescopes might emit radiation 10 000 times as great as all the radiant energy of earth's sun.

The other experiments on *Heao 1* were a scanning-modulation collimator to make parallel observations, using instruments designed by Harvard and MIT, to pinpoint locations of sources and determine their structure; a cosmic x-ray experiment designed by NASA and CalTech, to look for diffuse x-rays or cosmic rays; and another experiment from MIT and the Univ. of Calif. San Diego, to look for hard x-rays and low-energy gamma rays. NASA planned a guest observer program for this mission, participants to be selected from responses to a "Dear Colleague" letter issued by the Office of Space Sciences.

A malfunctioning rate gyro had delayed the planned June 30 launch; the Bendix-built instrument had been removed and returned to the manufacturer March 31, but further similar delays occurred [see July 26]. Early data from orbit showed spacecraft systems and experiments were operating normally. (NASA Releases 77-45, 77-56, 77-71, 77-83, 77-87; OSS rept S-832-77-01 [prelaunch] May 16/77, July 29/77; MOR S-832-77-01 [postlaunch] Aug 12/77, Apr 3/78; MSFC Releases 77-32, 77-64, 77-107, 77-142; KSC Releases 85-77, 140-77; Spacewarn SPX-284; *NYT,* Apr 10/77, 23; Aug 13/77, 33; NRL Release 23-8-77S)

• The Shuttle orbiter Enterprise made its first free flight at Edwards Air Force Base in Calif. with a "flawless landing" on a runway in the Mojave desert 53min 51sec after taking off at 8am local time (11am EDT) and 5min 22sec after release from its 747 carrier. Astronauts Fred W. Haise, Jr., and Charles Gordon Fullerton were the Enterprise pilots; veteran test pilots Thomas C. McMurtry and Fitzhugh L. Fulton, Jr., flew the Boeing 747.

Spectators, estimated to number about 70 000, had begun arriving the previous night (Thursday, Aug. 11) to cheer the landing just after 9am local time with the desert temperature nearing 100°F. The orbiter had separated from its carrier 3min late because the heat of the air had slowed the climb. The 747 had "strained its engines" to carry the 73ton Shuttle as high as possible; it had climbed to 27 000ft (8100m) before it started downward, in order to gain speed. The 122ft-long orbiter, released at 22 100ft (6738m) altitude, had soared one way, the 747 another, to ensure separation. Pilots of the T-38 chase planes following the flight had assured the crews of the 747 and the orbiter that they were clear of each other. "Thanks for the lift," Haise told the 747 crew as the craft separated.

Haise had rehearsed the free flight the previous day in a small twin-engine jet modified to behave like the Shuttle. Actual separation, which could not be rehearsed in advance, was the only way to prove that the Shuttle would lift cleanly over the tall vertical tail fin of the 747.

Separation had occurred without difficulty, however, and both pilots reported themselves delighted with the orbiter, which had handled in un-

Space Shuttle orbiter Enterprise separates from its Boeing 747 carrier in first free flight, August 12, 1977. (NASA 77-H-552)

powered flight as easily as design engineers had predicted. It was "a lot like flying a Concorde," according to Haise, and Fullerton described it as a "very crisp, very stable airplane." On the orbiter's way down, Haise had pulled up in three inflight "flare" maneuvers to test its landing characteristics. One of the identical redundant computers on the orbiter had failed at separation, cutting off flight-control accelerometers, without affecting the flight.

The Enterprise landed like a glider, without engines, relying on its battery of computers and "the most advanced control system ever installed in a spacecraft." It had approached the ground at an 11° angle, more than 3 times as steep as a commercial jetliner's approach, and at a speed of 338kph (210mph) compared to a typical 747 landing speed of 170mph. The orbiter had touched down a mile in front of the aiming point, because Haise underestimated its glide capability; it had rolled another 2mi before stoppping.

For this first free flight, the orbiter wore a tail cone over its three aft rockets; on the fourth flight, scheduled for Oct., NASA would remove the cone, causing the orbiter to land more steeply (as it would upon returning from space). In regular use, the Shuttle as a weight-saving measure would have no power after firing its rockets to deorbit, no air-breathing engines for maneuvering in the atmosphere; it would be slowed by air friction only and would glide down at 200mph or more to a landing. For the actual spaceflight early in 1979, the Shuttle would ride into orbit pushed by solid-fuel rocket motors on a giant external tank and by the liquid-fuel rockets on its tail.

Purpose of the free flight exercises was to verify the orbiter's subsonic airworthiness, the operation of the integrated system, and the pilot-guided vs. automatic approach and landing capabilities. The ability to land at an airport instead of parachuting into the ocean gave the Shuttle the "versatility and economy required to make space operations routine in the next decade," according to the *W. Star.* (NASA Release 77-160; *NASA Actv.,* Aug/77, 2; *NYT,* Aug 13/77, 1; *W Post,* Aug 13/77, A1; *W Star,* Aug 13/77, A-3; *B Sun,* Aug 12/77, 7; *C Trib,* Aug 13/77, 1-1; *Time,* Aug 22/77, 64)

• MSFC announced that a young student assistant under the first NASA research grant awarded Alabama A&M Univ. 12yr ago, now a nationally known organic mass spectrometrist, had become a NASA researcher again, representing Meharry Medical College and the Natl. Inst. of Health as a summer faculty fellow. Dr. Stanley L. Evans, one of 23 researchers selected for the Aeronautics and Space Research Faculty Program sponsored by NASA and the Am. Society of Engineering Education, would study separation and preserving of cells in zero gravity. (MSFC Release 77-144)

August 13: The *W. Post* reported that the U.S. Naval Observatory had honored the memory of its astronomer Asaph Hall with a 2:30am party attended by Hall's descendants ("a fine crop of teen-agers and young adults") on the 100th anniversary of his discovery of the moon of Mars, which he had named Deimos (flight) and Phobos (fear) after the horses that drew the war god's chariot. "It now turns out," the *Post* added, that the little moons would crash into Mars within the next 70 million yr [see During June]. Other Naval Observatory observances honoring Hall included a reception, a group of technical meetings, a dinner, and a 100th anniversary memorial entry in the logbook of the observatory's 26in refracting telescope. (*W Post,* Aug 13/77, B1)

August 15: LaRC reported it had held a luncheon Aug. 11 to recognize the contributions of 26 inventors on the center staff to NASA's patent program. Director Donald P. Hearth and Patent Counsel Howard J. Osborn presented awards, including one to Richard T. Whitcomb as 1976 coinventor of the yr for his patent "Airfoil Shape for Flight at Subsonic Speeds," shared with Robert T. Jones of ARC. (LaRC Release 77-38)

August 16: NASA announced it had chosen, from more than 1000 respondents to invitations, 59 scientists (47 from the United States and 12 from the United Kingdom) to participate in the second flight of Spacelab scheduled for launch in 1981. The Solar Terrestrial Division in NASA's

Office of Space Sciences, responsible for managing Spacelab 2's payload, would choose experiments to provide new data on very high energy cosmic rays; x-ray emissions in galactic clusters; diffuse cool objects in earth's galaxy; and relationships between solar activity and magnetic fields. (NASA Release 77-172; MSFC Release 77-148)

• NASA announced that Kenneth S. Kleinknecht, assistant manager of the orbiter project at JSC, would begin an assignment in Paris Aug. 28 as deputy associate administrator for spaceflight (European operations) acting as senior NASA adviser to ESA on Spacelab. (NASA anno Aug 16/77)

August 17: MSFC announced it had begun the first of a series of tests on the external propellant tank of the Shuttle to verify its ability to stand the structural stresses of launch. The proportions of the tank required conduct of the test in phases; the first tests would check out the intertank structure joining the two main segments of the external tank and containing the forward attach points for the two solid-fuel rocket boosters carried by the Shuttle. The heaviest loading would occur at these points, both while the Shuttle was on its pad and while the boosters were thrusting during flight.

The tests would apply upward loads as great as 4.35 million lb to the intertank cylinder subjected to opposing downward forces. The loads would come from computer-actuated hydraulic jacks, 28 on the forward end of the test article and as many as 10 in the center section. Chuck Verschoore, MSFC test engineer, said the forces exerted by the jacks could simulate bending and twisting effects as well as straight up-and-down loading, and could go as high as 140% of the tank's design limits. Test engineers would receive data from as many as 2800 measurement channels. Follow-on tests would check the external tank segments containing the liquid oxygen and liquid hydrogen propellants. (MSFC Release 77-149)

• The *NY Times* reported an announcement by Hughes Aircraft Company that Telesat Canada had chosen it to negotiate a contract for a new generation of Anik comsats operating at frequencies of 12 and 14 billion cycles per second, to replace in the 1980s the 4- and 6-billion-cps now the standard for national and international communications systems. The higher frequencies would ease radiocommunications crowding in 22 300 mile altitude equatorial orbits, and would expand the number of satellite-communications channels to approach the number available on terrestrial microwave and cable links.

Telesat Canada, which had operated that country's domestic comsat system since 1972, said that two rival companies—RCA's Astroelectronics Div. and GE's Space Division—remained in the running with Hughes for award of the contract, which could go as high as $50 million. RCA had built the Satcoms used by the U.S., and GE had built a direct-broadcast comsat for Japan. The *NYT* said the electronics industry had been watching the competition closely because the same three companies had been vying to build 3 data-transmission satellites for Satellite Business Systems in the U.S. (*NYT,* Aug 17/77, 49)

- INTELSAT reported that the Peoples Republic of China on Aug. 16 had become the 98th member of the organization. Peking's Administration of Long Distance Telecommunications had signed the operating agreement, making the PRC the 28th Asian nation to join INTELSAT. China, which had used INTELSAT since 1972, currently had 3 ground stations, 2 near Peking and 1 near Shanghai. It currently operated 23 full-time satellite circuits on the INTELSAT system and was using TV capacity on an as-needed basis. (INTELSAT Release 77-20-I)

- The Natl. Space Club announced that the guest speaker at its Aug. luncheon would be Col. W.J. Mellors, head of the Washington, D.C. office of the European Space Agency (ESA) since 1973. Col. Mellors had begun his career with the British army in 1942 and had become commander of the Royal Electrical and Mechanical Engineers in the Middle East by 1965. After joining ESA in 1970, he worked until 1973 with ESTEC, the technology center in the Netherlands. ESA had united the programs of its predecessor organizations, the European Space Research Organization (ESRO) and the European Organization for Development and Construction of Space-Vehicle Launchers (ELDO); its members nations were Belgium, Denmark, France, West Germany, Italy, the Netherlands, Spain, Sweden, Switzerland, and the United Kingdom. (NSC bulletin Aug 17/77)

August 18: MSFC reported that a new pattern of painting Shuttle solid-fuel rocket boosters for orbital flight tests in 1979 and 1980 could save the agency an estimated $300 000. The center's system dynamics laboratory had changed the 6 black stripes running the length of the giant motors to 4 large rectangles in a spiral pattern that would let engineers measure roll and pitch rates of the boosters on reentry. (MSFC Release 77-150)

August 20: Today newspaper reported that two USAF fighter planes had dropped about 20 bombs July 8 across a lake bed runway used by aircraft from the Jet Propulsion Laboratory, which operates for NASA the deep-space network antennas at Goldstone. NASA and the Department of

Defense had been at odds over the Mojave desert airspace, and "the military is winning," the paper said.

Twice in the past 6mo a B-1 bomber had narrowly missed the multimillion-dollar tracking antennas, and over the past yr USAF planes with electronic devices had been jamming space signals to and from Goldstone. In the July 8 incident, all bombs from the USAF planes had fallen no more than half a mile from the 5 big-dish antennas serving as the sole links to satellites in orbit around the earth, the sun, and the planet Mars and to one spacecraft headed for the planet Saturn.

The commandant of the Marine base at El Toro, where the planes came from, had grounded the pilots for 2wk, blaming the incident on "navigation error" and misreading of the target. However, on July 22, the B-1 bomber that had previously buzzed the Goldstone station flew with its chase plane low over the heads of station technicians to check the sky for radio interference from other military craft. According to JPL officials, the B-1 was "far off the course it normally flies on a test run."

Fighter planes from the Georgia air force base had interfered "no fewer than 60 times" with Goldstone radio links to its space vehicles. Last Oct. a fighter pilot had "accidentally" turned on an electronic device and disrupted contact with the Viking orbiter at Mars for more than an hour. Had the spacecraft been in a maneuver or an emergency, the mission would have been lost, the Viking scientists said. (*Today*, Aug 20/77, 1A)

August 20-29: Two yr to the day after launching the Viking mission to Mars, NASA launched *Voyager 2* at 10:29am EDT from ETR on a Titan IIIE-Centaur. Countdown was smooth except for a brief hold to check the status of a valve. Minutes after launch, problems began to appear: a suspected gyro failure, incomplete data transmission, and failure of the boom holding the science platform to deploy.

NASA later reported the gyro was working and data transmission was satisfactory. The boom, which was supposed to deploy 53min into the flight, had not extended and locked. To determine its position, controllers 12hr after launch had activated a plasma instrument on the scan platform and combined its readings relative to a known axis and the direction of the solar wind (data supplied by GSFC) to show that the boom had extended to within 2° of full deployment.

JPL reported Aug. 22 that the spacecraft had been stable since 3pm EDT Aug. 20, except for a pitch and yaw disturbance at 5am Aug. 21; flight controllers were investigating. It reported Aug. 25 that the craft had undergone another disturbance like that of Aug. 21; controllers had ruled out the propulsion module's having bumped the spacecraft. On the morning of Aug. 29, JPL calibrated the sun sensors and removed the dust cover on the infrared interferometer spectrometer. It had tried Aug.

26 to move the science boom to the locked position, but the computer system had aborted the command, showing that a problem existed. Controllers tried jettisoning the dust cover while reorienting the spacecraft, to jolt the assembly enough to open the boom hinge and lock it into position, but this was unsuccessful. JPL had decided to put *Voyager 2* on hold to allow controllers to concentrate on the launch of the other Voyager Sept. 5. (Voyager mission status bulletins 3 through 6)

August 23: MSFC announced that all systems aboard the *Heao 1* observatory launched Aug. 12 had been turned on and were operating as planned. Dr. Fred Speer, project manager for HEAO, said the observatory was already acquiring x-ray sources. Experimenters using *Heao 1* reported they were pleased with the early performance of their instruments. (MSFC Release 77-153)

August 23-26: The *NY Times* reported JSC had announced it could no longer afford to monitor instruments left on the moon by Apollo astronauts. "I sure hate to see them go, but NASA headquarters has notified us there will be no funds for operations beyond Sept. 30," said Wilbert F. Eichman, chief of JSC's payload requirements and operations branch.

Longest-lived lunar instruments were seismometers recording moonquakes and hits on the lunar surface by outer space objects; four seismometers were still operating, and scientists said some could send data for up to 4yr more. Data from other instruments, monitored in a small control room at JSC, had included measurements of the moon's field of gravity, heat inside the lunar surface, and solar wind and lunar ionosphere samples. Roy Kelly, a senior engineer for Bendix (manufacturer of the instrument packages), said the transmitters would be left on with no operational support available.

An Aug. 26 *NYT* editorial said NASA's decision to abandon monitoring was unfortunate because, if it continued the program "for just another year, there will be that much more data on lunar seismography and structure." Apollo had cost the nation close to $30 billion; shutting down to save a few hundred thousand dollars seemed to be "a repudiation of one of the finest chapters in American exploration," the paper said (*NYT*, Aug 23/77, 16; Aug 26/77, A-20)

August 25: NASA launched Italy's first experimental domestic comsat SIRIO from ETR at 7:50pm EDT on a Delta into a transfer orbit with 37 670km apogee, 229.7km perigee, and 23° inclination. At 10:57am EDT on Aug. 27, its apogee boost motor fired it into a near-circular orbit where it would drift by Sept. 8 to a station on the equator above the west coast of Africa at about 15°W. NASA would transfer operation to the Telespazio VHF/SHF control center at Fucino, Italy, about 45 days after launch. Italy's National Research Council (CNR) would reimburse NASA for the cost of the launch vehicle and support services, estimated at $15.7 million, as agreed in March 1975.

The Satellite Italiano Richerche Industriale Orientata (industrial research-oriented Italian satellite), 2m (6.5ft) high and 1.4m (4.6ft) in diameter, weighing 398kg (878lb) at liftoff, would transmit TV and voice data at superhigh frequencies (SHF) from Europe to North America in adverse weather conditions. The project had begun in 1969 as part of an ELDO program, specifically to carry out a communications experiment in the 12 and 18GHz bands proposed by Prof. Francesco Carassa of Milan Polytechnic Inst. The idea of using higher frequency ranges (30 to 60GHz) because of overcrowding on conventional frequencies had offered the advantages of greatly increased traffic in communications, less interference with present ground and satellite communications, and the use of small ground terminals. The Compagnia Industriale Aerospaziale (CIA) had designed and developed the spacecraft and was responsible for system management and orbit operations.

Defense/Space Daily said the original Aug. 17 launch date had slipped because of possible interference between the SIRIO antenna and the third-stage system that might hinder separation. The *NY Times* said that "adverse wind conditions" Aug. 16 that delayed a test launch of the Navy's Trident missile had automatically postponed the SIRIO launch. *AvWk* said that the SIRIO launch was important to Italy not only technically but also politically: it said NASA officials believed major changes would have occurred in the Italian government if the spacecraft had failed. (NASA Release 77-152; KSC Release 140-77; MOR 492-209-77-01 [prelaunch] Aug 9/77, [postlaunch] Nov 9/77; *D/SD*, Aug 22/77, 275; *NYT*, Aug 17/77, A-11; *AVWk*, Sept 5/77, 23)

• NASA announced that a third group of 20 astronaut applicants, all in the mission specialist category, would report to JSC Aug. 29 for a week of individual interviews and physical examinations. Eight of this group were women; all 20 had Ph.D. or medical degrees, or both, and one had a degree in veterinary medicine. (NASA Release 77-176)

• MSFC announced plans for a 3-day meeting beginning Aug. 31 for the 59 investigators selected earlier this mo [see Aug. 16] to collaborate on 11 experiments for the second Spacelab mission. Dr. Eugene W. Urban of MSFC's space sciences laboratory, mission scientist for Spacelab 2, would chair the meeting of the Investigators Working Group to brief the participants on engineering and management requirements. (MSFC Release 77-155)

August 26: LaRC announced selection of RCA Service Company, Camden, N.J., to negotiate a $3.5 million contract for electronics fabrication services to the center. Work on the contract, covering a 2-yr period with options for three 1-yr extensions, would begin in Jan. 1978. (LaRC Release 77-40)

• The *NY Times* reported that "observers in the U.S. and abroad" were puzzling over a Soviet space vehicle called *Cosmos 929* that had been orbiting for a mo and was large enough to be manned, but apparently carried no passengers. Orbital and radio telemetry characteristics of the vehicle were like those usually related to a manned mission; speculation was that the vehicle might be the first unit of a large space station to be assembled in orbit.

Reports from Europe that the USSR would launch a space vehicle during the summer had assumed that the mission would be another Salyut, five of which had previously been orbited, the first early in 1971. *Cosmos 929,* launched July 17, had drawn attention because of its size and its unusual telemetry, apparently two distinct systems, as though it were two objects joined together. On Aug. 18, after a major change in orbit, one of the two radio transmissions had also changed "significantly," and part of the vehicle might have been detached at that time. Members of the Kettering Group, an international group of nongovernment space watchers, had confirmed that the vehicle was as large as a Salyut and that its 51.6° orbital inclination had usually, though not exclusively, been used for manned flights as had its telemetry signals. *Cosmos 929* had produced no voices, however, and the group was "virtually certain" that no crew was aboard. (*NYT*, Aug 26/77, A10)

August 28: The *W. Post* reported that the leader of a USSR expedition on the first surface ship to reach the North Pole had told a news conference that his major problem was deciding when the ship had reached its destination. Soviet Minister of Merchant Marine Timofey B. Guzhenko said that, as the ship neared the pole Aug. 17, fog had made sun readings impossible; the polar daylight obscured the stars; and magnetic compasses were worthless when they "began to jump at the pole."

The vessel's officers had to spend several hr determining whether they had reached the pole. They finally relied on a satellite navigation system accurate within about 800yd, and made two passes through the ice floes to make sure they had reached their goal. The crew of the 13-day 3582mi expedition on the nuclear-powered icebreaker Arktika had planted a Soviet flag and capsule containing a copy of the USSR constitution to mark their accomplishment. (*W Post*, Aug 28/77, A24)

August 29: NASA announced it had tentatively chosen 114 scientists to participate in the Jupiter-orbiter probe mission scheduled for 1981-1982. More than 500 scientists had submitted proposals for research in response to a 1976 invitation from NASA. Thirteen selectees would have interdisciplinary jobs, helping with mission plans and coordinating research and instrumentation to ensure best results from the mission. NASA would assign the others to 17 individual-experiment areas; principal investigators for 3 of those areas would be from nations other than the U.S. JPL would manage the Jupiter-orbiter project, and ARC would manage the probe system. (NASA Release 77-177)

• NASA reported that LeRC had been operating a power generator for 30,000hr without a break and was aiming for 50 000hr of continuous operation, hoping to use the system in deep space missions. The Brayton space power generator, a closed-cycle gas turbine, had been running for more than 3yr under rigorous space environment conditions of temperature, pressure, and speed.

For the test, ordinary electric lab heaters had provided energy to the generator; in space, the generator would draw energy from the sun or from a nuclear source. The Brayton engine had dispelled the "myth that only static nonrotating systems have the endurance potential for long space voyages," said Jack A. Heller, LeRC engineer in charge of the advanced engine work. The system, he said, "not only is showing reliability but it is performing at a high energy-efficiency level of 25% or better without any signs of degradation." (NASA Release 77-178)

• The *W Post* reported conflict between the USAF and NASA not only over use of the Mojave desert area [see Aug. 20] but also over use of radio frequencies that NASA considered the key to contacting alien intelligences, and that the Air Force wanted for a satellite global-navigation system. The USAF wanted to use a band scientists called the "waterhole," in which hydrogen and oxygen radiate natural radio signals into space (the combination of the two into water had given the band the name), thought to be the area in which alien civilizations might conduct a dialogue in space. The USAF had selected these frequencies for navigation signals as carrying less interference. Should the USAF have a navigation-satellite network in place by 1984, its signals would drown out any other signals on those frequencies. (*W Post*, Aug 29/77, A1)

August 30: ARC announced that 7 U.S. biological experiments launched Aug. 3 on the USSR's *Cosmos 936* had landed by parachute in Siberia Aug. 22. The Soviet satellite also carried experiments from France, Czechoslovakia, Poland, Romania, Bulgaria, Hungary, and East

Germany. Five of the U.S. experiments had followed up those on the first cooperative U.S.-USSR biological satellite *Cosmos 782* launched in Nov. 1975; 4 of them had Soviet coinvestigators, and all of them used some Soviet scientific or technical assistance. The last U.S. biosat was *Biosatellite 3,* launched in 1969; without the Soviet flight opportunity, the U.S. biology experiments would have to wait for Shuttle flights in the 1980s. (ARC Release 77-37; NASA Release 77-169)

• MSFC reported that astronaut Paul Weitz had arrived on center to check out a procedure developed for jettisoning a jammed-open outer door of the Shuttle scientific airlock while on a space mission. Weitz would wear a pressurized space suit for the task, to be carried out in MSFC's neutral buoyancy simulator. The 75ft-diameter tank filled with water 40ft deep would duplicate movement of people or hardware almost as if weightless in space. Engineers and technicians could design instruments and procedures to handle almost any foreseeable problems before actual spaceflight, and unforeseen problems (such as solar array failure on Skylab) could be solved during a space mission.

Center workers had been devising a way to remove the outer door of the airlock in case it could not be closed at the end of a mission, using a full-scale mockup of the Shuttle cargo bay in the simulator. Underwater TV cameras would record the astronaut's movements on control room monitors, and both video and voice communications would be recorded for later study. Scuba divers in the simulator would help as needed. (MSFC Release 77-158, 77-166)

• NASA announced that the largest Space Shuttle component—its external propellant tank—would roll off the assembly line in New Orleans Sept. 9. The huge tank would then go by barge to the Natl. Space Technology Laboratories in Miss. for the first static firing of the Shuttle main engines.

The external tank consisted of 2 propellant tanks and an intertank structural connector; the forward tank would contain liquid oxygen, the larger aft tank liquid hydrogen. It would be the main structural member carrying the orbiter and the two solid-fuel rocket boosters. At launch, the tank would feed cryogenic propellants to the three main engines on the orbiter; the solid-fuel rocket boosters on each side of the external tank would use their own fuel and would separate about 2min into flight for later recovery. The mated orbiter and external tank would thrust for another 6min, separating just before the Shuttle reached orbital velocity. The complete external tank was 47m (154ft) long; the Shuttle orbiter it would carry was 37.2m (122)ft long. (NASA Release 77-179)

August 31: NASA reported it would join NOAA in a 5mo test to detect turbulence in the flight paths of 2 planes in the skies over Colorado, "a built-in turbulence laboratory," according to Dr. Peter M. Kuhn. An earlier study by Dr. Kuhn and others from NOAA's atmospheric physics laboratory had used infrared radiometers to detect fluctuations in atmospheric water vapor, indicating turbulent areas ahead of a plane with 81% reliability [see *A&A 76*, Nov 11]. Successful tests could produce an inexpensive turbulence-warning instrument for a plane's control panel. (NASA Release 77-180)

• GSFC announced "a comprehensive programmatic and institutional review" of its FY 1978-1979 program to define the skills it needed to do its job "within its allotted resources." The review would result in fewer Civil Service employees, the announcement said, and a work force adjustment would require a center-wide review and validation of position descriptions to match program redirection and "changes in manpower levels." GSFC Director Robert S. Cooper called on all center employees to cooperate fully "in the updating of our position descriptions." (GSFC anno Aug 31/77)

• The *NY Times* reported that the Voyager spacecraft being launched toward Jupiter and Saturn would use a little-known power source that had made possible much of the dramatic space activity of the last decade: a capsule of plutonium 238 producing heat convertible into electricity. In localities far from the sun, or on planetary bodies with drastic temperature swings between day and night, the plutonium capsule had proved far superior to solar cells.

Early missions such as Skylab and Salyut space stations; the *Mariner 9* and *10* that surveyed Mars, Venus, and Mercury; and communications, weather, and navigation satellites had all used solar cells. But in 1959 President Eisenhower had revealed the possibility of nuclear power for spacecraft, and in June 1961 a grapefruit-sized 2.7watt generator had gone into orbit on the Navy's *Transit 3A* navsat. NASA had used a larger system called SNAP 19 on *Nimbus 3*. Later models powered the still-active *Pioneer 10* and *11* launched in 1972 and 1973 that reached Jupiter in 1973 and 1974, as well as the Viking landers that touched down on Mars in July and Sept. 1976 to search for evidence of present or past life.

Another model called SNAP 27 had powered the still-active stations left on the moon by Apollo missions between 1969 and 1972; the *Apollo 12* station had lasted nearly 8yr, the length of time the late version nuclear capsule on the Voyagers would have to last to permit observing Uranus in 1985. The long-lived moon stations had attracted notice recently [see Aug. 23-26] when NASA announced it would stop monitoring them in order to save "something less than $1 million a yr," the article said.

The plutonium power sources, known as SNAP (system for nuclear auxiliary power) or RTG (radioisotope thermoelectric generator), had been a byproduct of U.S. production of power for nuclear weapons. The Energy Research and Development Administration (ERDA), destined to be part of a new federal Department of Energy, had undertaken a $30-million-per-yr development of even larger nuclear power systems for space, including selenide thermoelectric units with plutonium 238 heat sources for an orbiter with probe scheduled for launch toward Jupiter in 1982, and a system generating between 1000 and 2000 watts being developed for DOD. (*NYT*, Aug 31/77, 45)

During August: NASA announced it had detailed astronaut Russell L. (Rusty) Schweikart to the state of Calif. effective Sept. 1 as assistant for science and technology to the governor. The assignment would be for 1yr under the Intergovernmental Personnel Act.

Schweikart, one of the third group of astronauts, had come to NASA in 1963 and had flown the first lunar module in 1969 on *Apollo 9*. He had been backup commander for the first Skylab mission in 1973. In May 1974 he had joined the NASA Hq office of applications, and in Sept. 1976 became assistant for payload operations to the Assistant Administrator for Planning and Program Integration, working on policies and plans for payloads in the Shuttle era. In the new postion, Schweikart would coordinate the application of emerging technology to the state's needs and become informed about the problems and challenges facing state governments. (NASA anno Aug 24/77)

• The Natl. Aeronautic Association announced that 127 planes had finished in the last Powder Puff Derby on the 30th anniversary of the transcontinental air race for women. Pilots represented 36 of the United States as well as Australia, West Germany, and the Bahamas; 3 were flying their 24th Powder Puff Derby. NAA had decided in 1976 to make this run the last because of shortages of fuel and financing. The 30th flight was not a race but a "sentimental journey," the NAA said (NAA newsletter, Aug/77, 1)

• NASA announced it had established a fare structure for Space Shuttle payloads on flights scheduled to begin in 1980. Depending on the weight and volume of cargo, the price would run from less than $10 000 to more than $21 million. Other factors would be the optional services desired and whether the payload would go on a reserved basis or as a standby.

Costing $10 000 or less would be the "getaway special" (GAS), a small research package weighing less than 200lb, self-contained, with a volume of less than 5ft^3, flying on a space-available basis. The $21-million fare would be for a payload using the full capacity of the Shuttle on a reservation basis by non-U.S. government customers. Lesser charges would apply to customers sharing flights with other customers; those willing to fly on a standby basis; and those having made a substantial investment in Space Transportation System development (e.g., The European Space Agency, its member nations, and Canada).

Chester M. Lee, STS operations director, said NASA had set the prices to encourage Shuttle use by making costs attractive but still recovering operating expenses. Commercial and foreign users would pay fees covering depreciation of facilities and equipment, and amortization of orbiters. The pricing plan would undergo review in 3yr to make any necessary adjustments, Lee said; it contained provisions for cost increases due to inflation. NASA prices would include space in the Shuttle, with standard services such as design review, safety review, flight plans, transmission of payload data, and services of a 3-man crew. Optional services, costing more, would include special training, upper stages and services, special mission kits, revisit or retrieval of satellites, or any special equipment. (NASA Release 77-173)

- NASA announced it had identified about 40 payloads to be carried on 11 Space Shuttle flights in its first yr of operation, beginning in 1980. Three civilian firms had deposited "earnest money" with NASA for payloads on 8 flights; NASA had plans for 5 payloads, and DOD had plans for one.

Chester M. Lee, STS operations director, said NASA also had forecast 12 non-NASA civilian payloads, 10 NASA payloads, and 4 DOD payloads for launch during 1980 and 1981. Allocating the payloads to specific flights would not be firm until about a yr before launch, he noted. A working model for Shuttle traffic in the first 12yr, totaling 560 Shuttle flights, would show NASA flying about half the missions on its own; 20% would be DOD missions; 6%, other U.S. government agencies; 12%, commercial firms; and 12%, foreign customers. (NASA Release 77-175)

- NASA announced it had awarded Boeing Commercial Airplane Co. a contract to demonstrate advanced technology for fabricating composite horizontal stabilizers to be used on B-737 transport aircraft. Managed by LaRC, work under the contract would be done over a 4yr period on a cost-sharing basis, NASA contributing about $11.7 million and Boeing contributing about $1.3 million. Boeing would devise advanced manufacturing methods suitable for producing composite-material structures, and would carry out all analysis and testing needed to meet FAA requirements. Boeing Commercial Airplane Co. had previously received a similar contract to fabricate and devise techniques for using composite materials in elevator shipsets for B-727 aircraft [see July 7]. (NASA Release 77-174)

- Researchers from the Soviet ship Akademik Korolev, anchored off the coast of Virginia near Wallops Flight Center for joint U.S.-USSR rocket launches, had discovered the charms of the Eastern Shore and were enjoying their contacts with the natives, according to reports in the *Washington Star* and *Washington Post*. The meteorological soundings had been completed by the last week of August, the newspapers said, and the 134 Russians waiting on the ship for results to be correlated had time to visit the mainland on trips planned by local officials in cooperation with NASA.

Coming ashore in groups of about 50, the visitors (mostly young) had seen Pocomoke City, Md., a small U.S. city; the village of Chincoteague, Va.; the Assateague wildlife refuge; and a tour of Accomack county that included a visit to a private home, a farm, and a chicken-processing plant. Through NASA interpreters, the visitors asked scores of questions on topics from labor relations at the plant to the history of local buildings. They had joined in a Frisbee game with local youngsters, eaten largely of the local seafood, and bought out the entire beer supply of a local general store.

Capt. Oleg Rostovsev, commanding officer of the 409ft research ship, said he had found the visit "a very interesting experience." His assistant for scientific affairs, Evgeny Nelepov, said that many of the researchers had visited U.S. cities before, but that major cities everywhere had become more and more alike, "losing the individual characteristics and cultural tradition . . . To find these national characteristics one must go to the small villages and rural areas where traditions and local culture endures. That is why this visit here has meant so much to us." The Russians had held a press tour and reception on the Akademik Korolev as a thank you before the ship departed for Baltimore. (*W Star*, Aug 28/77, D-1; *W Post*, Sept 1/77, A-1)

September

September 1: The Space Age might not yet be out of its teens, the *Chicago Tribune* reported, but the two decades that would end Oct. 4 since the launch of *Sputnik 1* had brought many wonders. The newspaper cited worldwide television; photographs from space of the blue-white earth in the vastness of space; color pictures of the surface of Mars; close-ups of the planet Jupiter; men walking—and riding—on the surface of the moon; astronauts living aboard a space station for 84 days; and men "walking" in space, outside their ships. Satellites had changed human life as few other objects (the wheel, the printing press, gunpowder) had changed it, and had affected everything from the educational system to politics, from ideas about ecology to ideas about humanity itself and about its future.

Predictions of things to come in the next 20yr of the Space Age included weather modification, energy transmission from space, and permanent outposts in orbit and on the moon. People would use satellites to monitor air and water pollution, find minerals, chart water currents, predict earthquakes, map land use, keep tab on crops, look for fish, and make worldwide live television commonplace, the paper said. (*C Trib,* Sept 1/77, 1-7)

- The *Washington Post* said NASA was looking for "someplace else to land the space shuttle." Its second free flight had been postponed when Hurricane Doreen turned the planned landing strip (a dry lakebed at Edwards AFB) into a "mud puddle." A NASA source said officials had investigated other sites for the flight, tentatively rescheduled to Sept. 7. (*W Post,* Sept 1/77, A-27)

- Johnson Space Center confirmed that NASA would turn off science instruments on the lunar surface and would dismantle the JSC control center for Apollo lunar-surface experiment packages (ALSEP) as of Sept. 30. The five stations still operating were put there by *Apollo* missions *12, 14, 15, 16,* and *17*. In July 1969 the original moonwalkers of *Apollo 11* had left behind a prototype station with a design life of only 14 days; it had survived for 45 days and quit when its power supply failed. Specifications for the remaining five had included a 1yr design life for 4, and 2yr for the *Apollo 17* station; the *Apollo 12* station was "well into its eighth year," NASA noted.

Built to record and transmit long-term lunar-surface data, the stations with their long life had provided researchers with a real bonus: instead of data from only one seismometer at a time, for instance, the group had operated as a sensor network, greatly enhancing available information on the moon's internal temperature and magnetic field, charged particles in its environment, and especially moonquakes and meteoroid impacts. Although the experiments would be terminated, the transmitters would continue to operate, serving earth as an astronomy reference point. The Jet Propulsion Laboratory would use the signals in deep space work such as astrometrics and spacecraft navigation or monitoring the motion of the lunar orbit against a background of extragalactic stars to test gravity theories.

The lunar stations had had problems: The *Apollo 14* ALSEP ran for 4yr, quit for 2 days in March 1975, started up again, and had repeated this 6 times since. Understanding the trouble (temperature fluctuations resulting from the sun's position had caused short circuits) would help in designing future science stations. (JSC Release 77-47; NASA Release 77-203)

September 2: Ames Research Center announced that NASA had chosen 29 experimenters for the atmosphere probe portion of the Jupiter mission [see Aug. 29]. The probe would carry 6 experiments into Jupiter's atmosphere; 5 of the experimenters would be interdisciplinary scientists, and 24 would be assigned to the 6 experiments: helium interferometer, Dr. Ulf von Zahn, Federal Republic of Germany, principal investigator; mass spectrometer, Dr. H.B. Niemann, Goddard Space Flight Center; atmospheric structure, Dr. Alvin Sieff, ARC; nephelometer, Dr. Boris Ragent, ARC; net-flux radiometer, Dr. R.W. Boese, ARC; sferics receiver, Dr. L.J. Lanzerotti, Bell Laboratories.

The probe would separate from the orbiter about 55 days away from the planet, continuing on a separate flight path for about 30min of high-speed descent while relaying information to earth through the orbiter. After the entry probe concluded its work, the orbiter would fly close to Jupiter's moon Ganymede and near the large moons Io, Europa, and Callisto.

ARC's Dr. Lawrence Colin would be project scientist for the Jupiter probe; JPL would manage the Jupiter orbiter/probe project. (ARC Release 77-40)

• ESA announced selection of 53 candidates from 12 European countries for the single ESA opening on the first Spacelab mission scheduled for late 1980. By the end of 1977, ESA would pick 6 from the 49 candidates chosen by their respective countries and the 4 from ESA's staff, to undergo further tests and evaluation by NASA between Jan. and April 1978. By mid-1978, 3 candidates would be appointed as European experiment specialists; several months before the mission, one would be the first West European scheduled to travel and work in earth orbit, the other two serving as backup. (ESA release Sept 2/77)

September 4: FBIS reported a Tass announcement that the USSR had installed its first cameras designed to photograph bolides (celestial bodies invading the earth's atmosphere) in a 30-unit network in the Ukraine. The cameras would permit astronomers to calculate the trajectory and place of fall of the bolides, the release said. (FBIS, Tass in English, Sept 2/77)

• FBIS carried the official report by L.A. Aleksandrov, chairman of the USSR-U.S. group on space meteorology, that the joint rocket launches over the Atlantic to compare instrument results [see July 1] had been a success. (FBIS, Krasnaya Zvezda in Russian, Sept 4/77)

September 5-7: NASA launched the 1820lb *Voyager 1* from Cape Canaveral on a Titan/Centaur vehicle at 8:56am EDT Sept. 5 in a firing described as flawless by project manager John Casani. A second burn of the Centaur at 9:49am boosted the spacecraft out of earth orbit; firing and jettison of the onboard propulsion module sent *Voyager 1* on a trajectory toward Jupiter, where it should arrive 4mo before its sister ship launched Aug. 20. JPL reported Sept. 6 that *Voyager 1* was proceeding "without any of the problems that plagued Voyager 2." NASA had postponed the *Voyager 1* launch to permit installation of extra springs to lock the second spacecraft's science boom in place. The *Voyager 1* launch would be the last time NASA would use the Titan/Centaur, although the USAF would continue to use its Titan for future missions, most of them secret. (Voyager mission status bulletin 6; *W Star*, Sept 6/77, A-2; Sept 7/77, A-14; *NYT*, Sept 6/77, 1, 22; Sept 7/77, A-18; *C Trib*, Sept 6/77, 1-1, 1-24; *W Post*, Sept 6/77, A-2; *P Ing*, Sept 6/77; *LA Times*, Sept 6/77; *Today*, Sept 6/77, 1A; *D/SD*, Sept 7/77, 8)

September 7: NASA reported that Kenneth Souza of ARC and Dr. Eugene Benton of the Univ. of San Francisco had returned to the U.S. from Moscow Sept. 2 with the biological samples carried on the USSR biosat *Cosmos 936* [see Aug. 30]. This mission was the first to subject laboratory rats to artificial gravity in the weightlessness of space. NASA would send the samples (tissue from the rats and live fruit flies, processed and packed in dry ice at the mobile landing site in Siberia) with radiation records to U.S. laboratories at the Univ. of Calif., Berkeley; Univ. of Wash., Seattle; Univ. of San Francisco; Univ. of Southern Calif.; and ARC, for analysis. (NASA Release 77-182)

- NASA announced that JSC's space and life sciences directorate payload to be managed by the Hq Office of Applications on the second flight of Shuttle orbiter 102 (sister ship to the Enterprise used in recent approach-and-landing tests at Dryden Flight Research Center) in 1979 would investigate earth resources, environmental quality, and severe storms. The instrumentation and other hardware carried on the second flight would be reusable, to permit adjusting or modifying it for later flights at relatively low cost. The first orbital flight would mainly record performance and monitor space immediately around the orbiter for contamination; the monitor would fly on all 6 orbital flight tests in 1979 and 1980. NASA had estimated the cost of payload hardware, science and technical integration, and mission support at about $10 million. The 6 orbital test launches from Kennedy Space Center would aim at preparing the Shuttle for regular operations by May 1980. (NASA Release 77-181; JSC Release 77-48)

- ESA announced that NASA would launch the orbital test satellite OTS, ESA's first comsat, from Cape Canaveral Sept. 13 or 14 on a Delta. The European launcher Ariane would launch the operational European comsat system to follow OTS between 1981 and 1990, carrying domestic telephone, telegraph, and telex traffic besides relaying Eurovision TV. OTS would operate in the 11- and 14-GHz bands, an advantage over the European 4- and 6-GHz links subject to radio interference in regional service. OTS's original June launch date had slipped because a motor prematurely separated from the vehicle sitting on a launch stand in May. NASA had redesigned the bolt that failed, and had replaced the damaged motor and vehicle booster. (ESA Release Sept 7/77; MOR M-492-210-77-01 [prelaunch] Sept 8/77; NASA Release 77-168)

September 8: NASA announced that GSFC and Marshall Space Flight Center would join the USAF in solving a yrs old problem of comsat makers and users: flying in a synchronous orbit at about 34 500km (22 000mi) altitude, the comsats routinely encountered electrically charged particles from the magnetosphere that built up static electricity on the spacecraft surface, causing arcing that could damage solar-power cells and thermal insulation, interfere with signal transmission, and switch spacecraft functions erroneously. The joint project called SCATHA (satellite charging at high altitudes) would obtain specific data on the phenomenon. The USAF planned to launch a satellite in Jan. 1979 to study electrostatics at geostationary orbit. MSFC would provide 3 light ion-mass spectrometers to measure speed, temperature, and density of low-energy ions and detect their points of origin. GSFC would provide two instruments, one to measure space electric fields, the other to measure space magnetic fields. (NASA Release 77-183; MSFC Release 77-164)

• NASA announced plans to develop a prosthetic urinary sphincter for patients losing control of bladder functions because of injury or disease. Patients now used external collection devices, implanted catheters, or prosthetics difficult to operate. John Richardson of MSFC's technology utilization office said NASA's biomedical applications office had recommended MSFC for this job because of its expertise in design, manufacture, assembly, and operation of miniature valving. MSFC would have a contractor design and test the device and prepare documents for Food and Drug Administration approval, as well as manufacture the device commercially. A successful device would permit safe and quick emptying of the bladder and avoid infection, the prime source of fatal kidney failure among those without bladder control. (NASA Release 77-184; MSFC Release 77-163)

• NASA announced that the Office of Management and Budget had called for an agency ceiling of 23 237 permanent positions by the end of FY 1978, 500 fewer than current. Dr. Alan M. Lovelace, deputy administrator, said work force adjustments to meet the goal would include greater use of contractor R&D; consolidation of applications work at ARC, GSFC, and the Natl. Space Technology Laboratories; reduction of support manpower at JSC, KSC, and MSFC; consolidation of GSFC data management; transfer of helicopter research from Langley Research center to ARC; and reductions in support office and facility staff agencywide. MSFC would lose 150; Lewis Research Center, 108; JSC, 90; KSC, 80; LaRC, 77; GSFC, 74. DFRC, Hq, and Wallops Flight Center would lose 30, 20, and 5, respectively. The shift in programs would add 45 to ARC and 6 to NSTL. (NASA Release 77-185)

September 9: The quality of information from the Voyagers would greatly surpass that of earlier missions, said Philip Abelson in *Science* magazine. "A decade ago the level of knowledge concerning the planets and their satellites was such that almost any arm-waving explanation would fit the facts. This is no longer true." Citing the potential of 5 of the 11 instruments, Abelson noted the Voyager cameras could "achieve a resolution and observing time at Jupiter better by a factor of 40 than those of the Pioneer spacecraft. . . . The radio infrared, visible, and ultraviolet instruments can all participate in a mutually reinforcing fashion." (*Science,* Sept 9/77, 1858)

• ComSatCorp said the Federal Communications Commission had

asked for a status report on the escrow account the FCC ordered set up in 1976 to cover possible rate refunds to its customers. ComSatCorp said it had "meticulously complied" with the FCC order, but would respond in detail to the inquiry. (ComSatCorp Release 77-22)

September 10: The *Washington Star* reported that the first Space Shuttle fuel tank emerged from the Michoud assembly facility near New Orleans on its way to tests at the Natl. Space Technology Laboratory in Miss. NASA contractor Martin Marietta had built the 520 000gal tank, half as long as a football field, at a cost of about $2.5 million. If successful, as many as 500 similar tanks would be built during the next 15 to 20yr. The 153ft-long tank, almost 30ft in diameter, was built of aluminum alloy and was disposable, designed to fall away and disintegrate in the atmosphere just before the Shuttle reached orbit. Shuttle flights would cost about $10 million each, as compared to $150 million each for earlier launches. (*W Star,* Sept 10/77, A-3)

September 12: NASA announced it had scheduled the second free flight of the Space Shuttle orbiter Enterprise on or about Sept. 13 at DFRC with astronauts Joe Engle and Richard Truly as pilots. The free-flight test series called for 3 more flights before the orbiter would go to MSFC for vibration tests early in 1978. (NASA Release 77-186)

• MSFC announced that Roy Gibson, director general of the European Space Agency, had visited the center for the first time to discuss the NASA/ESA Spacelab and other NASA programs with MSFC director Dr. William R. Lucas and his staff. Thomas J. Lee, manager of MSFC's Spacelab program office, said the visit would familiarize the ESA official with NASA capabilities and programs "in which the Europeans might be interested as potential areas for future cooperative ventures." (MSFC Release 77-167)

• JPL announced that NASA had selected an ion-drive propulsion system rather than the solar sail [see June 21] for use by automated interplanetary shuttles in earth's solar system toward the end of this century. NASA scientists had studied the concepts for a yr before making the choice on the basis of lower risk and greater potential for growth. LeRC had orbited ion rockets, forerunners of the ion-drive spacecraft, in 1969 as part of SERT (space electric rocket tests). Those rockets were still operating in space. An ion-drive ship would carry a cluster of mercury-ion engines and generate power from solar-cell arrays converting sunlight to electricity. First use of the ion-drive system might be a Halley's Comet rendezvous in 1986 by a spacecraft launched from the Shuttle in 1981 or early 1982. (JPL release Sept 12/77; NASA Release 77-187; *W Star*, Sept 13/77, A-6)

• The Natl. Oceanic and Atmospheric Administration reported a new device to supply near-instantaneous information on hurricane growth and movement in the Caribbean and North Atlantic. The device, an airborne data collection platform on a weather research aircraft, would relay information on pressures, wind speeds, temperatures, and other hurricane data within 15sec from NOAA's *Goes 2* orbiting at 35 400km (22 000mi) to the Natl. Hurricane Center in Miami. Conventional transmission of voice radio from aircraft to a ground station would take 5 to 15min. Quick transit of information as hurricanes approached populated centers would allow time for alerts. As no Atlantic hurricanes had occurred this yr until the end of Aug., NOAA had tested the device in actual flight over a tropical depression and would install it in 2 other planes next yr. (NOAA Release 77-240)

September 13: The Shuttle orbiter Enterprise made its second free flight at Edwards AFB in Calif., a 5min unpowered glide from about 24 000ft altitude, during which it reached a speed of more than 330mph and performed several maneuvers to test its handling before air brakes slowed it to a landing at 250mph. Separation from the Boeing 747 carrier was delayed 3min because a power failure at DFRC made mission control in Houston lose radio contact; the delay caused no problems. For about 50min before separation, pilots Joe Engle and Richard Truly took turns at the controls and the computer to monitor the orbiter's behavior at various speeds. Pilots of the 747 were Fitzhugh L. Fulton, Jr., and Thomas C. McMurtry. The flight had been postponed from Aug. 30 because of a tropical storm [see Sept. 1]; a third test was set for late Sept. (Postflight rpt, free flt 2, SpSh orbiter ALT; *W Post*, Sept 14/77, A6)

September 13–November 9: NASA announced that the Delta vehicle on which it had launched ESA's orbital test satellite OTS at 7:21pm Sept. 13 from ETR had exploded 54sec after liftoff. The booster and 5 strap-on motors [see Sept. 7] had performed properly before the explosion. On Sept. 14 John F. Yardley, NASA's associate administrator for spaceflight, had appointed a failure review board whose investigation concluded from flight data analysis, photo records, and "substantial" vehicle debris recovered near the launch site that the no. 1 strap-on motor had failed from causes unknown. NASA had scheduled an OTS backup launch for April 1978. (MOR M-492-210-77-01 [postlaunch] Nov 9/77; NASA newsrm rept Sept 14/77; *W Post*, Sept 14/77, A9; NASA Release 77-193)

September 14: NASA reported that its *Nimbus 6* had helped rescue from the Atlantic off Iceland a 2-man balloon crew attempting a transatlantic flight. The balloonists—Maxie Anderson, 44, and Ben Abruzzo, 47—were unharmed after an after-sunset pickup Sept. 12 by a USAF helicopter. They had lifted off Sept. 9 near Bedford, Mass., carrying a compact satellite-tracking unit giving out signals received by the satellite and relayed to GSFC, which plotted the balloon's location and forwarded it to the control center at Bedford. The balloon ran into a snowstorm that blew it off course and broke radio contact; it splashed down 1930km (1200mi) short of its target in France. The crew owned the balloon, gondola, and tracking equipment, the latter purchased from a commercial firm. GSFC had cooperated earlier in search-and-rescue tests of the tracking unit. (NASA Release 77-190)

• The Natl. Aeronautic Assn. announced it would present the Wright Brothers Memorial Trophy for 1977 to Lt. Gen. (ret.) Ira C. Eaker, former chief of air staff and a central figure in U.S. military and civilian aviation. He would receive the trophy—a miniature silver replica of the original plane flown Dec. 7, 1903—at the annual dinner Dec. 9 sponsored by the Aero Club of Washington.

Gen. Eaker had been commissioned lieutenant in the U.S. Army in 1917 and received pilot rating at the age of 21. After World War I, he and other Army flyers set world flying records including the first transcontinental flight using refueling, and a transcontinental blind flight on instruments in 1936. In 1942 he had led the first attack by Army bombers in Europe. He had retired from the USAF in 1947; since then, he had been an executive for Hughes Tool Co., Douglas Aircraft, and the U.S. Strategic Institute, and had set up a program to hire physically handicapped persons in the aircraft industry. (NAA newsletter Sept 14/77)

September 15: LaRC announced it would open a major educational exhibit on energy Sept. 22 through 29. Assembled by the Energy Research and Development Administration, the exhibit in two 50-ft trailers would include films, animated sections, and visitor-operated consoles describing all types of energy now in use and possible future energy sources. Operating the displays for ERDA on the U.S. energy situation and what could be done about it would be Oak Ridge Associated Universities, a nonprofit educational and research organization of 45 colleges and universities in the southeast. (LaRC Release 77-41)

September 16: NASA announced that a fourth group of 20 Shuttle astronaut applicants would report to JSC Sept. 19 for a week of physical examinations and individual interviews. All of this group were pilot applicants, one a civilian. Of the others, 12 were USAF; 4, U.S Navy; 3, U.S Marine Corps. A final group would report to JSC in mid-Nov. (NASA Release 77-192)

• NASA announced that ARC had awarded two San Francisco Bay area firms a $2.5 million contract for managing construction of a major modification to the largest wind tunnel facility in the western world. Turner Construction Co. and Lord Electric Co. would modify the ARC 40-by 80-ft wind tunnel built in 1944 by adding a new 80- by 120-ft (24 by 36m) test section and increasing drive power from 36 000hp to 135 000hp. The modifications should enable ARC to handle new types of aircraft, especially large helicopters and vertical- and short-takeoff and landing aircraft. (NASA Release 77-191)

• Two balloonists attempting a transatlantic crossing from Bar Harbor, Maine, on or about Sept. 19 would carry a lightweight satellite-tracking beacon like the one that helped save Maxie Anderson and Ben Abruzzo when their balloon ditched near Iceland on Sept. 12. The FAA had required Charles E. Reinhard and Charles A. Stephenson to carry the device as an air traffic control aid and in case of a need for search and rescue. This type of beacon had previously gone around the world on a polar flight, and another rode a 33ft sailboat on a 600mi trip through the Bermuda Triangle east of Fla., both successfully tracked by a Nimbus satellite. The balloonists would cover the cost of the beacon; NASA would take no part except for use of the satellite link and ground systems, which would operate 24hr anyway. (NASA Release 77-194)

• NASA announced it had awarded to McDonnell Douglas Aeronautics Co. a contract to build spinning solid upper stages for the Space Transportation System. The contract, amounting to about $9 million, had resulted from previous NASA agreements with Boeing Co. and McDonnell Douglas that the firms could design, test, and market the stages independently. Bids had been limited to those two firms because they had agreed to develop the stages with private funds. The contract called for delivery by Dec. 1979 of the first SSUS with all hardware, logistics, and services necessary to demonstrate the craft and inject it into proper orbit. Three other stages would launch Intelsat Vs, advanced commercial comsats. Later deliveries would be assigned to other missions. (MSFC Release 77-170; NASA Release 77-196)

• FBIS reported that the USSR had signed an agreement with Intersputnik, an international organization formed by 9 socialist countries in 1971, that the group would work out of Moscow and create favorable conditions for its activity in the USSR. Organizers of Intersputnik to use satellites and ground stations for international communications were Bulgaria, Hungary, the German Democratic Republic, Cuba, Mongolia, Poland, Romania, the USSR, and Czechoslovakia. (FBIS, Tass in English, Sept 16/77)

• FBIS reported a Tass interview with Boris Petrov, chairman of the Intercosmos council, on the high points of the program, including cooperation with capitalist countries in exploration of space. France 11yr ago was the first capitalist country to cooperate; India's satellite *Aryabhata* had orbited on a Soviet booster 2yr ago; Soviet-Swedish cooperation was developing; and the U.S. and USSR would work together on space biology, medicine, meteorology, environment, and the moon and planets. (FBIS, Tass in English, Sept 16/77)

September 17: X-ray sensors on the U.K.'s *Ariel* satellite had detected an explosion of "catastrophic dimensions" in the constellation Ophiuchus [the Serpent-Bearer], the *NY Times* reported. Similar sensors on *Heao 1*, the U.S. high-energy astronomy observatory launched Aug. 12, had made the same observation. Dr. Daniel A. Schwartz of the Harvard-Smithsonian Center for Astrophysics said intensity of x-rays from that area increased a thousandfold in the past 2wk, making it the second strongest source in the sky (that in the constellation Scorpius, a continuous source of x-rays, was the strongest). No star of comparable brilliance appeared in previous photographs of the region. *Ariel* had first detected the x-rays Aug. 31; *Heao 1* began recording the source Sept. 7. An observatory in Australia that had received data on the explosion had also recorded it in visible light; Dr. Herbert Friedman of the Naval Research Laboratory said that a nova was rarely observed at widely separated wavelengths such as those of visible light and x-rays. (*NYT*, Sept 17/77, B8)

September 19: MSFC announced it would sponsor a data-management symposium Oct. 18-19 jointly with the Univ. of Ala. in Huntsville (UAH) to discuss use of space-acquired or remotely sensed data in the 1980s. MSFC deputy director Richard G. Smith would chair the session; Daniel J. Fink of General Electric's Space Division would deliver the keynote address. Representatives of NASA and other government agencies, industry, and universities would deliver 31 papers during the meeting. Leonard B. Jaffe, NASA deputy associate administrator for applications, would speak at a banquet Oct. 18. (MSFC Release 77-172)

• ESA announced that the U.S. delegation to an Aerosat Council meeting in Washington, D.C., Sept. 15-16, had notified it that lack of funds would prevent the Federal Aviation Administration from proceeding with the program as contemplated. The Aerosat memorandum of understanding had provided that ESA, the government of Canada, and a U.S. company (ComSatCorp was selected in 1974) would proceed with development, production, launch, and operation of the satellite, with ESA and the FAA (leasing from the U.S. firm) using the comsat capability in equal portions. (ESA Release Sept 19/77)

September 20: NASA announced it would sign a $5.6 million contract with General Electric Co. to build an atmospheric cloud physics laboratory for launch on the Shuttle in the early 1980s [see July 11]. NASA had already selected 8 of 26 experiments for the ACPL on early Spacelab missions. (NASA Release 77-197; MSFC Release 77-173)

• NASA announced that Dr. Robert A. Frosch, the agency administrator, had approved awards totaling $39 800 to 7 NASA and 2 contractor employees for their contributions to the agency's program. The awards covered 5 major efforts: a central timing system for tracking stations using microwave signals bounced off the moon; a new silica insulation material to protect the Shuttle during reentry into the atmosphere; a temperature-resistant coating to cover the Shuttle insulation; an economical and durable black-chrome coating for commercial-type solar collectors; and a new optical glass surface to improve Landsat images. Largest single award, $10 000, went to Mahlon F. Easterling of JPL for development of the timing system. (NASA Release 77-195)

• NASA announced it had appointed Dr. Anthony Calio as Hq associate administrator for applications as of Oct. 1. Dr. Calio had served as deputy associate administrator for space science since Nov. 1975. He had begun working for NASA in 1963 at the electronics research center in Cambridge, Mass., then at NASA Hq. In 1968 he transferred to JSC and in 1969 became its director for science and applications. He had previously been a nuclear physicist with Westinghouse and the American Machine and Foundry Co., and was in the U.S. Army Chemical Corps from 1954 to 1956. Among his honors were NASA medals for exceptional service, 1969; exceptional scientific achievement, 1971; distinguished service, 1973. (NASA anno Sept 20/77; NASA Release 77-200)

September 21: DFRC announced that employees John G. McTigue and Stanley Markey would share a $5000 award for working out a way to truck the Shuttle orbiter overland from the Rockwell plant in Palmdale, Calif., to DFRC rather than use the Boeing 747 carrier, eliminating the need for a mate/demate device costing $2.53 million and saving the Shuttle program nearly $800 000. The two men used their knowledge of local terrain, geography, and manpower to set up a ground route with relatively minor problems. (DFRC Release 30/77)

• NOAA reported that airline passengers between Miami and Latin America could enjoy smoother flights because of satellite weather images. Lineas Aereos Nicaragua (LANICA) and 8 other carriers were sharing *Goes 2* weather photos of the area from the southeastern U.S. into South America, including all of the Gulf of Mexico and the Caribbean Sea. A LANICA official said that, until now, the airlines had little idea of the location of storms; now they could avoid them, or use the windflow to save time and fuel. (NOAA Release 77-247)

• ERDA announced it had selected Windworks, Inc., of Mukwonago, Wisc., to negotiate a contract for design and production of advanced small windmills for farms, rural homes, and other uses. The $388 000 contract covered designing a machine to produce 8kw of electric power from 29ft-diameter blades in a 20mph wind, and delivering it for testing by prime contractor Rockwell Intl. at ERDA's test facility in Colo. (ERDA Release 77-168)

September 22: NASA announced that Thomas C. Duxbury, a guest investigator at JPL, would use the shadow of Martian moon Phobos on Sept. 20, 24, and 28 to pinpoint the exact location of the *Viking 1* lander on the planet's surface.

JPL explained that the spacecraft had landed within a known area but was too small for detection by orbiter cameras, which had been unable to tie in photographs of the site to the lander's position. Duxbury had programmed both the lander and the orbiter to take a picture as the shadow of Phobos passed over the landing site; the time of the photograph and the exact position of the shadow on 3 occasions would permit calculation of the lander's position.

Duxbury was one of 20 chosen to participate in the Viking guest investigator program set up by NASA to offer experimenter opportunities beyond those originally planned for the mission. (NASA Release 77-199)

• The USSR announced it had launched a 910kg *Prognoz 6* to continue study of galactic energetic particles and of solar activity impact on the magnetosphere and the interplanetary medium. (FBIS, Tass Intl Svc in Russian, Sept 22/77)

• The USSR announced that Aeroflot's IL-62 jet passenger flagship had set 2 world's records. Piloted by a crew of 6 women, the IL-62 set a world distance record along a closed route of 10 388km, and a world record for average speed along a closed 10 000km route of 804kmph. (FBIS, Tass in English, Sept 22/77)

September 23: The *W Post* said that the U.S. had "quietly decided" to build a device for destroying enemy satellites in space: a small (ft-long, 8in-dia) heat-sensing vehicle "resembling a tomato can" without explosives but relying on collision at thousands of mph to accomplish its purpose. The *Post* said the USAF had "carefully camouflaged" this step in a Sept. 8 release announcing award to Vought Corp. of a $58.7 million contract "for development and test of hardware in support of the Space Defense Technology Program."

A 1967 treaty between the U.S and USSR had prohibited both nations from orbiting any object "carrying nuclear weapons or any other kinds of weapons of mass destruction. . ."; both sides had apparently decided this did not outlaw shooting at their own target satellites. U.S. military officials deduced that certain USSR launches were tests of hunter-killer satellites with non-nuclear charges to blow up on approach to a target. William J. Perry, DOD research chief, told the *Post* that the U.S. could either build antisatellite weapons also, or provide more protection for U.S. satellites. The DOD had begun on defensive measures such as making satellite wiring less vulnerable, but had also decided on offensive measures as shown in the Vought contract.

The small highly maneuverable U.S. device would have sensors allowing it to home in on slight temperature variations between a target spacecraft and its surroundings; it would be so light that a single launch vehicle could carry several into effective orbit. (*W Post*, Sept 23/77, A1)

• NASA announced it had joined with Rockwell Intl. Corp. in granting funds for a new solid-state electronics facility at Howard Univ., Washington, D.C., dedicated Sept. 20 at ceremonies with NASA Administrator Dr. Robert A. Frosch as speaker, attended by more than 200 persons from NASA, Rockwell, Cornell Univ., ERDA, Bell Laboratories, and other public and private groups. Cornell Univ., a leading school in electronics research, had helped set up the Howard project at the request of Rockwell. (NASA Release 77-201)

• LaRC announced NASA had awarded Lockheed Aircraft Corp. a contract to produce advanced-composite ailerons to decrease weight of the L-1011 transport aircraft. Lockheed, working with Avco Corp. in Nashville, Tenn., would demonstrate techniques of designing and fabricating 10 shipsets of composite ailerons for FAA approval. Costs of the 4yr contract, like those with other firms in NASA's aircraft energy-efficiency program (ACEE) to reduce fuel needs of civil-transport aircraft by 50% [see May 5, July 7], would be shared, $5.85 million to be contributed by NASA and $650 000 by Lockheed. LaRC would manage the contract. (LaRC Release 77-43; NASA Release 77-204)

• WFC announced it would officially open a new management education center Sept. 30 with Dr. Robert A. Frosch, NASA administrator, presiding. A unit of the Hq office of professional development, the center would offer short-term residential programs (3-day seminars to 2-wk courses) designed to train future agency leaders. (WFC Release 77-12; NASA Release 77-202)

• MSFC announced it was developing a coal mining machine using technology borrowed from the lunar rover to make mining safer, more productive, and more economical. The Interior Dept.'s Bureau of Mines had asked NASA to automate the mechanics of mining and eliminate the need for human operators in the unsafe and unhealthy environment of the cutting operation.

The longwall-shearing machine in current use would grind coal from the face of a seam in a linear direction, sending the coal out on a conveyor and advancing with its roof supports as it went deeper in the seam. Operators could not mine coal to the edge of a seam at roof or floor without breaking into surrounding shale, so that much coal was left unmined. An operator would find guidance difficult and progress slow because of poor visibility from dust.

An automated guidance and control system would allow removal of nearly all the coal at a much faster rate, recovering more of the coal left by other methods, in a purer state with less rock and other matter. The system would also reduce wear on the longwall machine's cutters and minimize downtime. The contractor to be selected would work with results of a 2yr MSFC study of a system using sensors such as gamma rays, radar, impact devices, and reflected light to guide a shear's cutting drums. (MSFC Release 77-175)

• FBIS reported that the USSR had opened a television station in Duki, town of Khabarov Krai on the Pacific coast of Siberia near Bering Strait and the Sea of Okhotsk, permitting reception of both central and local programs in all villages of the area. (FBIS, Moscow Dom Svc in Russian, Sept 23/77)

September 24: The press reported that the third free flight of the Shuttle orbiter Enterprise over Edwards AFB had successfully tested the microwave scanning-beam landing system originally developed for the Navy to land planes on ships at sea. Pilots Fred W. Haise, Jr., and USAF Lt.Col. Charles Gordon Fullerton turned over orbiter control to the system for part of its steep glide down, and ground transmitters kept the craft in line with the runway during the 5min 34sec free flight. Installed at Edwards earlier in 1977, the system would also be installed at KSC to handle Shuttle landings there.

Despite radio-transmission problems between the orbiter and Houston, the launch went off on schedule. At 24 000ft in a shallow dive at 287mph the orbiter separated from its Boeing 747 carrier and reached a maximum speed of more than 333mph during its landing approach. (*NYT,* Sept 24/77, 24; *W Star,* Sept 24/77, A-5; *W Post,* Sept 26, A-4)

September 26: JSC reported that another 20 Shuttle astronaut applicants would report for individual interviews and physical examinations; the group included 17 pilot and 3 mission specialist applicants. Eight of the applicants were from the USAF; 6, Navy; 2, Marines; 4, civilians. Of the more than 8000 applicants, about 200 would go to JSC for screening, the final group due in mid-Nov. and selections in Dec. (JSC Release 77-54; NASA Release 77-205)

• MSFC announced that solar energy would provide heat for some of the center this winter. A system already completed would supply 50% of the heat for one office building; solar collectors being installed on another would supply hot water. A flat-plate solar collector on the 10-story center headquarters building would be supplying hot water by July 1978. (MSFC Release 77-178)

September 27: MSFC reported on the dedication of the world's largest solar-energy cooling system at a hotel on St. Thomas, U.S. Virgin Islands. The hotel, surrounded by the ocean on three sides, had used diesel fuel for energy. The $554 000 system, part of ERDA's solar-energy demonstration program, used 13 000ft^2 of solar collectors to air condition the 300-rm hotel, saving an estimated $100 000 per yr over the cost of diesel fuel.

MSFC engineer Bob Middleton, design and installation manager, explained that the system was unique in not storing energy: it provided air conditioning during the day only, and the oil-fired system took over at night, as the most economical arrangement available. MSFC had worked on installations at a total of 59 sites under ERDA's demonstration program; ERDA had shared costs of this system with the hotel owners. (MSFC Release 77-179; NASA Release 77-207)

• INTELSAT announced that the People's Republic of Angola had become the 99th member of the organization on Sept. 23, and the 25th African nation to join. Angola's Empresa Publica de Telecomunicacoes

(EPTEL) had signed the operating agreement. (INTELSAT Release 77-25-I)

September 28: NASA reported that LeRC research under the Aircraft Energy Efficiency Program (ACEE) might bring back the use of propellers, once considered obsolete, in fast commercial aircraft service. Advances in composite-material technology had led to design of small-diameter 8-blade propellers that might save 20 to 40% in fuel over current turbofan engines. (MSFC Release 77-206)

September 29: NASA launched Intelsat IVA F-5 from Cape Canaveral on an Atlas Centaur that exploded less than a minute after launch, dropping the 3332lb spacecraft into the Atlantic Ocean. This Intelsat and an identical craft scheduled for launch in Nov. were to replace two *Intelsat IV*s now in orbit over the Indian Ocean and provide 6000 telephone channels and 2 television channels for use by 40 countries, compared to 4000 telephone and 2 TV channels offered by the *Intelsat IV* spacecraft.

At 60sec into the flight a range safety officer destroyed the secondary Centaur stage which had separated prematurely from the Atlas. Failure of the $49.4 mission was similar to the Sept. 13 explosion of a Delta rocket carrying a $42 million orbital test satellite (OTS) NASA was launching for ESA. (NASA Release 77-214; *W Post,* Sept 30/77, A-3)

An investigation concluded that a leak in the high-pressure gas generator system had ignited combustibles in the Atlas 35sec after launch, damaging engine components and terminating the flight. (MOR E-491-633-77-04 [postlaunch] Jan 5/78)

• FBIS reported that the USSR had launched space station *Salyut 6* into an orbit with 275km apogee, 219km perigee, 89.1min period, 51.6° inclination. A control center near Moscow would process information from the station aided by tracking stations on Soviet territory and on Soviet Academy of Sciences ships in the Atlantic. Onboard systems were functioning normally. (FBIS, Tass in English, Sept 29/77)

• Rockwell Intl. reported that George W. Jeffs, its president for space operations, had told the House subcommittee on space science and applications that the U.S. needed "bolder and more imaginative steps" toward space industrialization, and called on the federal government to demonstrate an aggressive "make it happen" program with a theme of Power from Space.

Space industrialization funding in the 1980 budget, Jeffs said, should approach $500 million for building-block phases leading to manufacturing, communications, or space power centers operational by the yr 2000.

Such funding would speed up the goal by at least 5yr and would encourage increased private investment. "As dramatic and technically rewarding as our space missions have been," he concluded, "they were more significant for creating enormously promising opportunity." (Rockwell Release SP-5)

September 30: MSFC announced that NASA had selected Martin Marietta Corp. and Ball Brothers Research Corp. for negotiations of fixed-price contracts for parallel $290 000 definition studies of a Shuttle tethered satellite system [see May 27]. MSFC would manage the studies for the NASA Hq Office of Space Flight. (MSFC Release 77-181)

• The Natl. Aeronautic Assn. announced that the Federation Aeronautique Internationale (FAI) would present the gold space medal, given yearly as the world's highest award for spaceflight, to former astronaut Michael Collins, director of the Natl. Air and Space Museum, at ceremonies in Rome Oct. 3. The award would recognize Collins's total contribution to manned spaceflight; he had been command module pilot during the *Apollo 11* mission that first landed humans on the moon. The U.S. would also receive 11 other international aviation awards at the world aviation body's 70th annual conference: 3 USAF pilots had set records last yr, giving the U.S. 5 of 6 possible absolute world records. Capt. Eldon W. Joersz achieved top speed over a straight course of 2193mph; Maj. Adolphus H. Bledsoe achieved speed over a closed circuit of 2092mph; and Capt. Robert C. Helt set an altitude record in horizontal flight of 85 068ft, all using a Lockheed SR-71 twin-engine aircraft. (The USSR held the altitude record of 118 897ft.)

Other U.S. award recipients would include Paul E. Yost of S.D. for a world-record balloon flight of 2475mi from Maine to the Azores, lasting 107hr, and Bruce Comstock of Mich. for being the only person to win the U.S. national hot-air balloon championship twice. Associate NASA Administrator Dr. Noel W. Hinners would receive a group diploma of honor awarded to the Viking project team, LaRC, JPL, and Martin Marietta for the first significant scientific information on Mars and its environment.

The FAI, formed in 1905, was the world governing body for aviation records; NAA was its U.S. representative. (NAA release Sept 30/77)

During September: INTELSAT announced it would award Hughes Aircraft Co.'s electron dynamics division a $126 099 contract to identify traveling-wave tube structures in the 4GHz frequency range with high reliability and uniform performance for use in future INTELSAT comsats. (Intelsat Release 77-24-M)

• ESA announced it had received a delegation of 9 telecommunications experts from the PRC's Electronics Society visiting Europe Sept. 12–Oct. 20 to see the principal national space facilities and industrial firms of ESA member nations. The visitors, headed by Lei Hung of the Electronics Society council, would discuss all aspects of communications-satellite programs and would give ESA and European industrialists their first opportunity to meet with people from PRC in the space business. (ESA Release Sept 13/77)

• The Natl. Aeronautic Assn. reported Dr. Paul MacCready, U.S. aeronautical engineer famous for his manpowered Gossamer flying machines, had claimed the Kremer Competition prize offered through the Royal Aeronautical Society of 50 000 British pounds (about $86 000 U.S.) for having completed on Aug. 23 the required course of clearing a start/finish line 10ft high, flying a figure 8 around 2 pylons a half-mile apart, and completing the flight over the same 10ft-high finish line.

The Gossamer Condor, a craft made of corrugated cardboard, balsa wood, aluminum, piano wire, styrofoam, transparent tape, and cellophane, with a 96ft wingspan but weighing only 77lb, was powered by Bryan Allen, an experienced hang glider and bicycle racer, who transferred energy through bicycle pedals to a propeller. The Kremer prize was the largest ever offered for a single aeronautical achievement. Dr. McCready's claim was subject to RAS approval upon submission of flight documentation. Although the NAA, as RAS contact in the U.S., had distributed numerous copies of the Kremer rules, only 2 other vehicles had been constructed and actually flown in the U.S., one a solo manpowered craft like the Gossamer Condor, the other a multi-manpowered triplane built by engineering students at MIT. (NAA newsletter Sept 77)

• The Natl. Aeronautic Assn. reported that Pan American Airways would mark the 50th anniversary of its first flight Oct. 28 by attempting a world-record round-the-world, over-the-poles flight in less than 48hr with a Boeing 747 jumbo jet. Carrying an official observer to certify record performance, the 747 would leave San Francisco, fly over the North Pole to London, then to Cape Town, South Africa, over the South Pole to Auckland, New Zealand, and back to San Francisco, a distance of 26 642 statute mi. A Boeing 707 had flown the polar route in 1965 in a record 62hr 27min 35sec. In May 1976 Pan Am's 747 had set an east-west record, a bicentennial flight covering 23,137mi over a New York-Delhi-Tokyo-New York route in 46hr 50sec. That flight had carried 98 passengers, many of whom would be aboard the polar flight in Oct. Pan Am's first flight had taken place Oct. 28, 1927, when a Fokker F-7 trimotor lifted off a dirt runway at Key West, Fla., for an hr-long flight over 90 miles of ocean to Havana, Cuba. (NAA newsletter Sept 77)

• NASA announced its yrs-old promise that scientists would operate their own equipment in earth orbit would become a reality, with selection of payload specialists for the first Spacelab mission in 1980 not by NASA but by scientists participating in the mission. One of the two payload specialists would be a representative of the European Space Agency (ESA), developer of Spacelab. A special panel from an international Investigators Working Group had interviewed 19 U.S. candidates at MSFC and would soon interview candidates in Europe. The entire working group would select four trainees for a two-yr program at MSFC; one pair would fly on Spacelab 1. (NASA Release 77-189)

• *NASA Activities* reported on international cooperation in space. Dr. Alan M. Lovelace last year had signed an agreement with the USSR Academy of Sciences, effective May 11 when signed by academy president A.P. Aleksandrov, to study joint operation of Shuttle and Soyuz-Salyut spacecraft in scientific and applications "objectives of merit," and the potential uses of a space platform and developing such a platform bilaterally or multilaterally. By the end of 1978, within a yr of beginning work on space-platform proposals, two study groups would bring in recommendations. The publication cited U.S.-USSR cooperation in rocket instrument tests during Aug., the U.S. biological experiments carried on a Soviet launch, and the arrangement for delivery to NASA of 7 lunar soil samples from the Soviet *Luna 24* mission.

NASA's mandate called for "cooperation by the United States with other nations and groups of nations . . . and in the peaceful application of the results of that cooperation." Best known current example would be Spacelab, funded by 10 member nations of ESA, and the concurrent development by Canada of the Shuttle's remote manipulator system. NASA's Landsat program had made data available to 130 countries, 7 of which had agreed to set up ground stations for direct reception of Landsat information. A joint NASA-AID project based on *Ats 6* broadcasts of instructional TV to some 5000 remote villages in India had sent special films and live discussions to 27 countries in Asia, Africa, the Middle East, and Latin America. Other joint efforts included solar-terrestrial studies, solar probes, infrared astronomy, the Space Telescope, and a satellite-assisted global search-and-rescue project. A major benefit to the U.S from such cooperation would be a favorable balance of payments resulting from foreign purchases of data from reimbursable launches. (*NASA Activities,* Sept 77, 14)

- The *Naval Research Reviews* reported on satellite use in the oceanic biology program of the Office of Naval Research, covering biodeterioration (growth of fouling and boring organisms), sound in the ocean (marine mammals), dangerous marine organisms, and special projects. Sound in the sea, second major research area in the program, had centered on whales and seals, especially sounds produced by whales. Navy scientists had implanted radio transmitters in large whales such as blues, fins, and grays, to track them by boat, plane, or satellite. Knowledge of whale behavior and movements would be valuable to designers and operators of naval equipment. (*NR Rev,* Sept 77, 6)

- FBIS reported on the 28th congress of the Intl. Astronautical Federation in Prague, celebrating the 20th anniversary of the launch of *Sputnik 1*. Chairman Marcel Barrere of France reviewed at a press conference the "tempestuous impact of space exploration on the planet's industry," mentioning the joint U.S.-USSR approach to a manned flight project using the Shuttle and the Salyut orbital station. Presentations at the congress had included papers on the search for extraterrestrial civilizations, the U.S. scientists reporting on the Voyager mission, and the USSR on remote probes of planetary atmospheres for signs of gas exchange between soil and atmosphere. (FBIS, Tass in English, Sept 24/77, Sept 28/77)

October

October 1: The new federal Dept. of Energy opened its doors with Secretary James R. Schlesinger promising "to resolve our energy problems . . . through the effective work of all the people of this department." Proposed by President Carter March 1, the department had become a reality Aug. 4 upon signing of its organization act and Schlesinger's confirmation as secretary the following day. The department inherited nearly 20 000 employees and had a first-yr budget of about $10.4 billion. Programs it would administer included conservation, resource development and production, research and development, data management, environment, and regulation. (DOE Release R-77-001)

• Reporting on the end of the Intl. Aeronautical Federation's 28th conference in Prague, FBIS quoted "American scientist O'Neal" [Gerard K. O'Neill] on whether "his project of big settlements on near-terrestrial orbits was not too remote from reality": "Mankind will never give up the idea of settlements in space . . . The question is only how fast will space industry and rocket technology reach the necessary level to enable people of the earth to start building the first cities in orbit."

The congress discussed for the first time the efficiency of space research, concluding that spacecraft created in different countries would inevitably have a uniform appearance, depending on their use. The idea of merging engineering ideas was "a dominant one." Marcel Barrere of France was reelected president of the organization. (FBIS, Tass in English, Oct 1/77)

October 3: Johnson Space Center announced that a sixth group of 20 Shuttle astronaut applicants would report this week for individual interviews and physical examinations. This group, all mission specialist applicants, included one woman. (JSC Release 77-56; NASA Release 77-215)

• Kennedy Space Center announced award to Intl. Business Machines Corp. of a $3 203 666 contract for a system to test Shuttle cargoes for compatibility with the orbiter before loading into the payload bay. The system, called CITE (cargo integration test equipment), would simulate the orbiter's electrical and electronic environment, stimulate the payloads, monitor their response, and record the results. Keeping up with the fast turnaround time planned for Shuttle missions required payloads ready to plug into the orbiters without extensive onboard testing. The CITE would be mobile for use either in the assembly building to check out cargoes in vertical positions or in the operations building to check out horizontal payloads. (KSC Release 156-77)

October 4: Postlaunch reports on the Voyagers said both were functioning as expected. The one launched first, called *Voyager 2* because it would arrive second at Jupiter and Saturn, had had difficulty with attitude control as well as with its science-platform boom; its data showed that the craft was merely obeying faulty software, which ground control had corrected. The boom was stable enough to capture satisfactory starfield images. *Voyager 1,* launched later, had not achieved the proper velocity during two trajectory corrections; Jet Propulsion Laboratory engineers said rocket exhaust was apparently pushing against the craft, and had changed the procedure accordingly. (NASA Release 77-209; MOR S-802-77-01/02 [postlaunch] Oct 6/77)

• NASA announced that its administrator, Dr. Robert A. Frosch, and Chairman A.B. Wolff of the Netherlands Agency for Aerospace Programs had signed a memorandum of understanding on the Infrared Astronomical Satellite (IRAS) project, a cooperative undertaking in which the United Kingdom would also share under a separate agreement. An earth-orbiting observatory built by the Netherlands and scheduled for launch in 1981 would carry a cryogenically cooled IR telescope furnished by the U.S. and a package furnished by the U.K. for command, control, and data acquisition. Nearly 500 scientists, engineers, and technicians from the participating nations would join in the first survey of the entire sky at infrared wavelengths undetectable by earth-based telescopes obscured by the atmosphere. The mission would have a 1yr lifetime. Ames Research Center would manage the telescope, built by Ball Bros. Research Corp.; JPL would have U.S. project management responsibility. (NASA Release 77-210)

• Marshall Space Flight Center announced award to Intl. Business Machines of a $1 344 950 contract for 134 site data-acquisition subsystems to monitor performance of ERDA's solar heating and cooling demonstration program at sites throughout the U.S. The IBM machines would provide data on solar energy at each site, amounts used for heating, cooling, and hot water, and other areas of performance, efficiency, and fuel saving. Each site would feed data to a central computer in Huntsville, Ala., for compilation of reports to business, industry, and the general public on the most economical and practical systems tested. (MSFC Release 77-184)

• The *Washington Post* reported on *Sputnik 1*'s 20th anniversary as observed in the Soviet Union. Sputnik had set off a "search for nationhood in the United States even as it seemed to confirm it in the Soviet Union . . . Ask any Soviet citizen above the age of 24 and each can tell you exactly where they were and what they were doing when they heard . . . a Soviet satellite was in orbit."

The U.S. search ended "in a symbolic and real sense" with the moon landing 8.5yr ago. However, the certainty of power that Sputnik gave the USSR "has never quite materialized out of the void of space." The A-2, early main launch vehicle for the USSR "civilian" program, was still that today. U.S. experts in the joint Apollo-Soyuz project had been "dismayed by the crudeness of Soviet space hardware": one compared the Soyuz craft to a "hot water radiator . . . it works, but that's about all you can say for it." An aviation weekly, interviewing the experts, said they rated the Soyuz with the Mercury capsule, first U.S. manned spacecraft. "Their scientists are every bit as good as ours," said one U.S. physicist, "but the technical gap is still huge."

The article noted that the U.S. had dropped its space medicine program in the early 1970s but the USSR had kept on with its, and had maintained large space stations in orbit since 1971 manned with crews up to 63 days at a time. The Soviet space program had had less trouble than the U.S. program "because of the closed nature of the system . . . no citizens groups or congressional critics to question the need for a manned space program." Competition for limited resources was the same in both countries, but "lack of public debate allowed managers luxuries of continuity not available to their counterparts in a democracy." (*W Post*, Oct 4/77, A-1)

October 6: NASA announced that its administrator, Dr. Robert A. Frosch, and Hans Matthoefer, minister for research and technology for the Federal Republic of Germany, had signed a memorandum of understanding for participation in a Jupiter orbiter/probe mission to be launched by the Space Shuttle in 1982.

The mission would send an orbiter and a probe on the most detailed investigation to date of the environment and moons of Jupiter, the first opportunity for 114 investigators to measure Jupiter directly as well as remotely from various orbital positions over a long period of time. The orbiter, carrying 10 instruments, would circle Jupiter for at least 20mo; the probe, carrying 6, would plunge into the atmosphere. West Germany would provide a retropropulsion module to inject the orbiter into its path around Jupiter, as well as the scientific instruments and the services of 14 German investigators in preparing the two spacecraft and analyzing their data. JPL would manage the U.S. project, ARC the probe, JSC the Space Shuttle. An office in West Germany would manage FRG's share. (NASA Release 77-211)

• MSFC reported that a highly sensitive gamma ray telescope carried to 40km (25mi) altitude by a giant helium-filled balloon on a 40-hr flight had studied an unusual binary-star system (AM Herculis in the constellation Hercules) in which material from the larger star was falling onto the dwarf companion.

Dr. Thomas A. Parnell of MSFC's space sciences laboratory said the flight was seeking discrete gamma-ray emissions from particular celestial bodies to identify the chemical elements and isotopes existing there; AM Herculis was one of 6 unusual objects sought by the 0.75m-diameter (2.5ft) telescope loaned by Rice Institute to MSFC in a joint astronomy project. The balloon instrumentation was similar to part of the payload on *Heao 1,* launched Aug. 12 to make high-energy astronomy surveys. Detection of gamma rays in objects by *Heao 1* would be confirmed by the balloon flights, the two sets of information being complementary. Dr. Parnell noted that balloons had carried models of HEAO instruments in preliminary tests; another MSFC balloon launched last week had carried a cosmic-ray detector like one to be flown on HEAO-C in 1979. (MSFC Release 77-186)

• ARC reported that the city of Genoa, Italy, had awarded it the Columbus gold medal for the multibillion-mile flights of two Pioneers to Jupiter. The medal, presented annually during Columbus Day celebrations in the explorer's birthplace, would be accepted Oct. 12 by acting ARC director Clarence A. Syvertson, who would be in Europe at the time. (ARC Release 77-42)

• Langley Research Center announced it would dedicate Oct. 11 the site of Project RECOUP (refuse-consuming utility plant), first jointly funded federal and municipal project of its kind, to generate steam from burning refuse [see Jan. 19]. Construction would begin in 1978 and be completed in 1980; NASA would share the $8.4 million cost of the plant with the US Air Force and the city of Hampton, Va. The plant would burn more than 80% of the refuse from Hampton, LaRC, the Langley Air Force Base, the Army's Fort Monroe, and the Veterans Administration hospital; in a 24-hr operation, it would burn about 200 tons a day, doubling the capacity of Hampton's landfill. LaRC would use the steam generated by the plant to augment its present heating system, saving about 2.4 million gallons of oil per yr. (LaRC Release 77-46)

October 7: ESA announced that NASA Administrator Dr. Robert A. Frosch and Roy Gibson, director general of ESA, had signed at a meeting in ESA headquarters a memorandum of understanding on cooperation in NASA's space telescope program.

The European contribution would include a faint-object camera for high-resolution imagery in the ultraviolet, visible, and near-infrared portions of the spectrum, with associated photon-counting detector, to be left in orbit as long as considered scientifically useful; the solar array to power the scope; and support of a scientific operations center to be established by NASA for managing the observatory. ESA would get 15% of the observing time for the duration of the program in return for its participation, and would share data with astronomers in its member states and the international community.

NASA would develop and integrate the telescope, launch it into orbit on the Shuttle, supply tracking and data-acquisition services, and maintain the telescope throughout the program. MSFC would manage the space telescope, Goddard Space Flight Center would manage instrument development and observatory operation, and the European Space Technology Center (ESTEC) would manage the ESA effort. (ESA Release Oct 7/77; NASA Release 77-212)

• LaRC announced that Sir Robert Mark, former director of New Scotland Yard (his official title was Commissioner of Police of the Metropolis), would be guest speaker at a center colloquium on terrorism and law enforcement as part of the public lecture series, "Our Future in the Cosmos." The lecture would contrast philosophy and methodology of British and U.S. criminal justice systems, including the areas of capital punishment, terrorism, and political influence on police matters, reflecting the speaker's feeling that a free society is governed by consent rather than by force. (LaRC Release 77-45)

October 9-12: Newspapers reported the launch Oct. 9 of *Soyuz 25* carrying two rookie cosmonauts—Vladimir Kovalenok and Valery Ryumin—toward a linkup with new space station *Salyut 6* launched Sept. 29. Western observers said *Soyuz 25*—first manned Soviet mission since Feb., when Col. Viktor Gorbatko and Lt.Col. Yuri Glazkov had spent 18 days aboard *Salyut 5*—was probably connected with the *Sputnik 1* anniversary and the upcoming 60th anniversary celebration Nov. 7 of the Russian revolution. The observance might include either an attempt at a new record of more than 63 days in orbit or the launch of another space station to link with *Salyut 6* and form a huge orbiting laboratory.

Officials had canceled *Soyuz 25* only 26hr into the mission because of unspecified trouble, although the craft had come within 130yd of the *Salyut 6*. The Moscow domestic service in Russian said only that the linkup was called off "due to deviations from the planned procedure for docking." The word that the flight was canceled came after 24hr of silence, which had caused fears for the safety of the crew. Tass said Oct. 12 that the capsule had returned safely, landing in Kazakhstan at 6:25am local time Oct. 11.

U.S. officials speculated that, like *Soyuz 15* in Aug. 1974 and *Soyuz 23* in Oct. 1976, this mission failed because the cosmonauts had overshot the docking point and could not make another attempt because their battery power allowed insufficient time to correct the navigational error. When a crew last overshot its Salyut, 4mo elapsed before another crew made a docking attempt. (*W Star,* Oct 10/77, A-1; Oct 11/77, A-6; *W Post,* Oct 11/77, A-1; Oct 12/77, A-21; FBIS, Tass in English, Oct 9/77, Oct 10/77, Oct 11/77)

October 11: ARC announced that a team of its scientists had made a "major breakthrough" in explaining the origin of life. ARC's Dr. James Lawless, team leader, presented a report on the work to the Pacific Conference on Chemistry meeting in Anaheim, Calif. Chemical evolution experiments in the past several yrs had applied electrical discharges to ammonia, methane, and water to produce basic life molecules, including amino acids and nucleotides, but with no indication how the molecules formed organic building blocks in ever more complex groups until a group appeared that could replicate itself.

The new experiments had used metal clays normally present on primordial earth and ocean shores. Mixing amino acid solutions with everyday metal clays showed that the clays attracted the thousand varying aminos out of solution. One nickel-containing clay preferentially attracted the 20 amino acids making up protein, main ingredient of living cells; of 8 clays tried, only the nickel clay did this. The other metal clays destroyed nonprotein-forming amino acids faster than protein acids, indicating how the life-forming acids were selected and concentrated. Experiments simulating tidal action on the clays produced chains of amino acids, as many as 8 so far; time would eventually produce the far longer chains found in life. A zinc clay, only one of 9 metal clays tried, had a similar effect on DNA building blocks; the presence of metals in life systems today had resulted from prebiological chemistry, Dr. Lawless said. (ARC Release 77-43; NASA Release 77-220)

• JPL announced appointment of Dr. Rochus E. Vogt, professor of physics at CalTech and principal investigator for the cosmic ray experiments on *Voyagers 1* and *2,* as chief scientist. (JPL anno Oct 11/77)

October 12: NASA's Space Shuttle orbiter Enterprise successfully went through its fourth free flight with its streamlined 5700lb tailcone off, increasing drag and reducing the altitude achievable by its 747 carrier (to about 18 000ft) as well as the distance it could glide after separation. Three previous free flights, tailcone on, had lasted about 5.5min each; this time the Enterprise, fitted with 16 000lb of dummy engines with nozzles 8ft in diameter, wide enough for a man to stand upright in, would have a glide time of only about 3 min.

Pilots Joe Engle and Richard Truly, alternating at the controls of the orbiter, brought it to a stop on a runway at Edwards AFB after recording a touchdown speed of 212mph and 5000ft of landing roll. NASA's goal had been no more than 9000ft of landing roll, showing that the orbiter could land at any major airport in the world. Fitzhugh L. Fulton, pilot of the 747 carrier plane, said the aerodynamic resistance created by the dummy engines had caused "heavy shaking" on both the orbiter and the 747 during the hr-long ascent to launch altitude. Tom McMurtry, copilot of the 747, said the turbulence resembled "what you might get in a commercial airliner . . . only we couldn't change altitudes to get away from it as an airline pilot does."

NASA officials said the last free flight, scheduled for Oct. 26, would end with the microwave landing system controlling the touchdown. (DFRC Release 31-77; postflt rept, free flt 4, SpSh orbiter ALT; *NYT,* Oct 13/77, A18; *W Post,* Oct 13/77, C-11; *LA Times,* Oct 13/77, 1; JSC *Roundup,* Oct 14/77, 1; *W Star,* Oct 13/77, A-10; *C Trib,* Oct 13/77, 1-13)

• JSC announced that another set of 20 Shuttle astronaut candidates would report to the center Oct. 17 for physical examinations and individual interviews. All of this group would be mission specialist applicants, 8 of them women. Applicants screened at JSC so far had totaled 140, 77 of them pilots and 63 mission specialists; the 17 women were all mission specialist applicants. (JSC Release 77-59; NASA Release 77-218)

• NASA announced that Dr. David R. Scott, director of Dryden Flight Research Center, would leave the agency Oct. 30 to form with others a firm in Los Angeles specializing in technology transfer. Deputy Director Isaac T. Gillam would be acting director until a replacement was selected. Before becoming director of DFRC in 1975, Dr. Scott had been deputy director since 1973. He had left the astronaut corps in 1972, having flown on *Gemini 8, Apollo 9,* and as spacecraft commander of *Apollo 15* (fourth manned lunar landing, during which he was first to visit the moon's Hadley rille and Apennine mountains). He had also been technical assistant to the Apollo program manager at JSC and special assistant in the Apollo spacecraft program office. (NASA anno Oct 12/77; NASA Release 77-217; DFRC Release 34-77)

October 13: LaRC announced that NASA had awarded Boeing Commercial Aircraft Co. a $1.5 million contract to develop advanced aerodynamic and active-control concepts for civil transport aircraft. The contract included 6 tasks that would apply studies of aircraft systems to initial design processers. LaRC would manage the contract. (LaRC Release 77-48; NASA Release 77-219)

October 14: NASA announced plans for Noise-Con 77, third in a series of national conferences on noise control, to be held at LaRC Oct. 17-19. Sponsored jointly by the Institute of Noise Control Engineering and LaRC, the conference would center on transportation noise. Some 40 technical papers would discuss control of noise from railways, aircraft, motor vehicles, and highways, as well as community impact of transportation noise, standards and measurements, and federal programs for noise control. (NASA Release 77-221)

October 17: NASA announced it had selected 5 materials-processing experiments to be flown on an early orbital-test flight of the Space Shuttle beginning in 1979. The Materials Experiment Assembly (MEA) would offer scientists an interim opportunity to conduct studies before Spacelab missions began; it would be automated and unattended, containing its own power source and a minicomputer to collect data. Principal investigators and their experiments were Dr. J. Bruce Wagner, Jr., Ariz. State Univ.—solid electrolytes containing dispersed particles; Ralph A. Happe, Rockwell Intl.—containerless preparation of advanced optical glass; Dr. Herbert Wiedemeier, Rensselaer Polytechnic Inst.—vapor growth of alloy-type semiconductor crystals; Dr. John W. Vanderhoff, Sinclair Laboratories, Lehigh Univ.—large-particle-size monodisperse latexes; Dr. S.H. Gelles, Gelles Associates, Columbus, O.—liquid miscibility gap materials. (NASA Release 77-222)

• Wallops Flight Center announced its annual Woman's Day observance Nov. 2 would stress the need for equal opportunity for women to contribute to all phases of national life. Speakers would be Lillian Levy of NASA Hq and Jill Barbon of the Salisbury, Md., chamber of commerce. (WFC Release 77-13)

• The Dept. of Transportation announced it had awarded contracts to Perkin Elmer Corp. and United Technologies Inc. for devices to measure the effect of jet-engine emissions on the stratosphere. The first contract, valued at $385 000, would be for a system to measure nitrogen oxides in the stratosphere between 50 000 and 115 000ft altitude and predict jet-emission effects. The second contract, funded jointly at $458 000 by FAA, NASA, EPA, USAF, and the U.S. Navy, would document and explain the differences in reported measurements of nitrogen oxides in jet-engine exhausts, comparing optical measurements with the sampling technique. (DOT Release 98-77)

October 18: NASA announced it had reached an agreement with the Navajo Tribal Council in Ariz. for a 2-yr project to use Landsat data for an automated resource inventory of the 16-million-acre reservation in the southwest U.S.

The tribe had asked NASA whether satellite and aerial remote sensing could help solve inventory and multiple-use management problems including range rehabilitation, timber, agriculture, harvest prediction, and wildlife counts. Phase one of the project would demonstrate how Landsat data could apply to specific tribal needs and changes in resources; phase two would include training of tribal personnel and setting up an operational system at Navajo headquarters. The Navajo would assist other tribes to realize the benefits of satellite information systems. (NASA Release 77-223)

- DFRC announced that the final free flight of the Shuttle orbiter, scheduled for Oct. 26, would have Fred Haise and Gordon Fullerton as pilots, and that Prince Charles, Britain's Prince of Wales, would attend. (DFRC Release 35-77)

- The Dept. of Transportation reported it had asked for suggestions on solving the problem of ozone irritation encountered on high-altitude flights. Ozone, a colorless atmospheric gas found in increasingly higher concentration at higher altitudes, had been identified earlier in 1977 as the cause of eye, nose, and throat irritation in airline passengers and crew on high-altitude long-range flights. Higher concentrations were found closer to earth in late winter and early spring months, increasing the number of aircraft exposed.

The FAA had issued an advisory on avoiding encounters with ozone, but was now seeking a permanent solution and soliciting views of manufacturers, crew organizations, health groups, high-altitude researchers, and airline customers, to be submitted by Dec. 6. (FAA Release 100-77)

October 19: ESA announced that the tandem launch of a single Delta of the U.S. ISEE-A and the European ISEE-B, originally scheduled for Oct. 19 and postponed to Oct. 21, had been postponed again. NASA was making additional checks of the Delta launch vehicle. (ESA Release Oct 19/77)

- The Dept. of Commerce announced that 2 oceanographic research ships and 4 aircraft had scanned the Gulf of Mexico during Oct. to demonstrate measurement of ocean properties using satellite-borne instrumentation.

Dr. Warren Hovis, director of NOAA's satellite experiment laboratory, said the program was a preliminary to launch in 1978 of the Nimbus-G with a coastal-zone color scanner. The joint NASA-NOAA tests had used aircraft equipped with models of the scanner to record data that could be confirmed by surface measurements taken on the ships. (NOAA Release 77-265)

• The Natl. Science Foundation reported that federal funding for research and development had risen significantly between fiscal yrs 1974 and 1978, having shown only slight growth in the 1969–1974 period. Total federal R&D funding for FY 1978 was about $26.3 billion, $8.9 billion more than for FY 1974. For the 1974–1978 period, 6 functional areas accounted for nearly all the increase in federal support: defense, 44%; energy, 25%; space and health, 7% each; environment, 5%; science and technology base, 4%. (NSF Release PR77-93)

October 20: JSC reported the 8th group of Shuttle astronaut applicants would report to the center Oct. 25 for physical examinations and interviews. The 20 mission specialist applicants in this group were all military except for one civilian; they represented the USAF (13), the Army (3), the Navy (2), and the Marines (1). The group included one woman, USAF Capt. Jane L. Holley. Of the 160 applicants screened at JSC, 77 wanted to be pilots and 83 mission specialists. (JSC Release 77-66; NASA Release 77-225)

October 21: LaRC announced it would hold the 33rd annual completion exercises Oct. 28 for 34 engineering technicians trained at the center. James P. Mitchell, deputy administrator of the Dept. of Labor's Bureau of Apprenticeship and Training, would be keynote speaker. LaRC Director Donald P. Hearth would preside at the ceremony and give special recognition to honor graduates. Woodrow W. Midgette, Jr., top honor graduate, would be speaker for the class of 1977, consisting of 7 electrical engineering technicians; 7 electronics technicians; 2 engineering draftsmen; and 18 engineering technicians in the areas of aerospace models, facility operations, materials processing, mechanical development, and systems environments. (LaRC Release 77-51)

• NASA announced it had named Dr. James J. Kramer associate administrator for aeronautics and space technology, effective Oct. 23. He had been acting in that capacity since Dec. 1976. Starting at Lewis Research Center in 1951, Dr. Kramer had come to NASA Hq in 1971 as chief of OAST's noise and pollution reduction branch. (NASA anno Oct 21/77; NASA Release 77-227)

• The *Washington Post* reported Senate confirmation of 11 high-level nominees to the new Dept. of Energy. Among those confirmed was Dale D. Myers, former NASA associate administrator for space flight, as undersecretary of DOE. Myers, an aeronautical engineer with North American Rockwell since 1943, had joined NASA in Jan. 1970; he had been a vice president and manager of the Apollo program at Rockwell. (*W Post*, Oct 21/77, B-5)

October 22: NASA launched two international satellites, the U.S. ISEE-A and ESA's ISEE-B, on a single Delta rocket at 9:53am EDT (1353hr GMT) from ETR at Cape Canaveral into a highly elliptical orbit ranging from about 138,000km apogee to 280km perigee, 28.7° inclination, and 2.4-day period. The ISEEs had separated an hr after launch, the A craft (*Isee 1*) spin axis perpendicular to the ecliptic and the B spacecraft (*Isee 2*) spin axis parallel to the ecliptic to permit magnetometer calibration and boom extension. The satellites would measure earth's environment, assisted by a third spacecraft ISEE-C scheduled for launch in 1978. (NASA Release 77-123; ESA Release Oct. 24/77; MOR S-862-77-01/02 [prelaunch] Oct 11/77, [postlaunch] Oct 25/77; *W Post*, Oct 23/77, A13; *W Star*, Oct 23/77, A-2)

October 23: The *NY Times* reported that Capt. Edgar D. Mitchell, astronaut on the *Apollo 14* mission, had taken 55 special stamp covers on his flight and had just sold the first for $4200. The San Francisco firm that auctioned the stamp earlier in Oct. said the buyer was a major stamp dealer in Italy.

In 1972 NASA had reprimanded the *Apollo 15* astronauts for smuggling 400 first-day covers to the moon and selling 100; the 300 covers not sold were turned over to the U.S. for "safe storage in the archives." The *Apollo 14* crew had carried 200 silver medals on the 1971 moon flight, keeping 150 for "private use," but had not been reprimanded because none of the medals had been sold. The Calif. dealer said Mitchell, now a civilian living in Fla., had brought in the covers for appraisal and was advised to find the market value by selling one. NASA said it knew Mitchell had taken the covers in his "personal preference kit"; when it asked about them, he replied "It's none of your business," and a few days later had announced his retirement. Official NASA comment on the sale: "Unfortunate." (*NYT*, Oct 23/77, 45)

October 24: FBIS said the USSR in 1978 would install a second generation of laser long-distance satellite rangefinders usable in daylight, increasing the range to 40,000km with an accuracy of ±10cm. USSR engineers were working in Czechoslovakia on the new rangefinder. The report said Intercosmos engineers from Hungary, East Germany, Poland, Czechoslovakia, and the USSR had built a rangefinder measuring distances up to 3000km accurate up to ±5m; however, it could operate only at night. Ground stations in Poland, Egypt, Bolivia, India, and Cuba had used the device. (FBIS, Intl Serv in Russian, Oct 24/77)

October 25: NASA announced it had agreed with the St. Regis Paper Co. in N.Y. to use the Landsat for monitoring from 900km (560mi) commercial timberlands owned by the company over a 5-state area in Fla., Ga., Ala., Miss., and La. St. Regis would use the satellite data to manage its timberland. JSC would manage development of an automatic system for classifying forested areas, to identify kinds of trees, estimate timber volume, and detect changes in health or growth of the forests, combining the new system with existing data from aerial photography and ground surveys. Software and techniques from this first major undertaking between NASA and a private company would be in the public domain, to benefit other forest-resource managers. Past cooperative uses of Landsat had been between NASA and other U.S. or foreign government agencies. (NASA Release 77-226; JSC Release 77-73)

October 26: NASA reported that the fifth and final free flight of the Shuttle orbiter Enterprise, testing its braking on a concrete runway, had successfully made a landing on concrete at Edwards AFB—although it had a bumpy finale when the orbiter touched down and bounced twice before stopping.

At a postflight press conference, pilots Fred W. Haise and C. Gordon Fullerton said the craft had reached speeds higher than expected after separating from its Boeing 747 carrier. Shortly after 8am local time the 747 had taken off with the orbiter attached to the top of its fuselage, climbing during the next 50 minutes to about 20 000ft altitude, where it nosed over and began to dive. At about 17 000ft the orbiter separated and came straight down without the turns made in previous flights; the dive angle was 25-22° compared with a commercial jetliner's descent slope of about 3°. Touchdown, which came 1min 55sec after separation, overshot the 5000ft mark on the concrete runway by about 1000ft. The rear landing gear had touched, but the craft hopped back into the air, bounced, and tipped slightly before the pilots steadied it and stopped it with 2000ft of runway remaining. Approach speed had reached 330mph on the indicators, and the pilots described "oddities in the way the speed dropped" during the descent.

Donald K. Slayton, in charge of the landing tests, said engineers this week would check out the airspeed indicators and the vehicle's calculated energy profile, but further tests were not probable. Plans for future orbiter landings would change on the basis of data from the fifth test, modifying the aim point to allow for higher lift/drag coefficients with the tailcone off and the slower bleed-off of airspeed. (NASA Release 77-224; postflt rept free flight 5, SpSh orbiter ALT *NYT,* Oct 27/77, A-18; *LA Times,* Oct 27/77, 2; *W Post,* Oct 27/77, A10, B-3; JSC *Roundup,* Oct 28/77, 1; *AvWk,* Oct 31/77, 16)

• NASA Administrator Dr. Robert A. Frosch announced reorganization of the agency effective Nov. 8 "to strengthen NASA organization and improve our effectiveness . . . we will reduce the number of staff offices reporting directly to the Administrator." Headquarters operations would consist of 3 functions: the office of the Administrator, for general management; the chief scientist, chief engineer, and 5 program offices for program planning, management, and review; and 8 staff offices to give program, staff, and functional support. Each of the 5 program offices would have an associate administrator to manage aeronautics and space technology; space sciences; space and terrestrial applications (formerly applications); space transportation systems (formerly space flight); and space tracking and data systems (formerly tracking and data acquisition). (NASA Release 77-228)

October 27: NASA launched the Navy's *Transat* spacecraft at 9:52pm PDT from the Western Test Range on a Scout vehicle into an orbit with 600.7nmi apogee, 572.6nmi perigee, 89.9° inclination, and 106.9min period. The spacecraft, a modified Transit navigation satellite, would join 5 others in a system being used by U.S. Navy and other ships for global navigation, as the first to carry 2 modified transponders with a capability called Satrack to support Trident (fleet ballistic missile) development.

Built by Johns Hopkins Univ. Applied Physics Laboratory, *Transat* would receive, translate, and relay NavStar global-positioning signals simulating a Trident missile in flight, to evaluate Satrack procedures before using them in Trident test flights and to check out Satrack processing at JHU/APL. NASA had agreed with DOD in Jan. 1970 to launch Scouts, carrying 15 Transit satellites and 3 Transit Improvement Program (TIP) satellites so far, the Navy to reimburse NASA for the launch vehicles, launch services, and mission support. The Transit system normally operated 5 satellites, with 2 on standby. (*Langley Researcher,* Nov 28/77, 2; MOR M-490-601-77-03 [prelaunch] Sept 30/77)

NASA's launch participation was judged successful as of Nov. 8-10. (MOR M-490-601-77-03 [postlaunch] Nov 15/77)

• FBIS reported that the USSR had completed a series of rocket launches in cooperation with France to compare procedures and instruments used by the two countries in high-altitude atmosphere sounding. The rockets were launched simultaneously from the French space center at Kourou, French Guiana, and offshore from the Soviet research vessel Akademik Korolev recently at work with the U.S. off the Atantic coast. (FBIS, Tass in English, Oct 27/77)

• INTELSAT announced that the People's Republic of the Congo had become the 100th member of the organization Oct. 26, and the 26th African nation to join. Created in 1964 with an initial membership of 11 nations, INTELSAT was providing communications service through 152 stations in 86 countries. (INTELSAT Release 77-27-I)

October 28: MSFC announced plans for a teleoperator retrieval system (a retrievable reusable low-thrust stage for maneuvering payloads in low earth orbit) to fly on the Space Shuttle in the early 1980s, its first use probably a Skylab-orbit adjustment to permit revisiting or control its reentry over an open ocean area. Riding into orbit on the Shuttle, the TRS would be deployed and remotely operated by an astronaut aided by a television image from a camera mounted on the TRS. It could also maneuver to higher orbit and return, using its own guidance and computer systems, to be remotely controlled again for retrieval into the Shuttle when it came within range of the TV signal. MSFC would manage development of the TRS, and Martin Marietta would be responsible for vehicle integration. (MSFC Release 77-201)

• INTELSAT announced that Upper Volta had become the 101st member of the organization Oct. 27, and the 27th African nation to join. The Office des Postes et Telecommunications de Haute Volta had signed the operating agreement for Upper Volta. (INTELSAT Release 77-28-I)

October 30: The *Washington Post* reported that NASA would try to save the Skylab space station "which is moving closer to earth and may fall in bits and pieces over the next few years." The Space Shuttle would launch a "remote-controlled rocket" to Skylab to push the station to a new orbit, where it would be available for future astronaut visits.

The 118ft Skylab launched in 1973 had been circling the globe every 1.5hr at a 200mi altitude. It had been home for three 3-man crews for as long as 12wk at a time; its last 84-day manned mission in 1974 was still the world's endurance record for time in space. (*W Post*, Oct 30/77, A-10)

October 31: The *Washington Star* reported that U.S. and European scientists were planning a two-spacecraft mission to an unexplored area of space to get a first look at the sun and its solar system from a new perspective. No spacecraft had as yet ventured off the so-called plane of the ecliptic (the solar equatorial plane) by more than 15° latitude relative to the sun. Using two spacecraft, one below the plane of the ecliptic and the other in the usual latitude relative to the sun, scientists could learn more about solar physics and conditions responsible for variations in earth's climate.

The mission would use two unmanned spacecraft weighing 650 to 800lb each, launched by a Space Shuttle. The report said NASA would request funds for the "solar polar" (or out-of-ecliptic, OOE) project in its next budget, giving engineers time to meet a favorable 10-day launch period in Feb. 1983. ESA, whose members included most western European nations, was said to be ready to share in the mission; it would build one of the vehicles, the U.S. the other. JPL would manage NASA's part of the mission, estimated to cost about $141 million. (*W Star*, Oct 31/77, A-6)

During October: The USAF reported that the first two-seater F-16 airplane, a fighter-trainer version known as F-16B, had made a successful first flight at General Dynamics Corp.'s Ft. Worth, Tex. facility. The test flights were to evaluate the handling of this version, weighing 600lb less than the standard single-seat F-16 when fueled, and to use rear-cockpit controls to put it through takeoff and landing, formation flying, and roll and pitch.

The one- and two-seat versions of the F-16 were almost identical in size and appearance, each 48ft long with a 300ft^2 wing area, but the F-16B had a larger canopy to cover the second pilot and would carry 1100 fewer gallons of fuel. Development of the new plane, first fighter designed to withstand forces up to 9g, was managed by the Aeronautical Systems Division at Wright Patterson Air Force Base. (AFSC *Newsreview*, Oct 77, 1)

• ERDA announced it had signed a contract with the Ford Motor Co., to be managed by LeRC, for developing the Stirling passenger-car engine to the point of deciding in 8yr whether or not to produce it commercially. A Stirling engine would be at least 30% more fuel efficient than those of comparable size and performance in current cars. It differed from the conventional internal combustion engine in being powered by heat from an external source: a burner using any of a variety of fuels would heat hydrogen gas inside the engine whose expansion and contraction would move pistons to generate power.

Invented in the 1800s, the Stirling predated the internal-combustion engine and had been used to pump water for mines and quarries. By the 1930s the availability of gasoline-powered cars had overshadowed other development. N. V. Philips of the Netherlands and United Stirling of Sweden had pioneered and continued work on the Stirling engine and had licensed Ford to work on it in 1971. ERDA and Ford had become interested because of the engine's high fuel efficiency, flexibility in choice of fuels, low noise and pollution, and good driving characteristics. Cost of the 8-yr effort would be about $160 million, $110 million contributed by ERDA and $50 million by Ford; if the Stirling should go into production, Ford's investment might eventually amount to more than $500 million. (ERDA Release 77-179)

November

November 2: Marshall Space Flight Center director William R. Lucas announced that the teleoperator retrieval system project assigned by NASA to the center would be headed by John H. Harlow, who had been deputy manager of the Space Shuttle main engine project since 1976 and served previously as assistant manager in the Shuttle Projects Office. (MSFC Release 77-206)

November 3: NASA announced that biologists working for it and the Natl. Science Foundation had identified a "new" form of life that might date back to earth's first billion yr: a methane-producing organism representing what might be the oldest form of life as well as a line of evolution totally separate from the two traditionally recognized lines of bacteria, and animals and plants. A research team headed by Dr. Carl Woese of the Univ. of Ill.-Urbana, described as a world expert on the genetic code, said the organisms were a distinct class no more related to typical bacteria than to higher forms. Biologists working with a group of methane-producing organisms had assumed they were ordinary bacteria; however, analysis of ribosomal RNA had shown the genealogy of the new form to be distinct from those of both bacteria and the animals or plants. Dr. Woese said the study might shed light on a missing stage of evolution, the chemical stage that immediately preceded identifiable life on earth. (NASA Release 77-229)

November 4: Johnson Space Center announced that the ninth group of Shuttle astronaut applicants would report Nov. 7 for physicals and interviews. The group, 21 mission specialist applicants and 2 pilot applicants, included 12 civilians; 2 were women, bringing the number of female applicants interviewed to 20. Of the military applicants, 5 were Navy; 4, USAF; and 1 each from the Army and Marines. NASA had screened 183 applicants so far, 79 seeking to be pilots and 104 mission specialists. (JSC Release 77-70; NASA Release 77-232)

• JSC announced it had selected ILC Industries, Inc., of Fredrica, Md., to negotiate a $1.2-million contract for Shuttle crew equipment and stowage provisions. The cost-plus-fixed-fee contract running from Jan. 1978 to the end of Sept. 1980 would cover items such as crew clothing, an orbiter survival kit, and personal hygiene kits, as well as replacement, servicing, and maintenance, with special studies of the crew equipment. (JSC Release 77-71)

November 5: Today newspaper reported that two launch failures in Sept. at Cape Canaveral (explosions of an Altas Centaur and a Delta rocket) had altered the schedule to include launch of a Meteosat Nov. 17 on a Delta; an Intelsat Dec. 8 on an Altas Centaur; and the Japanese comsat (originally scheduled for launch Dec. 8) Dec. 14 on another Delta. ESA's Meteosat, originally scheduled to go on Nov. 15, had been put off 2 days for further checkout of the Delta; the Intelsat was put off from Nov. 10 to Dec. 8 for further checkout of the Atlas Centaur. The 15 missions scheduled for launch from the Cape next yr had risen to 17: a FltSatCom launch for the Navy originally scheduled for Dec. 15 had been pushed forward to 1978, and a launch of ESA's OTS backup had been added to replace the one that failed. (*Today,* Nov 5/77, 1A)

November 7: Dryden Flight Research Center announced it had scheduled four ferry-test flights of the Shuttle orbiter Enterprise mated to its Boeing 747 carrier for Nov. 14, 15, 17, and 18 to prepare for the orbiter's cross-country ferry flight in March 1978 from DFRC to MSFC for ground-vibration testing. Later flights would ferry orbiters from Rockwell Intl.'s facility in Calif. to Kennedy Space Center in Fla. for launch.

Besides determining the best speeds and altitudes for the ferry flights, the test would measure holding-pattern and engine-out performance in cruise and landing/takeoff. The 747 crew would be Fitzhugh Fulton, commander, and Tom McMurtry, pilot, with Victor Horton and Skip Guidry as flight engineers; the 4 had flown the 747 in recent approach and landing tests. (DFRC Release 37-77; NASA Release 77-233; JSC Release 77-74)

• MSFC reported its engineers had prepared electronic simulations as dress rehearsals of Shuttle orbital missions set for 1978, with computers in the roles of major hardware. The simulations, carried out at JSC, would check out Shuttle avionics (electrical and electronic systems controlling the Shuttle during flight and sending signals to pilots and ground controllers monitoring the mission). MSFC's engineers would deliver to JSC in Jan. 1978 the Marshall mated-elements system (MMES) simulating items for which MSFC was responsible: main engines, solid-fuel rocket boosters, and external tank. JSC would provide an orbiter simulator and a computer to simulate Shuttle dynamics, a system called SAIL (Shuttle avionics integration laboratory) able to simulate a Shuttle flight beginning 20min before launch under normal or abnormal conditions. The MMES could simulate failure of almost any MSFC element to see how the systems would react with one main engine out or if high temperatures or pressures should occur in vital areas. (MSFC Release 77-208)

• NASA announced changes in organizations or titles for operational purposes during the transition period of its reorganization (Nov. 8 to Dec. 4) in the Office of Management Operations and the Office of External Relations, as well as establishment of the position of special assistant to the Administrator to direct implementation of the reorganization. (NASA anno Nov 7/77)

• The Natl. Geographic Society announced that Gen. William J. Evans would be recipient for 1976 of the Gen. Thomas D. White Space Trophy given annually since 1961 to the USAF member, military or civilian, contributing most significantly to U.S. aerospace progress. Gen. Evans, commander of the Air Force Systems Command at Andrews AFB from Sept. 1975 to Dec. 1976, had played a key role in developing defense space systems. (NGS Release Nov 7/77)

November 8: Langley Research Center announced that Dr. Edgar M. Cortright, who had been director of the center 1968–1975 and was now a vice president of Owens-Illinois, Inc., would speak at the center coloquium Nov. 14 on recent developments in solar energy.

Dr. Cortright, in charge of overall technical direction at the large multinational corporation now interested in uses of solar energy, would discuss technical, economic, and legislative aspects of the newly developing industry that would depend for wide application upon being price-competitive with other energy sources in spite of being inexhaustible and nonpolluting. The next few years would see its first domestic applications in areas such as space heating and air conditioning. (LaRC Release 77-54)

• NASA Hq notified its employees of changes in operational control during the reorganization, including requests for communications services, travel services, parking permits and other transportation support, assignment and scheduling of senior duty officers, library services, and other administrative functions. Changes would be effective immediately. (NASA anno Nov 8/77)

November 9: The *Washington Post* reported that Charles Kowal of the Calif. Inst. of Technology had discovered "something out there in space that orbits the sun roughly 2 million miles out once every 115 yr or so," but had not "the slightest idea" what it was. "You might say it's a mini-planet," far too small to be a conventional planet but as big as some asteroids, about 100 to 400mi in diameter; "It really doesn't resemble anything else. It is definitely not a satellite of any planet or a comet. . . ."

Kowal, a veteran at discovering new objects in the solar system, had detected the 13th moon of Jupiter and another object that might be the 14th; he had located the latest unusual body while working with the 48in telescope at Mt. Palomar Oct. 18–19 and had been trying to establish its orbit so that other astronomers could check it out. Researchers would need 2 or 3wk to determine the orbit, lying beyond Saturn and possibly both inside and outside the orbit of Uranus, "maybe going out as far as Neptune," between 1.3 million and 2.7 million miles from the sun. If the object should be defined as a planet, Kowal by tradition would have the privilege of naming it. Final decision on whether or not it was a planet would be made by the Smithsonian Astrophysical Laboratory in Cambridge, Mass. (*W Post*, Nov 9/77, A-17)

November 10: The *Chicago Tribune* reported that astronaut Walt Cunningham had written "the first kiss-and-tell treatise about the space program," a book called *The All-American Boys*, intended as a "candid, close-up, even critical look" at NASA, the space program, and his fellow "partynauts," none of whom had dared to be as critical as Cunningham in his first "but probably not last" book. The review said the book was "about people—imperfect mortal men—and what happens when they suddenly find themselves cast as superheroes in a world that literally knows no bounds." (*C Trib*, Nov 10/77, 1-17)

November 11: Lewis Research Center's 2-yr program of sampling air around the world to detect substances detrimental to the environment had been unable to cover the polar regions, the *Lewis News* said, until last month when its air monitors rode the polar route on Pan American's 50th anniversary flight.

Known as GASP (global air sampling program), the LeRC effort had used instruments designed at the center to detect and measure small particles and pollution-related gases, as well as carbon monoxide, water vapor, ozone, and various other oxides, operating automatically and recording data on magnetic tape for analysis at NASA labs and distribution to scientists and engineers anywhere for air-quality evaluation. Sampling the upper air over the poles would show "what clean air is actually like," said GASP director Porter J. Perkins. "What kind and amount of manmade pollutants are up there, if any? Especially are we interested in fluorocarbons from aerosol cans . . . What we may find . . . may conceivably alter presumptions we've made about global distribution of certain airborne contaminants." (*Lewis News*, Nov 11/77, 3)

- KSC's *Spaceport News* reported on Center efforts to recover launch-vehicle and payload wreckage from the "unprecedented" back-to-back explosions of a Delta and an Atlas Centaur carrying an OTS and an Intelsat spacecraft.

In "the most comprehensive and painstaking recovery in the Space Center's history," land, sea, and air crews searched every foot of the 50mi^2 impact area. Two Patrick AFB helicopters and a private chopper worked to spot scattered wreckage from the air; the USAF crew made 19 flights in 2wk, with a total of 88hr in the air, covering a 12mi^2 area and locating about 95% of the wreckage recovered. Ground crews in armored personnel carriers and sea crews in Navy and private craft then took over to retrieve and identify each piece, check it for live ordnance, and catalog its recovery location. USAF ordnance disposal personnel often had to penetrate thick, snake-infested vegetation to reach Delta and Atlas Centaur fragments. Personnel from Port Everglades and the Navy's harbor-clearance unit using an Air Force landing craft-utility (LCU) boat retrieved the Delta's solid-fuel motor #1 from the water as prime suspect in the OTS launch failure.

The teams worked from early morning to dusk, even over weekends, and recovery work for the Delta was still under way. Hundreds of persons helped on the recovery, from directors to office clerks. Personnel from KSC had the aid of representatives from McDonnell Douglas, General Dynamics, Patrick AFB, and Pan American Airways operations in the extensive ground search. To reward the combined efforts, KSC would present a group achievement award to the Delta 134/Atlas Centaur 43 recovery team at an awards ceremony Nov. 22. (*Spaceport News*, Nov 11/77, 4)

- MSFC reported successful completion of structural tests on the Space Shuttle external tank's intertank section, carrying the attach points for the Shuttle's two solid-fuel rocket boosters and subject to heavy loading both on the launch pad and during flight while the boosters were thrusting.

MSFC engineers would now proceed with vibration tests on the forward portion of the external tank, the liquid-oxygen container, attached to the intertank. Structural tests would follow later in 1978. (MSFC Release 77-212)

- JSC announced that the 10th group of Shuttle astronaut applications would report to the center Nov. 14 for physical examinations and interviews. This group consisted of 24 mission specialist applicants and one pilot applicant; 17 were civilians, 1 a woman, and 8 were from the military—3 each from the USAF and Navy, 1 each from the Marines and

Coast Guard. This group would bring the total to 80 pilot applicants and 128 mission specialist (21 women) applicants screened at JSC. In Dec., NASA hoped to name as many as 20 in each category for 2 yrs' further training before final selection as astronauts. (JSC Release 77-75; NASA Release 77-236)

• JSC announced that NASA and the Soviet Academy of Science would send representatives to a meeting Nov. 14-17 in Moscow to discuss a joint program for the 1980s using the flexible delivery and large capacity of the Space Shuttle and the Salyut's longer stay time in orbit. Dr. Noel W. Hinners, NASA's associate administrator for space science, would head the U.S. delegation, and Dr. Boris Petrov, chairman of the academy's Intercosmos council, would head the Soviet delegation. (JSC Release 77-78; NASA Release 77-234; FBIS, Moscow Dom Sv in Rusn, Nov 18/77)

• NASA announced that Wallops Flight Center would host the 8th annual meeting of the NASA-Soviet Space Biology and Medicine working group Nov. 19-25, preceded by a workshop Nov. 16-18 on simulated weightlessness, as part of a continuing program under a 1971 agreement between NASA and the USSR Academy of Sciences. Meeting participants would discuss biomedical results of the *Cosmos 936* mission carrying U.S. experiments, the *Salyut 5/Soyuz 19* mission, and a JSC test demonstrating Spacelab missions, as well as research on space motion sickness and health in weightlessness. Dr. David Winter, NASA's director for life sciences, would head the U.S. delegation, and Dr. Rufus Hessberg, director of space medicine, would head U.S. workshop participants. Soviet leader at both meetings would be Dr. Nikolai Gurovsky of the Soviet Ministry of Health. (JSC Release 77-78; NASA Release 77-234; WFC Release 77-15)

• Communications Satellite Corporation (ComSatCorp) announced it would buy back 1.5 million shares of its common stock for $37 cash per share until Dec. 6. The organization had decided on the repurchase as a preliminary step in possible future acquisitions, and a restructuring of its accounts to show a substantial debt, as the FCC had indicated its rate making would be based on such a debt in the organization's capital structure. ComSatCorp emphasized that it was currently debt free. (ComSatCorp Release 77-28)

November 14: Despite NASA's failure to win FY 1979 funding for a new Mars mission, such a mission remained the major planning element in the agency's planetary programs office, *Aviation Week* reported. A Mars rover/orbiter mission had strong support from the Natl. Academy of Sciences, to lay the groundwork for a sample-return mission using either Martian-orbit rendezvous or direct ascent from earth to Mars. At a KSC briefing, A. Thomas Young, NASA's director of planetary programs, said the highest priority missions NASA might implement "in the near term" would include the Venus-orbiting imaging radar (VOIR), a lunar-polar orbiter, a Saturn orbiter probe, and a rendezvous with comet Halley or Encke, or both, or possibly with an asteroid.

"As we go through the budgetary process, on one side I wish Mars would go away," Young said. "But when you stop and think about rivers, volcanoes, enormous Mars icecaps, the chemistry potential for life, and the atmosphere, you realize that understanding Mars is so important toward understanding where earth has been and is going that you've got to treat it in a responsible manner." He said NASA planning should benefit from the 2.4m-diameter space telescope, aimed primarily at bodies outside the solar system but sure to provide significant data on objects nearer earth. (*AvWk*, Nov 14/77, 55)

• *Av Wk* reported the Intl. Telecommunications Union's frequency board had distributed to its 100-plus member nations the technical details of a USSR plan to launch 7 geostationary Volna satellites for global maritime and air communications service. Members questioning possible interference with existing or projected systems had until Jan. 9 to respond. The 7 spacecraft would work in separate systems: numbers 2, 4, and 6 would each operate in 4 frequency bands, 2 allocated exclusively for mobile maritime service and 2 for mobile aeronautical service. Numbers 1, 3, 5, and 7 would also use these bands, plus a 240-400mc band now used by the US Navy for its maritime service which the Volnas would use for land-mobile service. The Volna proposal did not include fixed shore terminals, although USSR officials had said the system would be a national rather than international system. (*AvWk*, Nov 14/77, 20)

November 15: JSC announced it had signed a supplemental agreement with Rockwell Intl. Corp. for engineering-change orders on the Space Shuttle orbiter. The supplement, valued at about $226 million, would bring the value of the Rockwell contract to about $3.2 billion. The supplement would cover changes such as addition of lightning protection, payload and systems integration, avionics changes, provision of spares, and support of the carrier aircraft during ALT. (JSC Release 77-76)

JSC announced it had signed a contract modification with Ford Aerospace covering hardware and software systems engineering, plus maintenance and operation of JSC's mission control center and other ground-based data systems. The modification, valued at more than $1.4 million, would bring the value of the Ford contract to about $51.3 billion. (JSC Release 77-77)

November 16: NASA announced it would join with Canada in a 3-day meeting Nov. 29-Dec. 1 to assess performance of *Cts*, world's most powerful comsat, now in orbit for nearly 2yr. The meeting would hear experimenters discuss their results and the social, technical, and economic aspects of CTS-type comsats. Canada and the U.S. had shared time on *Cts*, which used a LeRC-developed transmitter tube to operate in a new frequency band at power levels 10 to 20 times higher than those of other comsats, making it possible to use smaller and less expensive ground receivers. *Cts* had permitted 2-way TV and voice contact experiments for health, education, business, and similar purposes. Besides managing all U.S. experiments on *Cts*, LeRC had also used a portable earth terminal in an over-the-road van resembling a TV studio to connect the satellite with local events to be broadcast. (NASA Release 77-237)

• MSFC reported that DOD had lent NASA a crew of Navy divers for 2 yrs to support tests in the neutral buoyancy simulator studying weightlessness. All 5 had been fleet-salvage divers for the Navy with 5 to 17 yrs experience in scuba, hard hat, mixed-gas, and other types of diving apparatus. Selected for their ability to use special tools to assemble or dismantle structures or equipment underwater, the crew would help test hardware, instruments, and tools to be used in space and would work underwater in full-scale Shuttle or Spacelab mockups to aid engineers in planning contingency procedures, equipment operation, space construction, and other manned-mission activities.

Each trainee working underwater in a space suit would have 2 Navy divers responsible for his safety, with others assigned to prepare hardware and ensure the necessary tools were in place before the test started. During the test, divers would stand by to retrieve dropped tools or reset apparatus for repeat testing; between tests, the divers would dismantle, remove, or rearrange hardware under water or install new hardware. Chief Petty Officer Richard Wiebe, a veteran diver with 17 yrs in the Navy, said the crew had found work on the Shuttle program challenging. (*Marshall Star*, Nov 16/77, 4)

November 17: NASA reported that a team of researchers using ultrasensitive radio equipment aboard an Ames Research Center U-2 jet aircraft to measure the background microwave radiation in the cosmos had concluded that the event initiating the universe had begun not with a violent and uncontrolled "big bang" but with "an extremely smooth process," in which matter and energy were uniformly distributed and kept expanding at an equal rate in all directions.

Drs. Richard Muller and George Smoot, with graduate student Marc Gorenstein, of the Lawrence Laboratory and the Univ. of Calif. at Berkeley, had designed and operated the radio equipment for the project funded by NASA and the Dept. of Energy. The plane had flown at 65,000ft altitude above 90% of earth's atmosphere to make the sensitive experiments indicating that the Milky Way galaxy (including earth and its solar system) was flying through space toward the constellation Hydra at a uniform speed of more than 1 million mph—so uniform that "it provides a universal reference for measuring this motion," said Gorenstein. The large-scale regularity found in the expansion of the universe made the speed of the local motion discovered for the earth "all the more surprising," said Dr. Muller. Furthermore, the measurements "seem to show that there is no rotation of the universe," Dr. Smoot said, which was surprising because "we can see that everything within the universe is rotating—planets, stars, and galaxies. If there is rotation it has to be less than one hundred millionth of a rotation in the last billion years. . . . Either conditions before the beginning were very regular, or processes we don't yet know about worked to make the universe extremely uniform." The uniformity was greater than 1 part in 1000 for matter, 1 in 3000 for energy, and 1 in 10 000 for expansion. (NASA Release 77-235; ARC Release 77-45)

• NASA announced it had appointed Raymond A. Kline as associate administrator for management operations, effective Nov. 20. Coming to NASA Hq in 1968, he had been acting associate administrator for center operations from July of this year. Previously he had served for 6yr on Dr. Wernher von Braun's executive staff at MSFC. (NASA anno Nov 17/77)

November 21: NASA announced that the next launch of an Atlas Centaur rocket from KSC would be no earlier than Jan. 6, 1978, carrying an Intelsat IVA commercial comsat. Originally scheduled for launch in Nov., the mission was first delayed for investigation of a previous Atlas Centaur launch that failed Sept. 29. That investigation was almost complete; the associate administrator for transportation systems, John Yardley, was to receive a report in early Dec. The postponement to Jan. was not caused by the investigation, NASA said, but by discovery of faulty feedback transducers in the Atlas engine-control actuators during routine testing. The new date would allow removal and replacement of components from that manufacturing batch and revalidation of the engine systems. (NASA Release 77-241)

- NASA announced tentative selection of 18 scientists to participate in the Space Telescope project. The 10-ton spacecraft scheduled for Shuttle launch in 1983 to orbit at 500km (310mi) altitude would make observations deeper into space, and with more detail, than heretofore possible, covering about 350 times the volume of space now visible from earth-based telescopes. It had been designed for remote operation from the ground, including maintenance or retrieval by a space-suited astronaut for return to earth to be overhauled and relaunched. The telescope could accommodate 5 different instruments at its focal plane: both the faint-object camera (ESA) and the wide-field camera could cover ultraviolet and blue spectra, the latter the red and near-infrared regions as well; the two spectrographs would offer a range of resolution impossible with a single instrument; and the photometer could calibrate the other instrument and image the galactic background. (NASA Release 77-239)

- NASA announced that on Nov. 29 it would present first results of the scientific experiments on high-energy astronomy observatory *Heao 1* launched Aug. 12. Dr. Noel Hinners, associate administrator for space science; Dr. Frank B. McDonald of Goddard Space Flight Center, project scientist; and principal investigators for the experiments would report findings on pulsars, quasars, exploding galaxies, and black holes. Scientific instruments on the observatory offered highly sensitive detection and resolution of x-rays emitted by stellar sources throughout the universe; the spacecraft, rotating end over end, could survey the entire sky within 6mo. (NASA Release 77-242)

November 22: NASA launched the ESA *Meteosat* at 8:35pm EST from the Eastern Test Range on a Delta vehicle into a synchronous transfer orbit with 37 001km apogee, 147km perigee, and 27.5° inclination. The apogee boost motor fired at 1:19 pm EST Nov. 23 to begin maneuvering it to a position on the Greenwich meridian at 0° longitude above the equator, where it would provide long-range forecasting coverage to Europe, the Near East, and Africa for at least 3yr.

The cylindrical craft 3.2m long and 2.1m in diameter weighed 697kg at launch and 300kg in orbit; its payload was a telescope radiometer to observe earth and cloud formations, providing relay and transmission of meteorological data. The World Meteorological Organization had set up a chain of 5 such satellites, 1 European, 1 Soviet, 1 Japanese, and 2 U.S. Besides transmitting to user stations the data from a center at Darmstadt, West Germany, *Meteosat* would relay Atlantic Ocean images from the U.S. *Goes.* (NASA Release 77-230; MOR M-492-102-77-01 [prelaunch] Nov 16/77, [postlaunch] Feb 28/78; KSC *Spaceport News* Nov 11/77,1; ESA releases Nov 3/77, Nov 16/77, Nov 23/77)

November 22: MSFC announced it had awarded to Martin Marietta Corp. a $1.735 million contract for analysis and design activities needed for a preliminary design review in March 1978 of the teleoperator retrieval system(TRS).

MSFC had been managing development of the TRS to adjust the orbit of Skylab [see Nov. 2], based on in-house NASA studies and earlier work by Martin Marietta on the astronaut-maneuvering unit and a free-flying teleoperator system. The contractor would make wide use of already qualified hardware in preparing for TRS assembly, test, and integration. (MSFC Release 77-220)

November 23: The *Marshall Star* reported that, before the real Shuttle orbiter arrived in Ala. next year, a homemade orbiter constructed of scrap rocket parts and steel beams would stand in for it during practice lifts into the tower where the real thing would be tested. The stand-in would first make use of the device that would remove the Enterprise from its 747 carrier, then follow the route to be taken by the real orbiter, in order to check clearances on the sides. Its major tryout would come when MSFC engineers put it into the 430ft (131m) test stand to see what effect the wind would have on installation.

The simulator, basically a motor casing from a Titan solid-fuel rocket 10ft (3.04m) in diameter and 80ft (24.39m) long with added framing, collars, nose and tail structures, and wings plus aluminum sheeting for the outer skin, would closely match the Enterprise in size, shape, weight, and center of gravity. After its performance at MSFC, it would go by barge to KSC to take part in launch-procedure checkouts, and from there to Vandenberg AFB for DOD use in preparing Shuttle launch facilities. (*Marshall Star*, Nov 23/77, 4)

• MSFC announced it would conduct on Dec. 1 at the Natl. Parachute Test Range in Calif. the third airdrop test of a parachute system designed to recover the Shuttle's reusable solid-fuel rocket boosters. Royce Mitchell of MSFC said the test article, a 48 000lb (21 773kg) dummy booster carried under the wing of a B-52 aircraft, would encounter 30% more dynamic pressure than the system should experience during a real deployment.

To test the durability of the drogue chute that would open first to pull out the main chutes, fins added to the test article would increase its speed and reduce drag. The drogue would deploy at a speed of about 420mph (676kph) 20sec after the plane released the dummy booster at an estimated speed of 230mph (370kph) and an altitude of about 19 400ft (5913m). The 2 boosters to be used on each Shuttle launch would separate after burnout at an altitude of about 27mi (43.5km), descending slowly by parachute into the ocean for recovery and reuse. (MSFC Release 77-221)

• LaRC announced that NASA Administrator Robert A. Frosch would speak at the center's annual awards ceremony Nov. 29, presenting awards for outstanding leadership, exceptional service, group achievement, and 45-yr service. LaRC Director Donald Hearth would present 4 awards for 40-yr service and 83 for 35-yr service. An award for outstanding volunteer service would go to 4 employees, and 3 would receive the center's highest honor, the H.J.E. Reid Award, named for the man who had headed the center for 35yr, starting as engineer-in-charge in 1926 and retiring as director in 1961. The center would also make an award for technology utilization for the first time. (LaRC Release 77-55)

November 28: The Dept. of Transportation reported that the Federal Aviation Administration would demonstrate an advanced all-weather microwave landing system Dec. 5-9 at JFK International Airport using a NASA Boeing 737 to fly a series of curved approaches and automatic landings like those recently flown at other locations, most recently Buenos Aires. The Intl. Civil Aviation Organization had recently tested the equipment as a replacement for 40-yr-old instrument-landing systems. Beginning Feb. 22, the British government would demonstrate a different MLS on the same runway to obtain comparative data on the two systems; similar comparisons would proceed at other locations abroad. In March 1977 the ICAO all-weather operations panel had endorsed the MLS design proposed by the U.S. and Australia, to be considered by the full membership of ICAO meeting in 1978. (DOT Release FAA-118-77)

November 29: NASA held a press briefing on results from *Heao 1*, launched 3.5mo ago, described by the project manager Dr. Frank McDonald of GSFC as "probably the largest scientific satellite" flown by the U.S. On hand for the briefing were Dr. Noel Hinners, associate administrator for space sciences; Dr. McDonald and Dr. Elihu Boldt from GSFC; Dr. Herbert Friedman of NRL, principal investigator on the large x-ray survey; Dr. Gordon Garmire of CalTech, who with Dr. Boldt had worked on the cosmic ray experiment; Dr. Hale Bradt of MIT, principal investigator for the scanning-modulation collimator; and Dr. Hale Peterson of the Univ. of Calif. San Diego, principal investigator for the low-energy gamma-ray experiment. The scientists gave the press brief statements of their results and responded to questions. (Text, Nov 29/77)

November 30: LaRC reported it had selected Klate Holt Co. of Hampton, Va., to negotiate a $13 million contract for on-center support services for 3yr starting in March 1978, with two 1-yr extension options. The contract would cover refrigeration and air conditioning, electric systems (compressors, large valves, and pumps), building-trades services such as carpentry, painting, masonry work, roofing, and street repairs, plus engineering services including drawings and specifications. (LaRC Release 77-56)

- The *Marshall Star* reported that the Nov. 22 launch of *Meteosat* for ESA had encountered a delay because of a "weak but spurious" radio signal that had reached the command destruct system. Although officials considered the signal too weak to affect the launch, they delayed it anyway to trace its origin. They ran tests to try and pick up the signal again, tracing it finally to the advanced range-instrumentation ship Redstone, which had not sent the signal but had leaked it from onboard computer testing. Upon identifying the source, officials allowed the launch to proceed. (*Marshall Star*, Nov 30/77, 3)

- ComSatCorp announced that the FCC had decided to make it pay extra sums into an escrow account ordered by the agency pending court review of an FCC rate order in 1975. Since then ComSatCorp had continued to charge the rates in effect at the time of the order, putting in escrow the difference between the amount collected and the lower rates ordered by the commission. ComSatCorp said the FCC's new order would require it to put about $25 million more in escrow, and it would petition for reconsideration of the order. (ComSatCorp Release 77-30)

- The Natl. Aeronautic Assn. reported on a nationwide poll of its members asking what three problems aviation faced today: the government rated first, then fuel costs and air traffic control, including the quality of weather information at airports. Inconsistent policies and the need for regulatory reform were the main complaints. U.S. efforts to control business arrangements for foreign military sales displeased 71%, who said the effort harmed U.S. aircraft manufacturers. Members wanted the U.S. to build an SST and open its airports to the Concorde (by an 84% vote); on subcontracting construction of some aircraft to European nations, to standardize and render interchangeable NATO's military equipment, the members split 52 to 48% in favor of keeping the business at home. (NAA newsletter Nov 30/77)

During November: The USAF reported completion of 40mo of tests in a joint U.S.-Canadian project using an XC-8A Buffalo military cargo plane from Canada with a huge air-filled rubber doughnut for landing gear, to prove the feasibility of an air cushion landing system for large transport aircraft.

Tests mostly at Wright Patterson AFB consisted of 84 flights, 34 of them air cushion takeoffs and 39 landings on grass, snow, and hard surfaces, with 25mi of taxiing over the same surfaces and over obstacles. Wallace Buzzard, program manager for the USAF flight-dynamics laboratory, said that besides being able to operate on difficult surfaces the air-cushioned plane "also taxied smoothly over a 30ft-long ramp with a 9in dropoff, crafters 6ft in diameter, and ditches 2ft wide. An aircraft with a conventional landing gear simply can't do those things; it would be seriously damaged if not wrecked."

The tests included numerous inflight inflations and deflations of the air cushion, and tests of flying quality: the pilot found he could control the aircraft in crosswinds as high as 20mph, letting the pressure assist him in moving it to a particular point although taxiing sideways. Whereas most planes would descend for a soft landing at about 3ft per sec, the air cushion plane could soft-land at a rate of 8ft per sec with a 6° nose angle.

The XC-8A weighed 33,000lb during the tests without cargo; Buzzard said it could have tested the same fully loaded, probably becoming more efficient with increased weight. "Outfitting an aircraft weighing more than a million lb with conventional landing gear will be extremely difficult," he said. "An ACLS could handle aircraft as heavy as 3 million lb and be competitive in weight with conventional landing systems." (AFSC *Newsreview*, Nov 77, 1)

• NASA reported that JSC had produced a compact medical kit containing most of the instrumentation of a well-equipped physician's office, using techniques and tools developed to monitor astronauts' vital signs during spaceflight. The NASA "black bag," weighing less than 14kg and fitting into a case 18cm by 56cm by 36cm, would contain electronic equipment, drugs, bandages, and instruments needed to diagnose and treat many kinds of illnesses or accidents; electronic tools included an electrocardiograph and electroencephalograph, plus either a strip-chart recorder or a cassette tape recorder, with a coupler built in to transmit data over a standard telephone line. The package would improve the treatment possible on emergency or house calls and provide more complete data before hospital admission. *(NASA Actv*, Nov 77, 24)

• MSFC reported that the new manpower level required at the center would mean dropping some 70 employees in 1978 by "involuntary separation." In Sept., NASA Hq had announced that MSFC would probably lose about 150 positions, but had not said whether the loss would come through attrition or reduction in force. Although resignation, retirement, and other causes had produced some vacancies, MSFC director Dr. William R. Lucas had notified all employees that a reduction in force would take place by April 1978, and that NASA would ask the Civil Service Commission for "early-out" authority to let employees with certain combinations of age plus yrs of service retire early on reduced annuity, granted in some reductions in force. (*Marshall Star*, Nov 77, 4)

• GSFC reported that it and the Univ. of Maryland had designed an experiment called GUMPS (Goddard/Univ. of Md. particle study) for detecting cosmic ray proton energies up to 10Tev, 20 times greater than available from manmade accelerators, to clarify the interaction between particles accelerated from active supernova-explosion remnants and the magnetic field in Milky Way interstellar space. Comparing intensities at the top of and deep inside the atmosphere would reveal much about the fundamental properties of matter.

Borne in a giant spherical aluminum gondola, the instrumentation would include a calorimeter of large iron plates to interact with incoming cosmic rays, producing a cascade of particles in numbers proportional to the energy of the incoming particle; 8 sensitive detectors would measure the energies. A multiwire proportional-counter hodoscope produced by the Univ. of Md. would determine the trajectory of the mass of originating atoms. The experiment would fly on a balloon at 80 000ft altitude before its Shuttle launch. (*Goddard News*, Nov 77, 2)

• KSC reported that the 10th anniversary of the maiden flight of Saturn V, world's most powerful rocket, had passed without fanfare. *Apollo 4* had lifted off Nov. 9, 1967, from KSC's Complex 39A in a successful demonstration of Saturn V's flightworthiness and the Apollo spacecraft's ability to withstand the intense heat of reentry. (KSC *Spaceport News*, Nov 11/77, 2)

• MSFC announced selection of William A. Brooksbank, Jr., as manager of the solar heating and cooling project in the center's special projects office. MSFC director Dr. William R. Lucas said that Brooksbank, most recently deputy manager of the Spacelab program office, was well qualified for the position by his background in engineering and management. Brooksbank had come to the Redstone Arsenal at Huntsville in 1958 as atomic energy project engineer for the Army's ballistic missile agency; he had worked at MSFC since its establishment in 1960. (MSFC Release 77-218)

• JSC announced selection of John C. Stonesifer as head of the center's new life sciences experiments program office, which would manage planning, development, and operation activities at JSC, ARC, and KSC for all such experiments to be carried on the Space Shuttle and other missions in the future. Stonesifer would continue to act as chief of JSC's bioengineering systems division until designation of a replacement. He had joined NASA at LaRC in 1957 and transferred in 1962 during the Mercury project to the space task group; he had been chief of JSC's recovery branch before becoming head of bioengineering systems. (JSC Release 77-72)

• NASA reported that Dr. Lo I. Yin, x-ray researcher at GSFC, had invented a hand-held device producing an instant x-ray image from a small source of radioactive material. The lixiscope (low-intensity x-ray imaging scope) had resulted from research on space energy sources that converted their x-rays to visible images; Dr. Yin said the idea had not been feasible until the Army declassified an image intensifier produced at its Ft. Belvoir night vision laboratory. Cost of the unit, not yet on the market, might be less than $5000. Using no new technology, it contained off-the-shelf items such as the radioactive source and an x-ray phosphor screen besides the intensifier.

Pulling a trigger unshielded the source, sending through the object being examined a low x-ray dose which the phosphor screen converted to visible light. The unit's high intensification would allow use of a small radiation source of 10 to 20 millicuries resembling a pen-size battery; an attached camera could produce instant x-ray pictures with exposure 1000 times weaker than that of a conventional x-ray machine. Dr. Yin said a device invented for x-ray astronomy "where there is a scarcity of x-rays should [be] of obvious value in medical fluoroscopy where there are many x-rays."

Cooperating in evaluating the device would be the National Institute of Dental Research, Howard Univ.'s College of Dentistry, Howard's cancer research center, and the Duke Univ. Medical Center at Durham, N.C. NIDR researchers had already worked up a configuration of the lixiscope for use in dentistry, and the cancer research center would compare it with existing techniques for detecting soft-tissue tumors or foreign bodies and for looking at bone fractures. Prime advantage of the unit would be for emergency and other field use requiring quick fluoroscopy (NASA Release 77-238)

• Calspan Corp. of Buffalo, N.Y., reported it was continuing technical work on the XV-15 tilt-rotor research aircraft from Bell Helicopter Textron that made its maiden flight earlier in 1977 at the Bell facility in Arlington, Tex. Bell had awarded the company a $678 000 contract to design and fabricate the automatic flight-control system for the XV-15, and a $62 000 contract to support ground and flight checkout of the system. NASA's ARC had awarded Calspan a $117 000 contract to study modifications of future XV-15s to adapt the craft for research into its handling qualities. (*Calspan News*, Nov 77, 3)

December

December 1: The Communications Satellite Corporation (ComSatCorp) reported it would proceed with its plan to buy back its common stock [see Nov. 11], noting that the FCC had authorized it to borrow $25 million to finance part of the cost of the repurchase. A letter to stockholders said the corporation would contest FCC's attempt to make it put more money in an escrow fund for possible repayments to customers, because it said the FCC's action was "neither lawful nor appropriate." (ComSatCorp Releases 77-31, 77-32)

December 2: The *Washington Post* reported that NASA had asked Congress to let it use $100 million of Space Shuttle production funds to pay for higher costs and delays in developing the first two Shuttles. Dr. Robert A. Frosch, NASA Administrator, told the appropriations subcommittee headed by Sen. William Proxmire (D-Wis.) that $65 million would bring production up to schedule and the other $35 million would go into contingency reserves that had run between $100 and $150 million; they were "down near zero right now," Frosch said. Associate Administrator John F. Yardley said the rocket to put the Shuttle into orbit was 5mo behind schedule; also running behind were computer controls and fuselage parts. Asked why NASA requested a fund shift that might cause problems in the future, instead of a supplemental appropriation, Frosch noted that twice when NASA had asked Congress for supplemental budget increases, it had had to wait until the next fiscal year for approval. (*W Post*, Dec 2/77, A-2)

- Dryden Flight Research Center reported that Ralph "Buzz" Sawyer of its flight systems laboratory had made a "paper airplane" to test a theory of reducing elevator flutter and had come up with a new type of aircraft its builder named Skyjacker. Sawyer, working on radio-controlled models, decided to change their airfoils into booms but wanted to test the idea without risking the $600 radio systems. He made a foot-long cardboard model and "heaved [it] off the roof" of DFRC's heat facility; the "paper plane" had demonstrated amazing stability. Sawyer obtained a patent on the design and built a demonstrator 18ft long with an 18ft wingspan that he flew for the first time in Jan. 1975. With a 200hp fuel-injected engine and a 3-blade constant-speed propeller, the Skyjacker had "very good low-speed" handling, was easy to fly and land, and (according to the designer) would not stall or spin. The craft had "lift panels" 6ft wide instead of actual wings, but derived lift from its configuration just as DFRC lifting bodies did; the design had no compound curves, and its flat surfaces would make it inexpensive to produce. Possible applications would include crop dusting and firefighting. (*NASA X-Press*, Dec 2/77, 4)

• DFRC reported it had made further study of the effect of insects sticking to the leading edges of aircraft wings [see *A&A 76*, Nov. 23] by flying a small jet transport from four airports (Los Angeles, San Francisco, Sacramento, and San Diego) to study actual environments from which long-range commercial flights would be made. The program, conducted jointly with Langley Research Center, was part of NASA's aircraft energy-efficiency program to achieve 20 to 40% fuel savings in a 1985 long-range transport aircraft. Such a craft would probably use laminar flow-control technology, depending on smooth airflow over the leading edge of the wings; insects adhering to the wings had made the airflow turbulent, canceling the fuel savings achieved with the laminar flow.

NASA had equipped the wings of the test aircraft with a modified leading edge consisting of 5 panels of different materials, to demonstrate changes in the airflow caused by insects adhering to the different surfaces; it had also added a water cleaning system to study its ability to either wash off or keep off the insects encountered during takeoff and climb. It had then flown the plane over alfalfa fields and sewage ponds, and at Kennedy Space Center and Johnson Space Center, to encounter as many different quantities, kinds, and sizes of bugs as possible. Bob Baron, DFRC project manager, said results indicated a need for some sort of washing system to reduce effects of insect impacts on the laminar-flow surface. (DFRC Release 40-77)

December 4: The Dallas *Morning News* applauded "NASA official Dave Williamson" [assistant administrator in the Office of Special Projects] for rejecting a White House proposal to resume UFO investigations: he had termed it "not wise to do research on something that is not a measurable phenomenon." The paper said there was no reason to reopen the matter, except that President Carter had mentioned it in his campaign and that interest in flying saucers was up again. Such "detached reserve" should apply to any multimillion-dollar government proposal, the paper said. (*Mn News*, Dec 4/77, 15)

December 5: NASA announced that the Jet Propulsion Laboratory had negotiated contracts with 5 firms for enough silicon (photovoltaic) solar cells to supply a combined total of 190kw of power for use by the federal government. JPL would manage the contract for the Dept. of Energy (DOE), sponsor of the program to reduce the price of the cells below 50¢ per peak watt by the end of 1986, while raising production capability to 500,000kw annually. The DOE program, besides reducing costs, would study the use of solar-cell materials potentially less expensive than silicon crystals and of alternatives such as optical concentrators.

Firms receiving contracts were ARCO, 20kw, $321 950; Motorola, 50kw, $676 614; Sensor Technology, 40kw, $643 907; Solarex, 30kw, $559 454; and Solar Power, 50kw, $757 665. Panels of solar cells previously purchased under the program were powering an irrigation system in Nebr., a test system for residential use at MIT's Lincoln Laboratory in Mass., a dust-storm warning system in Ariz. [see May 23], automatic weather-reporting systems in 6 states, and other remote uses. (NASA Release 77-246)

• NASA reported it had certified as fully operational the general aviation airfoil design and analysis center at Ohio State Univ. airport near Columbus. The center, established under a 3-yr LaRC contract with the university's aeronautical and astronautical research laboratory, would provide directly to aircraft designers and manufacturers, on a fee basis, services such as analysis and design of two-dimensional airfoil shapes, high-lift devices and aerodynamic controls, and technical assistance and consultation on airfoil wind tunnel and flight testing. (NASA Release 77-247)

• NASA announced retirement effective Dec. 31 of Gerald M. Truszynski, associate administrator for space tracking and data systems, after 33yr of service to NASA and its predecessor, NACA. In 1944, he had begun working at NACA's Langley aeronautical laboratory as an engineer in instrument research and development, and had transferred 3yr later to the NACA station at Edwards, Calif., an instrument project engineer on the rocket-powered X-1, first airplane to surpass the speed of sound. In 1954 he had become chief of the instrumentation division at Edwards, heading development and operation of systems for NACA's jet- and rocket-powered craft (X-1, X-2, D-558, and X-15) that pioneered supersonic and hypersonic flight.

At NASA Hq after 1960, Truszynski occupied his present position in

1968, succeeding from the post of deputy that he had held since its creation in 1961. He had been responsible for planning, development, and operation of global tracking systems, networks, and facilities for communications and for data acquisition and processing in all NASA's spaceflight programs. In 1969 he had received NASA's highest award (the distinguished service medal) twice, for support of Apollo manned flights to lunar orbit and of the first moon landing. Upon his departure, Norman Pozinski would be acting associate administrator. (NASA anno Nov 5/77; NASA Release 77-248)

• The Natl. Space Club announced that Dr. John E. Naugle, appointed NASA chief scientist Nov. 8, would speak at its Dec. meeting and present awards to Washington-area high school students in American Univ.'s 19th summer research program. Dr. Naugle had been NASA associate administrator since 1975 in charge of much of the agency's R&D work. (NSC newsletter Dec 77)

December 6: NASA announced appointment of Andrew J. Stofan as deputy associate administrator for the Office of Space Sciences, effective Jan. 8, 1978, succeeding Dr. Anthony J. Calio who had become associate administrator for the Office of Space and Terrestrial Applications. Stofan had been director of launch vehicles at Lewis Research Center since July 1974; he had begun there in 1958 as a research engineer, assigned in 1962 to the Centaur project and becoming project manager of the Titan-Centaur vehicle in 1970. (NASA Release 77-245)

December 7: NASA announced it had negotiated 12-mo contracts worth about $4 million with 9 firms, 3 of them small businesses, for developing low-cost automated solar-cell fabrication processes that could be adapted to automated assembly and mass production of the cells. The firms were Lockheed Missiles and Space Co., $213 163; M.B. Associates, $230 000; Motorola, Inc., $433 870; RCA, $846 900; Sensor Technology, $395 000; Solarex Corp., $392 000; Spectrolab, $706 000; Texas Instruments, $537 460; and Westinghouse, $426 000. The small businesses were M.B. Associates, Sensor Technology, and Solarex. The contracts were part of DOE's low-cost silicon solar-array project managed by JPL. (NASA Release 77-249)

December 8: FAA predicted the number of passengers on U.S. scheduled airlines over the next 12yr would increase from the current 232.1 million to 418.4 million in 1989. Commuter airlines (carrying fewer than 30 passengers in at least 5 scheduled flights per wk) would double operations, triple fare miles, and fly 14.5 million passengers compared to the current 6.5 million. Size of the general (non-airline) fleet would increase 65%. (FAA Release 121-77)

December 9: KSC announced it had selected 2 companies to negotiate a contract for a 10-yr administrative telephone system at the center, including switching equipment, telephones, cabinets, switchboards and consoles, interior wire and cable, and other equipment needed for proper operation. The contract would contain an option to purchase. Companies remaining in competition were GTE Automatic Electric and Northern Telecom, Inc.; other firms submitting proposals were Southern Bell Telephone and Telegraph, and Independent Business Telephone Co. (KSC Release 207-77)

• LeRC announced it had awarded a $2 895 750 contract to Teledyne Industries, Inc., of Northridge, Calif., for management, engineering, and repair services to Centaur digital computers and remote multiplexers. The cost-plus-fixed-fee contract would run for 1yr beginning Nov. 1. The digital computer would work with the Centaur inertial guidance system to compute and adjust flight without ground command, permitting the Centaur to deliver its payload to a preselected orbit; the remote multiplexer would work with the digital computer to supply inflight data during launch. Centaur, first U.S. liquid hydrogen/liquid oxygen high-energy rocket, had been a second stage on both Atlas and Titan III. Recently it had assisted the launch of the Voyagers, and was scheduled during the coming yr to launch several commercial comsats and the Pioneer mission to Venus. LeRC would manage the Atlas-Centaur and Titan-Centaur through all phases of manufacturing, testing, and launch. Teledyne would perform its job as contractor both at Northridge and at KSC in Fla. (*Lewis News*, Dec 9/77, 2)

• JPL reported it had assisted DFRC with an engine problem on a possible Mars reconnaissance plane, a small remotely powered vehicle (RPV) called the Mini Sniffer developed by DFRC's Dale Reed for environmental research at 100 000ft above earth's surface. The 13ft aircraft with a 22ft wingspan, powered by an airless 30hp engine running on hydrazine, was the first propeller-driven plane designed to operate above 50 000ft. JPL had taken a hand when asked by DFRC to help with an engine-catalyst problem; Dr. Jose Chirivella, who solved the problem, saw the possibilities of using the miniplane on Mars.

Deployment of 16 mini RPVs, 8 from each of 2 carrier spacecraft, over Mars so that each could make 5 or more separate takeoffs and landings in different locations, could achieve a total of 80 possible missions. The carriers, orbiting about Mars, would serve as comsats to collect data from the mini RPVs for relay to earth.

The Martian RPVs would be instrumented to collect seismographic, meteorological, and geochemical data; they would conduct aerial magnetic and gravity surveys, take high-resolution photos, look for subsurface water, and study the speed and direction of winds and the pressure and density of the atmosphere. They might also collect surface samples from widely separated areas, storing them at a central location for return to earth by later missions. The small craft could be folded into 3m-long canisters protected by coverings that would fall away when parachutes unfolded to support the vehicles as they opened out to flight configuration for descent and cruise through the lower Martian atmosphere, landing like helicopters under power from variable-thrust retrorockets. (*JPL Universe*, Dec 9/77, 1)

• LaRC reported on a recent visit to Buenos Aires, Argentina, by its Boeing 737 research aircraft outfitted with a new microwave landing system being demonstrated for FAA by a NASA team. The trip was scheduled during an Inter-American Telecommunications Conference sponsored by the Organization of American States that had invited the FAA to demonstrate the landing system.

Part of the LaRC team flew to Buenos Aires on a C-5A packed with support equipment for the 737, including their own power supply; the rest of the team flew on the 737 with stops in Puerto Rico, Belem, and Rio de Janeiro. The 44-member team operated from 2 airports during its visit: the demonstration flights went from Ezeiza airport 21mi outside the city to the Aeroparque George Newbery in mid-town near the Rio de la Plata. The LaRC plane made 68 landings with the system, 56 of them automatic, and demonstrated it to more than 100 persons including officials from Argentina and other countries, and the press. Jack Reeder, chief of the terminal configured vehicle (TCV) program, said the team made the first automatic landing at the Buenos Aires airport and the first automatic landing by a commercial plane. (*Langley Researcher*, Dec 9/77, 3)

• JSC reported on the medical evaluations of astronaut applicants for which the center's Flight Medicine Clinic had been responsible. The evaluation had four parts: the medical history (illness, injuries, surgery, etc.); a thorough physical exam; specialty evaluations (neurology, otorhinolaryngology, ophthalmology, etc.); and special tests such as the treadmill, pulmonary function test, and audiometry and body chemistry examinations. (*JSC Roundup*, Dec 9/77, 1)

December 10–31: The USSR launched *Soyuz 26* from Baykonur at

4:19am Moscow time Dec. 10 in what Tass described as a mission "jointly with the *Salyut 6* space station." The 300ton "ferry vehicle" carried 2 cosmonauts, flight commander Lt.Col. Yuri Romanenko and flight engineer Georgy Grechko, into an orbit with 329km apogee, 267km perigee, 90.2min period and 51.6° inclination. Tass reported that the crew would rest for 12 to 21hr while the spacecraft was beyond radio contact with Soviet territory, during which time two research vessels would receive and relay telemetry from the satellite.

The *Soyuz 25* mission launched in Oct. carrying Vladimir Kovalenok and Valery Ryumin had failed to link with *Salyut 6*, although it had come within 393ft of the orbiting station [see Oct. 9-12]. The crew returned safely 48hr after liftoff. The "deviation" causing the cancellation was never explained.

Tass reported Dec. 11 that *Soyuz 26* had docked successfully with *Salyut 6* at 6:02am Moscow time. The news agency noted that *Salyut 6* had been equipped with 2 docking ports, 1 on the transfer compartment of the station and the other opposite it in the equipment bay, to allow 2 spacecraft to service the manned station. The *Soyuz 25* had made its unsuccessful approach from the transfer-compartment side, but *Soyuz 26* had docked with the unit on the other side. "The presence of two docking units on orbital stations considerably expands the opportunities and possibilities of future space flights," Tass said. "In particular, this allows two ships to dock with a station, which is important for replacing crews, for carrying out rescue operations, and delivering foodstuffs and equipment to an orbital scientific laboratory." Tass said the 2 crewmen had crossed into the space station.

Lt. Gen. Vladimir Shatalov, training chief for the program, said on TV that during the docking he had watched it "with greater excitement than my own, which I performed in 1969." Press reports called it "an important recovery" for the Soviet space program, after the failure of the previous docking attempt which was to have marked the 60th anniversary of the Russian revolution. Grechko's outside assignment would be to inspect the docking device on the "main entrance," and some press reports speculated that the cosmonauts might be preparing for the docking of a second spacecraft.

On the following days the cosmonauts "changed into light working clothes" and began to "mothball the onboard systems of the transport" *Soyuz 26*, according to Tass. They had also begun reactivating the space station, where they would live on a time pattern "as close as possible to that of Moscow." Tass emphasized the added comfort of being able to communicate through the research vessels instead of staying up to make contact while over their home territory, besides being more convenient for "the hundreds of specialists who support the flight here on earth." The cosmonauts were the first to call "ready" for the first communications session with the control center Dec. 12. The next day was a rest period to prepare for a "complex experiment" scheduled in the next few days.

Dr. Konstantin Feoktistov in an interview described "great changes" in both the external structure and the flight control and life support systems of the space station, aimed at increased reliability, length of service, and better conditions for the cosmonauts. Scientific equipment for astrophysics and earth monitoring had been improved on the basis of previous experience. Feoktistov emphasized the addition of another docking port as a safety device, citing the possibility of accident to the ferry craft (failure of the engine or some automatic device, or encounter with a meteorite) and the risk necessary in undocking it for inspection. Having two ports would permit sending up another craft for rescue and return to earth, as well as access for additional equipment or supplies. He also noted that the water-regeneration unit labeled experimental on *Salyut 4* was now standard equipment, and that the new design includes an "experimental place for washing," no easy matter under conditions of weightlessness: "If water is let into the chamber it will immediately spread all around and get into the respiratory organs."

Another improvement was incorporation of a "Delta" autonomous system for navigation that would permit the station to determine its orbit without participation of earth-based radio equipment. Other automatic instruments to monitor the area around the station had detected "micrometeor particles" in quantities confirmed by earth radar observation.

On Dec. 20 the cosmonauts donned semirigid full-pressure spacesuits and left the space station to check the transfer compartment and its docking unit and to carry out any necessary repairs. In addition to the special tools for the docking unit, Grechko carried a mobile TV camera that he used to send views of the docking elements to earth. No damage was visible, and Grechko reported the equipment in full working order. Tass said the men spent most of the 88min in space depressurizing and repressurizing the airlock leading to outer space, again suggesting the possibility of further visits.

Grechko's spacewalk was the first by a Soviet spaceman in 9yr. The new spacesuits were said to be more flexible than previous models, with a small instrument panel in front and a door-like entrance hatch in back. In a Dec. 20 recorded interview, Gen. Shatalov described the lengthy training in a special simulator apparently resembling the neutral-buoyancy tank at JSC, where the cosmonauts had practiced their extravehicular activity, and the orbital-station mockup on "flying laboratories" where they had learned to put on the new spacesuits in conditions of weightlessness.

Ensuing days saw performance of experiments in medical research, earth resources monitoring (during which they reported forest fires on the African continent), and distortions caused by the optical properties of porthole surfaces. On Dec. 24 Tass reported that tadpoles had hatched out of frog eggs brought from earth in the same vessel with tadpoles born on earth; the earth-born tadpoles had reacted to weightlessness by "swimming disorderly," not distinguishing top from bottom, but the newborns were swimming in spirals.

For the remainder of Dec., the crew proceeded with a schedule that each research center had helped to establish, with each center "looking forward to its hour." The cosmonauts reported a burning meteorite flashing past the station Dec. 27; one that struck a window had left a tiny scratch. On Dec. 28 Tass reported that so far the crew had not used any of the onboard recreation facilities (chess, a small library, a videotape recorder). The time allocated for leisure was occupied by Romanenko in checking the control systems, and by Grechko with camera and sketchbook near the transfer-compartment portholes. Tass reported a New Year tree was in place, "packed for the journey by friends from the Star Town together with toys."

At year's end the cosmonauts reported that onboard systems were working normally, they were feeling well, and the planned program of research and experiments was being "completely fulfilled." A motor on *Soyuz 26* had been used Dec. 29 to correct the *Salyut 6* flight trajectory. (*W Post*, Dec 11/77, A-36; Dec 20/77, A-7; Dec 21/77, A-28; *W Star*, Dec 10/77, A-4; Dec 12/77, A-4; FBIS, Tass in English, Dec 10, 12, 13, 20, 21, 23, 24, 28/77; Tass Intl Sv in Russ, Dec 11/77; Moscow in Engl to Afr, Dec 28/77; Mosc Dom Sv in Rus, Dec 13, 15, 19, 20, 30)

December 13: ESA announced that the Societe Europeenne de Propulsion (SEP) had successfully conducted the first test of the complete first stage of its Ariane launcher at the Vernon test center north of Paris. The stage was in its flight configuration for this test, and consisted of the propulsion bay (containing 4 Viking II engines), the flight-standard tanks, the forward skirt, and the intertank skirt. The stage functioned during the test for 111sec. The purpose of the test was to verify first-stage behavior, especially with regard to the hot-gas pressurization system; thermal environment of the engine bay, tanks, and skirts; and the dynamic aspect.

This test was a follow-up to previous tests of the propulsion bay that had accumulated 404.5sec of successful operation, making it possible to include flight-configuration tanks in this series of first-stage tests. ESA had planned 2 more development tests before May 1978, followed by 3 qualification tests of the first stage before final qualification. (ESA anno Dec 13/77)

December 14: In its final launch of 1977, NASA sent Japan's domestic comsat *Cs* (*Sakura*) into a synchronous transfer orbit from the Eastern Test Range on a Delta at 7:47pm EST. Transfer-orbit elements were: 35 936km apogee, 166km perigee, 28.82° inclination. The apogee boost motor fired at 10:27pm EST Dec. 15 to begin maneuvering the *Cs* over the South Pacific to a station at about 135°E above New Guinea on the equator south of Japan.

The 677kg hatbox-shaped spacecraft (including apogee motor) was 3.48m long and 2.18m in diameter, covered with solar cells, and would weigh 340kg in orbit. It would be the first comsat to carry both C-band (6 to 4GHz) and K-band (30 to 20GHz) frequency ranges: K-band coverage would be for the main islands, C-band for the remote islands. An antenna horn reflector extending 128.8cm above the solar array would provide 33db K-band coverage to the main islands and 25db to the remote islands. Ford Aerospace had built the spacecraft under contract from Mitsubishi Electric Co. of Nagoya, which had prime contract responsibility under Japan's Natl. Space Development Agency (NASDA). *Cs*, designed for both telephone and television communications technology experiments, relaying signals between fixed and mobile stations, would provide coverage to the Japanese islands for at least 3yr.

NASDA would reimburse NASA for providing launch support for this mission, at an estimated cost of $16 million. The NASA portion of the mission was adjudged successful on Feb. 23-24, 1978. (NASA Release 77-244; mission summary M-492-211-77-01 [prelaunch] Dec 14/77, MOR M-492-211-77-01 [postlaunch] Feb 28/78)

• NASA announced that *Pioneer 11* would fly just outside the rings of Saturn during its first encounter with the giant planet in 1979. The spacecraft would pass 30 000km (18 000mi) from the edge of the outermost ring and would swing in under the plane of the rings to a distance of 25 000km (15 000mi) from the planet's surface. Dr. Noel W. Hinners, NASA associate administrator for space science, and A. Thomas Young, director of planetary programs, had made the decision to fly the Pioneer outside rather than inside the rings mainly because NASA wanted to use it as a pathfinder for the two Voyager spacecraft headed for Saturn encounter in 1980-81. The Pioneer would cross the ring plane at about the same distance as the trajectory that would use Saturn's gravity to propel *Voyager 2* toward Uranus.

The uncertainties regarding a ring crossing even at 30 000km (18 000mi) from the outer edge of the rings called for NASA "to do everything we reasonably can to ensure Voyager's success," said Young. Should the Pioneer not survive the rings, NASA would have to revise its plan of sending *Voyager 2* on to Uranus; however, a successful pass by *Pioneer 11* would increase confidence in committing *Voyager 2* to the

Uranus option even if *Voyager 1* had not achieved all its objectives at Saturn. An outside pass offered much better odds for success than an inner one, which would bring the spacecraft as close as 6000km (3700mi) to Saturn's surface. The 1979 flyby by *Pioneer 11* would be the first close encounter with Saturn; until the rings of Uranus were discovered, Saturn was considered the only ringed planet in earth's solar system. The origin of the rings remained unknown, but their composition was defined by astronomers in 1970 as ordinary water ice, and radar readings in 1973 detected snowball-size chunks of solid material. (NASA Release 77-250)

- LaRC announced that Dr. J.H. Wilkinson of the U.K.'s National Physical Laboratory, author of textbooks on algebraic processes, would speak at the Dec. 19 colloquium. Wilkinson, who broke ground in floating-point and backward error analysis and in eigenvalue solutions, would describe the early days of electronic computer development, his experiences using one of the first working computers, and the beginnings of numerical analysis on an automatic digital computer. (LaRC Release 77-58)

LaRC announced that Ed Yost, who made the world's longest manned balloon flight, would speak at the center Dec. 20 on his attempt to cross the Atlantic that failed just short of the coast of Portugal but broke 8 world records, "most successful failure in ballooning history." (LaRC Release 77-59)

- Marshall Space Flight Center reported it had coped with the threat of a serious power shortage when maintenance and emergency shutdowns dropped the Tennessee Valley Authority's generating capacity by a third at the same time that severe cold weather moved into the area. On Friday, Dec. 9, TVA had asked NASA and the U.S. Army to reduce power use by half, as it expected the shortage to continue for some time. C. Horton Webb, director of MSFC's facilities office, said the plant maintenance division had worked around the clock turning heat on and off to keep buildings from freezing, and had asked employees to help in minimizing use of electrical equipment. Security guards helped by closing blinds and turning off lights and equipment wherever possible. Webb said the combined effort had reduced power use by 50% on Saturday and 65% by Sunday morning. (*Marshall Star*, Dec 14/77, 1)

December 15: NASA announced that *Voyager 1*, launched Sept. 5, had overtaken its sister ship *Voyager 2*, launched Aug. 20, about 77.5 million miles from earth on its way to encounters with Jupiter and Saturn. *Voyager 1* would arrive at Jupiter in 1979 in the lead by 4mo, and at Saturn in 1980 ahead by 9mo. If all went well by the time *Voyager 2* reached Saturn in Aug. 1981, NASA might send it on toward Uranus where it should arrive by Jan. 1986. Both spacecraft would eventually escape earth's solar system. (NASA Release 77-252)

• JSC reported it had modified its mission control center to use solar power for dehumidifying the building. Solar panels installed on the roof would provide heat 60% of the time, according to weather calculations for the Houston area, and would save about 1.5 million ft^3 of natural gas per yr at a construction cost of about $240 000. The modification was part of a joint NASA/DOE solar-energy demonstration project. (JSC Release 77-86)

December 19: MSFC reported on the first meeting of a science working group for a proposed earth-orbiting system called Advanced X-Ray Astrophysics Facility (AXAF), to study stellar structure and evolution, large-scale galactic phenomena, active galaxies and galactic clusters, and cosmology. Chaired by Prof. R. Giacconi of Harvard Univ. and representing universities and government laboratories working on x-ray astronomy in the U.S. and Europe, the group would define the scientific requirements. The 12.8m (42ft) craft weighing about 9000kg (20 000lb) would be designed for Space Shuttle launch and repair or retrieval in orbit. (MSFC Release 77-227)

• NASA reported that its announcement on selection of Shuttle astronauts, set for December, would not be forthcoming until January. Dr. Robert A. Frosch, agency administrator, had met Dec. 12 with officials of the selection program but, because of current NASA budget activity, had not reviewed the data presented. On Dec. 16 Dr. Frosch had notified JSC director Dr. Christopher C. Kraft, who had supervised the selection program, that the review would proceed after the end of the year. (NASA Release 77-254)

• ComSatCorp reported its shareholders had tendered 5 827 678 shares in response to its recent offer of repurchase at $37 per share [see Nov. 11]. Those holding 10 or fewer shares would be reimbursed first; the remaining owners would be included in the repurchase on a prorated basis at about 34% for a total buy-back of 2 million shares. (ComSatCorp Release 77-36)

December 20: NASA reported KSC was preparing its facilities for next yr's Space Shuttle arrival and prelaunch work, more than 3yr after it last launched a Saturn rocket from Complex 39. In Oct. 1978, Shuttle orbiter 102 would fly in from Calif. on its Boeing 747 carrier, with the main engines, external tank, and solid-fuel rocket boosters due in Nov. and Dec. for assembly and test before launch March 1979. KSC's orbiter landing strip had been completed in Aug.; support facilities to be completed by April 1978 would include a device to unload the orbiter, plus a microwave landing system to guide it to an automatic landing. An orbiter-processing facility to store fuels and handle payloads was the only new construction except the landing strip, the rest being modifications of Apollo facilities. (NASA Release 77-253)

December 21: NASA reported it had scheduled 25 launches in 1978, 11 on Deltas and 8 on Atlas Centaurs. It would provide support for 3 Atlas-F launches from Vandenberg AFB, Calif. KSC would launch 10 of the Deltas and all the Atlas Centaurs; one Delta would go from the Western Test Range at VAB. Paying customers would subsidize 15 launches: ESA, ComSatCorp, the U.S. Navy, Japan, NOAA, the U.K., and Canada.

Whereas most of its 1977 launches were in the area of applications (communications, environmental, navigation, meteorological, earth resources, or geodetic), its 1978 mission would divide between applications and scientific missions almost equally. Scheduled in Jan. would be Intelsat IV-A F-3 for ComSatCorp, FltSatCom-A for the Navy, and its own IUE; the Intelsat IV-A F-6 would follow in Feb. March would see launch of NASA's Landsat-C and Japan's experimental broadcast satellite BSE. In April NASA would launch its own heat-capacity mapping mission (HCMM) and Comstar-C, as well as the backup OTS for ESA. May would see 4 launches: GOES-C, and NASA's TIROS-N, Pioneer Venus-A, and SeaSat-A. Scheduled in June would be GEOS-B for ESA and a backup satellite for Japan. In July NASA would launch from Wallops on a Scout the UK-6 to measure radiation particles, and from KSC its own ISEE-C. August would see launch of NASA's Nimbus-G and Pioneer Venus-B. In Sept. NASA would launch NOAA-A; a navsat for the Navy; and comsat NATO-IIIC. NASA would launch its own HEAO-B in Oct. Two launches set for Nov. would be Canada's Telesat-D and FltSatCom-B for the Navy. George F. Page, KSC's director for expendable vehicles, noted that the center's workload would be heavy with a schedule calling for 3 launches in Jan. and 2 each in some other months. (NASA Release 77-256; KSC Release 211-77)

• MSFC reported that Mack Vinson of its personnel office had been one of only 25 people, and the only one from Alabama, selected for a 2-wk study tour in March 1978 of the Peoples Republic of China. The group, a delegation from the American Society for Public Administration, would view firsthand the ways in which PRC officials solved public policy and administrative issues. Other objectives would be to set up a 2-way interchange of administrative ideas and techniques, obtain information for use in university courses, and broaden understanding of China and its people. Vinson said the tour was a matter of personal interest and would be made at his own expense. He had been a member of ASPA since 1965 and was president of its northern Alabama section in 1975-76. (*Marshall Star*, Dec 21/77, 2)

December 22: NASA and ESA announced that 6 U.S. and 4 European candidates were finalists in the competition for 2 payload specialist positions, 1 U.S. and 1 European, on the first flight of ESA's Spacelab scheduled to go on the Space Shuttle in 1980. For the U.S., Byron Lichtenberg of Natick, Mass., Ph.D. candidate at MIT, and MSFC's Ann Whitaker, M.S. at the Univ. of Ala., were the only non-Calif. finalists. The others were Craig L. Fischer, M.D., of Indian Wells; Michael L. Lampton, Ph.D., of Berkeley; and Robert T. Menzies and Richard J. Terrile, both Ph.D.s from CalTech (Terrile employed there and Menzies at JPL) and both from Pasadena. ESA's 4 were electronics engineer Franco Malerba of Italy; physicist Ulf Merbold of West Germany; and 2 from the Netherlands, astronomer Claude Nicollier and physicist Wubbo Ockels. (NASA Release 77-255; MSFC Release 77-229; ESA Release Dec 22/77)

• FBIS reported a Tass claim that Soviet physicists at the USSR's joint institute for nuclear research at Dubna near Moscow had synthesized the 106th element with a mass number of 259, by bombarding a lead target with ions of chromium (accelerated heavy particles). Academician Georgy Flerov said his group had been working for yrs to synthesize trans-uranium elements (those with atomic nuclei heavier than uranium) and had already obtained the 102nd through 105th elements on the periodic table. The U.S. and the USSR (both countries having "sufficiently powerful heavy-ion accelerators") had begun work on the 106th element about the same time. After the group at Dubna reported success, the U.S. scientists said they had found another isotope of the 106th element, with mass number 263. (FBIS, Tass in English, Dec 22/77)

December 23: MSFC reported that its homemade Shuttle orbiter [see Nov. 23] had made its practice run through the center and had played stand-in during practice hoists into the 430ft (131m) test stand where the Enterprise would be mated to the other Shuttle components for the first time. Werner H. Rubel, supervisor in MSFC's product planning branch, said the idea and design for the stand-in orbiter had originated in his group, where Jerry B. Bennich and John L. Ransburgh had done the engineering work; the mockup and prototype assembly branch had performed final assembly. (MSFC Release 77-232)

MSFC announced successful completion of the first loading test of the

Space Shuttle external tank to ensure that the test article, test facility, and hardware could handle the super cooled propellants (liquid hydrogen and liquid oxygen) to be used by the Shuttle main engine. (MSFC Release 77-234)

• KSC reported successful conclusion of the first open-sea tests on Shuttle retrieval equipment staged Dec. 5-16 from Port Everglades, in strong winds and seas from 3 to 6ft that had hampered operations both wks. Test manager Bob Everette said the hugh waves created conditions vital to the test although "it was hard even to stand up on the boat" and the aft section of the 158-ft offshore-support vessel was often under a ft of water.

The first group of tests simulated parachute recovery; retrieval of the 6 main parachutes and 2 drogues would use 2 support vessels, each responsible for recovering 3 main chutes and a drogue. The second series of tests was to verify performance of the nozzle plug in open waters. The plug was designed to "swim out" from the ship to a booster casing floating vertically, dive to the bottom of the casing, secure itself to the aft portion and pump out the water. The booster, rotating to a horizontal position, would be ready for tow to the Spaceport for refurbishing. The tests were also to confirm the deck layout of the support vessels and how they kept position while the plug was docking with the booster. Everette said retrieval gear would be modified according to the test results.

More ocean tests would follow about 1mo before the first launch of the Shuttle, pre-mission exercises to train personnel who would work on the recovery. Everette praised the teamwork of the support agencies taking part in the test: staff from KSC and MSFC; Tracor Marine, operator of the support vessel; Everglades Towing Co.; the Naval Ocean Systems Center; U.S. Booster, Inc.; Battelle Memorial Institute; Pioneer Parachute Co.; the U.S. Navy supervisor of salvage; and Martin Marietta. (*Spaceport News*, Dec 23/77, 4)

• In its year-end issue, the *Lewis News* described the relationship between U.S. Army and NASA activities at the center. Civilian employees of the Army Propulsion Laboratory had been working among LeRC employees since 1970, as 1 of 3 Army air-mobility research and technology labs working jointly with NASA centers: Ames Research Center housed the aeromechanics laboratory and the research and technology lab headquarters, and LaRC included the Army's structures laboratory. The only AMRT lab complex not on a NASA installation was the applied technology lab at Ft. Eustis, Va.

LeRC and the Army pursued R&D activities under a unique agreement, with the Army providing funds for programs of mutual interest and adding about 40 persons (mostly engineers, scientists, and technicians) to the LeRC staff, all of them supervised by Lewis personnel. Army interests included the joint aeronautical research group, working on compressors, turbines, bearings, seals, and gears, with special attention to Army needs in small-engine technology; the technical support group of scientists and skilled technicians, set up to help with the inhouse workload created by Army activities at the center; and a small staff of engineers called the Army aeronautical research group, using LeRC facilities to pursue investigations of interest to the Army specifically. An instance was the engine test cell in Bldg. 14 recently activated and primarily operated by the Army, with help from LeRC experts and resources, said Propulsion Laboratory director John Acurio.

Aiding Acurio in lab operations and efforts to improve aircraft propulsion were Curtis L. Walker, chief of the Army aeronautical research group, and administrative officer Gus Gold. (*Lewis News*, Dec 23/77, 4)

• JPL reported that Dr. M. Kudret Selcuk of its energy-conversion systems section had developed a fixed solar collector using a vee-trough concentrator and vacuum-tube receivers to produce temperatures above 350°F for generating steam or organic vapor, or electricity to drive turbines. The system, on which NASA had obtained a patent, had proved "far superior to flat-plate systems," Dr. Selcuk said.

The new system had no moving parts, which made it unique. Located on the roof of JPL's Bldg. 248, its torpedo-like glass tubes and piping would remain on a tilted rack, and the sheet-metal troughs in an asymmetrical V shape could be reversed at the spring and autumn equinoxes to take best advantage of the sun's rays. Heat collected could be used directly for shaft power to drive conventional air conditioning systems, for space heating, or for other uses such as supplying process heat. (*JPL Universe*, Dec 23/77, 2)

• JSC reported that NASA administrator Dr. Robert A. Frosch had approved development of a teleoperator retrieval system to be flown no later than Feb. 1980 as a Skylab-revisit mission. JSC would be responsible for integrating the TRS into the Shuttle and for planning and operations. Harold E. Gartrell, assistant manager of the Shuttle payload integration and development office at JSC, would manage JSC activities for the Skylab revisit and would work with the Shuttle program office on integrating that mission with the Shuttle orbital-flight test planners. His office would be responsible for interfaces outside JSC in preparing and conducting the mission. (*JSC Roundup*, Dec 23/77, 1)

• MSFC reported that Corning Glass Works, Canton, N.Y., had begun work on a huge primary mirror blank for NASA's Space Telescope, designed to give astronomers a view to the edge of the universe. Corning would work under a $1.74 million subcontract from Perkin-Elmer Corp., which would be responsible for assembling the optical telescope. Lockheed Missiles and Space Co. would build the spacecraft and support system and integrate the telescope into it.

Scientists expected the space telescope to detect light from objects 14 billion light-years from earth; it might, therefore, be able to provide views of the first galaxies at the time they were formed. The Hale telescope at the Mt. Palomar observatory in Calif. could penetrate only about 2 billion light-years because of the optical degradation from earth's atmosphere. The space telescope would be carried by the Shuttle to an orbit more than 300mi above earth to avoid that interference.

Corning would deliver the first of two 8ft discs to Perkin-Elmer within a yr and the spare by April 1979, for the process of grinding the blanks to an optical curvature. The blank would be 1ft thick with a 2ft-diameter center hole, and would consist of solid inch-thick top and bottom plates separated by a weight-saving structure of open cells 10in long and 4in square. It would carry a thin metal film constituting the mirror or reflective surface; glass had long served as best mirror material because its shape would change only slightly with changes in temperature, so that it maintained image quality. More than 40yr ago Corning had made the 200in blank for the Hale telescope out of borosilicate glass, having an expansion of 32.5 on a scale used by physicists; the material it would use for the space telescope blank would have an expansion of zero. Corning had chosen the new material, a titanium silicate called ULE (ultra-low expansion) first formulated in 1967, for ease of handling and ultrasonic testing. (MSFC Release 77-233)

• LaRC reported that its plans for a trash-burning steam plant called Project RECOUP [see Oct. 6] had hit a snag when the center opened construction bids Dec. 7. All four bids were higher than expected, the lowest being about $1.4 million more than the government's estimate, which was about $8 million. Funds available to the center for the plant, including contingencies, totaled no more than $8.4 million. Leo Daspit, LaRC project manager, said he would have to meet with his counterparts in the city of Hampton and at Langley AFB to "explore our options." (*Langley Researcher*, Dec 23/77, 3)

• LaRC announced it had selected a firm called Mercury, of Tustin, Calif., to negotiate a 3yr $2.4-million contract for support services to the center's steam and air-compression facilities. The contract, beginning Jan. 1, 1978, would have two 1yr unpriced options for extension. (LaRC Release 77-60)

- A year-end summary of NASA activities included 16 launch efforts and completion of two series of manned Shuttle orbiter tests. Three of the 16 launches had failed, 2 destroying launcher and spacecraft immediately after liftoff; the third resulted from upper-stage failure and did not put the spacecraft into proper orbit. (NASA Release 77-257; KSC Release 212-77)

December 27: NASA announced it had awarded RCA Astro-Electronics Division of Princeton, N.J., a $10.5 million fixed-price contract for a closed-circuit television system to support the Space Shuttle program. The contract, running from Jan. 1978 through June 1982, would cover design, development, production, qualification, and delivery of CCTV systems for training, certification, and flight. It would include hardware for the first orbital test vehicle design, with options for TV equipment on additional orbiters and for Canada-built remote manipulator systems. Cameras and monitors would fly in the crew compartment and the payload bay, and on the arm of the remote manipulator. JSC would provide technical direction. (NASA Release 77-259)

- FBIS reported that "the world's northernmost New Year's tree" had been lighted 500km from the geographical North Pole, carried there by members of the USSR's high-latitude North 29 air expedition with "New Year gifts and mail" for the drifting station North Pole 23. Tass said the arctic "Fathers Frost" then visited an icebound island, base of the North Pole 22 scientific station, near the Canadian Arctic archipelago in the part of the Arctic Ocean most remote from the Soviet Union. (FBIS, Tass in English, Dec 27/77)

December 29: KSC announced that the Thunderstorm Research International Program (TRIP) combining efforts of atmospheric physicists and lightning researchers from the U.S., Europe, and Africa would continue in 1978 as they had for the past 2 summers. At the recent meeting of the American Geophysical Union in San Francisco, KSC had invited the researchers to continue their studies of electrical properties, origins, and effects of thunderstorms.

The KSC Spaceport would benefit from the studies by applying the findings in Shuttle launch and landing operations; to have many investigators in the same area studying the same storms and exchanging data would offer more chances of gaining new knowledge than might come from isolated studies. The researchers had funded their own programs, KSC providing meteorological instrumentation, use of the local National Weather Service office, and normal support services such as power, communications, and file processing.

KSC, hosting the program for a third consecutive yr, had unique meteorological facilities accumulated during the Apollo and Skylab programs, plus a large number of summer thunderstorms occurring normally in the area. Although KSC would not host the group after next yr, the 1979 study site for continuations of TRIP would be the Langmuir Laboratories at Socorro, N.M., for studies of mountain storms of the southwest. (KSC Release 217-77)

• ARC reported that the Galileo memorial scholarship program established in 1973 by the San Francisco section of the American Institute of Aeronautics and Astronautics and ARC would award a $750 scholarship and other prizes. The program was open to high school seniors planning a career in physical or natural sciences, engineering, or mathematics, residents of the area or children of employees or retirees at ARC or of Galileo crewmembers. The program was a memorial to the men who died April 12, 1973, in an accident involving the Galileo I (a modified Convair operated by ARC as a flying laboratory for research in aeronautics, astronautics, astronomy, and earth observation). (ARC *Astrogram*, Dec 29/77, 4)

During December: NASA reported that Fitzhugh L. Fulton, Jr., civilian test pilot for DFRC since 1966 and a 23-yr veteran of the USAF, had received from the Society of Experimental Test Pilots the Iven C. Kincheloe award as test pilot of the year for his work in the Space Shuttle approach-and-landing test program in which he piloted the orbiter on 13 test flights beginning in Feb. and ending in mid-Nov. 1977. Fulton was also flying the "triplesonic" YF-12 aircraft for NASA; in the past, he had flown the XB-70 and was launch pilot for the X-15 and other research vehicles.

This award was the second top-level honor for Fulton, who won the Harmon Trophy in 1962 as an Air Force pilot in the B-58 Hustler test program at Edwards AFB that set an international altitude record. In his work for the Air Force Fulton had received 3 Distinguished Flying Crosses, plus another DFC and 5 air medals for 55 missions in Korea and scores of humanitarian missions during the Berlin Airlift of 1948-49. (*NASA Actv* Dec 77, 8)

• The USAF announced it had issued contracts to Martin Marietta Aerospace at Denver and General Dynamics Corp.'s Convair Divison at San Diego for design of a large space structure to be carried into orbit on the Space Shuttle and erected and deployed in space. The contracts, each for $750,000, called for conceptual design of a flight-demonstration model of an antenna-like structure to be assembled and deployed in a form exceeding the size of the orbiter cargo bay (60ft long by 15ft in diameter). Plans called for automated assembly in space with manned supervision. The flight article would also serve to verify the Shuttle's ability to deploy other DOD craft that might require assembly in orbit. (AFSC *Newsreview*, Dec 77, 4)

Appendix A

SATELLITES, SPACE PROBES, AND MANNED SPACE FLIGHTS, 1977

World space activity in 1977 approached the level of previous years: 124 recorded launches, compared with 128 in 1976 and 125 in 1975. Of the total, the U.S. had 26 launches, 24 of them successful, as described in Appendix B. The USSR had 98 launches with 111 possible payloads, including a single eightfold comsat launch like those of earlier years, plus two probes: a Vertikal rocket launched to a height of about 500km on Aug. 30, 1977, and another launched to a height of about 1500km on Oct. 25. The 79 launches (accounting for 86 payloads) in the Cosmos series, besides the usual photo-reconnaissance, radar-calibration, communications, electronic-intelligence, navigation, and test vehicles, included 4 antisatellite spacecraft and 3 targets; the ASATs (*Cosmos 910, 918, 961,* and *970*) either plunged into the Pacific Ocean or exploded in orbit. The ocean-surveillance satellite *Cosmos 954* exhibited difficulties with its control system 2mo after launch and finally crashed in Canada's Northwest Territory in January 1978 with its nuclear reactor aboard. The USSR also launched 6 spacecraft in the Molniya series; 4 in the Meteor series; 3 manned Soyuz vehicles; and one each in the Raduga, Ekran, Prognoz, Intercosmos, and Salyut series. The remaining launch was the cooperative French-Soviet spacecraft *Signe 3* equipped for gamma-ray studies.

Two other launches during 1977 were Japan's test vehicles, *Tansei 3* launched Feb. 19 from the Uchinoura Space Center, and *Kiku 2* launched Feb. 23 from the Tanegashima Space Center.

Sources of the data in the accompanying table include the United Nations Registry of Space Flights; the *Satellite Situation Report* compiled by Goddard Space Flight Center; and press releases of NASA, the Department of Defense, National Oceanic and Atmospheric Administration, and other government agencies, as well as the Communications Satellite Corporation. Soviet data derive from statements in the Soviet press, international news service reports, and announcements and briefings by Soviet officials. Data on satellites of other nations come from announcements of their governments and international news services.

SATELLITES, SPACE PROBES, AND MANNED SPACE FLIGHTS, 1977

Launch Date	Spacecraft, Country, Int'l Designation, Vehicle, Launch Site	Payload Data	Apogee (km)	Perigee (km)	Period (min)	Inclination (degrees)	Remarks
Jan. 6	Cosmos 888 USSR 1977-1A A-2 Baykonur-Tyuratam	Total weight: Unavailable. Objective: "Investigation of upper atmosphere and outer space." Description: Unavailable.	314	168	89.3	65.0	Probable military photo-reconnaissance satellite; reentered Jan. 19, 1977.
Jan. 7	Meteor 2-02 USSR 1977-2A A-1 Plesetsk	Total weight: Unavailable. Objective: Meteorology. Description: Unavailable.	904	888	103	81.3	Still in orbit.
Jan. 20	Cosmos 889 USSR 1977-3A A-2 Baykonur-Tyuratam	Total weight: Unavailable. Objective: "Investigation of upper atmosphere and outer space." Description: Unavailable.	353	210	89.8	71.4	Probable military photo-reconnaissance satellite; reentered Feb. 1, 1977.
	Cosmos 890 USSR 1977-4A C-1 Plesetsk	Total weight: Unavailable. Objective: "Investigation of upper atmosphere and outer space." Description: Unavailable.	1017	981	105.1	83.0	Probable navigation satellite; second-generation A-type. Still in orbit.
Jan. 28	Nato III-B U.S.-NATO 1977-5A Thorad-Delta ETR	Total weight: 670kg. Objective: To launch spacecraft into synchronous equatorial orbit for NATO use. Description: Drum 2.2m dia., 2.23m long (3.1m including antennas).	35 962	35 463	1432.3	2.8	Second of 3 planned comsats launched by NASA for NATO. Apogee boost motor fired Jan. 30 put satellite on station at 15°W above equator. Still in orbit.

Date	Spacecraft	Details				Remarks
Feb. 2	*Cosmos 891* USSR 1977-6A C-1 Plesetsk	Total weight: Unavailable. Objective: "Investigation of upper atmosphere and outer space." Description: Unavailable.	518	466	94.4	65.8	Probable radar-calibration satellite; reentered Feb. 4, 1981.
Feb. 6	DOD spacecraft U.S. 1977-7A Titan IIIC WTR	Total weight: 820kg. Objective: "Develop spaceflight techniques and technology." Description: Unavailable.	35 755	35 532	1436.0	0.1	Probable early-warning satellite (IMEWS-7, integrated missile/early warning satellite) stationed over Indian Ocean, carrying infrared sensors to detect Asia launches; still in orbit.
Feb. 7	*Soyuz 24* USSR 1977-8A A-2 Baykonur-Tyuratam	Total weight: Unavailable. Objective: Joint experiments with *Salyut 5* orbiting space station. Description: Unavailable.	281	218	89.2	51.6	Third attempt to place crew aboard *Salyut 5* orbiting laboratory, after 49-day flight of *Soyuz 21* and docking failure of *Soyuz 23*. *Soyuz 24* docked successfully Feb. 8; crew, Col. Victor Gorbatko and Lt. Col. Yuri Glazkov, transferred Feb. 9, undocked Feb. 25; recovered Feb. 25, 1977.
Feb. 9	*Cosmos 892* USSR 1977-9A A-2 Plesetsk	Total weight: Unavailable. Objective: "Investigation of upper atmosphere and outer space." Description: Unavailable.	454	170	90.4	72.9	Probable military photo-reconnaissance satellite; recovered Feb. 22, 1977.
Feb. 11	*Molniya 2-17* USSR 1977-10A A-2-e Plesetsk	Total weight: Unavailable. Objective: Transmission of television programs and multichannel radio to stations throughout USSR. Description: Unavailable.	40 753	468	735.4	62.8	Still in orbit.
Feb. 15	*Cosmos 893* USSR 1977-11A C-1 Plesetsk	Total weight: Unavailable. Objective: "Investigation of upper atmosphere and outer space." Description: Unavailable.	1678	331	105.2	74.0	Probable geophysics satellite, possibly a failed Intercosmos; still in orbit.
Feb. 19	*Tansei 3* Japan 1977-12A Mu-3H Uchinoura Space Center	Total weight: 134kg. Objective: Launch of test satellite. Description: Polyhedral cylinder 1m long, 1m in diameter.	3814	795	134.2	65.8	Still in orbit.

Launch Date	Spacecraft, Country, Int'l Designation, Vehicle, Launch Site	Payload Data	Apogee (km)	Perigee (km)	Period (min)	Inclination (degrees)	Remarks
Feb. 21	Cosmos 894 USSR 1977-13A C-1 Plesetsk	Total weight: Unavailable. Objective: "Investigation of upper atmosphere and outer space." Description: Unavailable.	1012	971	104.9	82.9	Probable navigation satellite; third-generation B-type. Still in orbit.
Feb. 23	Kiku 2 Japan 1977-14A Nu Tanegashima Space Center	Total weight: 130kg. (empty). Objective: Engineering test satellite (ETS-2) to be put in synchronous orbit. Description: Polyhedral cylinder 1.4m long, 1.8m in diameter.	35 785	35 784	1435.9	0.1	Japanese engineering test satellite launched by Natl. Space Development Agency, located above lat. 130°E; still in orbit.
Feb. 27	Cosmos 895 USSR 1977-15A A-1 Plesetsk	Total weight: Unavailable. Objective: "Investigation of upper atmosphere and outer space." Description: Unavailable.	634	609	97.1	81.2	Probable electronic ferret; still in orbit.
March 3	Cosmos 896 USSR 1977-16A A-2 Plesetsk	Total weight: Unavailable. Objective: "Investigation of upper atmosphere and outer space." Description: Unavailable.	209	195	89.6	72.9	Probable military photo-reconnaissance satellite; recovered March 16, 1977.
March 10	Cosmos 897 USSR 1977-17A A-2 Plesetsk	Total weight: Unavailable. Objective: "Investigation of upper atmosphere and outer space." Description: Unavailable.	328	168	89.5	72.8	Probable military photo-reconnaissance satellite; recovered May 26, 1977.
	Palapa 2 Indonesia 1977-18A Thorad-Delta ETR	Total weight: 574kg. Objective: Launch into synchronous transfer orbit of satellite to transmit TV, voice, other data throughout Indonesia. Description: Spin-stabilized cylinder 1.8m in diameter, 3.3m high, topped by earth-oriented antenna assembly.	36 250	35 915	1451.2	0.1	Hughes-built comsat to supplement Palapa 1 (1976-66A), launched by NASA into transfer orbit; apogee boost motor fired Mar. 12 put s/c into equatorial synchronous orbit at longitude 77°E; became operational Apr. 15. Still in orbit.

ASTRONAUTICS AND AERONAUTICS, 1977 — APPENDIX A

Date	Spacecraft	Details				Remarks
March 13	DOD spacecraft U.S. 1977-19A Titan IIIB-Agena D WTR	Total weight: 3000kg. Objective: "Develop spaceflight techniques and technology". Description: Unavailable.	329	141	89.1	96.4	USAF satellite; decayed from orbit May 26, 1977.
March 17	Cosmos 898 USSR 1977-20A A-2 Plesetsk	Total weight: Unavailable. Objective: "Investigation of upper atmosphere and outer space." Description: Unavailable.	225	214	88.9	81.3	Probable military photo-reconnaissance satellite; recovered March 30, 1977.
March 24	Molniya 1-36 USSR 1977-21A A-2e Plesetsk	Total weight: Unavailable. Objective: Transmission of television programs and multichannel radio to stations throughout USSR. Description: Unavailable.	40 818	451	736.4	62.8	Still in orbit.
	Cosmos 899 USSR 1977-22A C-1 Plesetsk	Total weight: Unavailable. Objective: "Investigation of upper atmosphere and outer space." Description: Unavailable.	547	502	95.1	74.0	Probable electronic ferret; decayed from orbit Oct. 19, 1980.
March 29	Cosmos 900 USSR 1977-23A C-1 Plesetsk	Total weight: Unavailable. Objective: "Investigation of upper atmosphere and outer space." Description: Unavailable.	521	455	94.4	82.9	Probable geophysics satellite for research on magnetosphere, esp. auroras; known to carry Czech. and E. German experiments in payload; decayed from orbit Oct 11, 1979.
April 5	Meteor 27 USSR 1977-24A A-1 Plesetsk	Total weight: Unavailable. Objective: Meteorology. Description: Unavailable.	895	853	102.4	81.3	Still in orbit.
	Cosmos 901 USSR 1977-25A A-1 Plesetsk	Total weight: Unavailable. Objective: "Investigation of upper atmosphere and outer space." Description: Unavailable.	819	268	95.5	71.0	Probable radar-calibration satellite; decayed from orbit June 28, 1978.
April 7	Cosmos 902 USSR 1977-26A A-2 Plesetsk	Total weight: Unavailable. Objective: "Investigation of upper atmosphere and outer space." Description: Unavailable.	262	171	88.8	81.4	Probable military photo-reconnaissance satellite; recovered April 20, 1977.

Launch Date	Spacecraft, Country, Int'l Designation, Vehicle, Launch Site	Payload Data	Apogee (km)	Perigee (km)	Period (min)	Inclination (degrees)	Remarks
April 11	Cosmos 903 USSR 1977-27A A-2e Plesetsk	Total weight: Unavailable. Objective: "Investigation of upper atmosphere and outer space." Description: Unavailable.	40 157	598	725.9	62.8	Probable early-warning satellite; still in orbit.
April 20	Cosmos 904 USSR 1977-28A A-2 Baykonur-Tyuratam	Total weight: Unavailable. Objective: "Investigation of upper atmosphere and outer space." Description: Unavailable.	326	202	89.4	71.2	Probable military photo-reconnaissance satellite; recovered May 4, 1977.
	Geos U.S.-ESA 1977-29A Delta ETR	Total weight: 573kg. (fueled). Objective: Launch into synchronous transfer orbit of s/c to provide data on electric and magnetic fields 36 000km above earth. Description: Cylinder 1.3m long, 1.6m in diameter, carrying 4 radial, 4 axial booms.	11 710 (38 270)	241 (2106)	227.8 (718.2)	26.0 (26.4)	European-built geodynamic experimental ocean satellite launched by NASA for ESA; booster malfunction put s/c into unsatisfactory transfer orbit; apogee boost motor fired April 27 put s/c into most desirable orbit; all experiments returning useful data; still in orbit.
April 26	Cosmos 905 USSR 1977-30A A-2 Plesetsk	Total weight: Unavailable. Objective: "Investigation of upper atmosphere and outer space." Description: Unavailable.	337	169	89.6	67.1	Probable military photo-reconnaissance satellite; recovered May 26, 1977.
April 27	Cosmos 906 USSR 1977-31A C-1 Kapustin Yar	Total weight: Unavailable. Objective: "Investigation of upper atmosphere and outer space." Description: Unavailable.	515	464	94.4	50.7	Probable geophysics satellite, which apparently remained attached to upper stage of launch vehicle; decayed from orbit Mar. 23, 1980.
April 28	Molniya 3-7 USSR 1977-32A A-2e Plesetsk	Total weight: Unavailable. Objective: Transmission of television programs and multichannel radio to stations throughout USSR. Description: Unavailable.	39 963	437	718.7	62.8	Still in orbit.

Date	Spacecraft	Details				Remarks
May 5	Cosmos 907 USSR 1977-33A A-2 Plesetsk	Total weight: Unavailable. Objective: "Investigation of upper atmosphere and outer space." Description: Unavailable.	301	167	89.2	62.8	Probable military photo-reconnaissance satellite; recovered May 16, 1977.
May 12	Dscs II-7 U.S. 1977-34A Titan IIIC ETR	Total weight: Unavailable. Objective: Communications. Description: Unavailable.	35 762	35 438	1426.7	2.44	Military comsat, part of Defense Satellite Communications System, USAF; still in orbit.
	Dscs II-8 U.S. 1977-34B Titan IIIC ETR						Still in orbit.
May 17	Cosmos 908 USSR 1977-35A A-2 Baykonur-Tyuratam	Total weight: Unavailable. Objective: "Investigation of upper atmosphere and outer space." Description: Unavailable.	285	174	89.1	51.8	Probable military photo-reconnaissance satellite; recovered May 31, 1977.
May 19	Cosmos 909 USSR 1977-36A C-1 Plesetsk	Total weight: Unavailable. Objective: "Investigation of upper atmosphere and outer space." Description: Unavailable.	2109	988	117.1	65.9	Probable target for satellite-interception tests (see *Cosmos 910*); still in orbit.
May 23	Cosmos 910 USSR 1977-37A F-1M Plesetsk	Total weight: Unavailable. Objective: "Investigation of upper atmosphere and outer space." Description: Unavailable.	560	149	91	65.1	Probable test interceptor of *Cosmos 909*; test unsuccessful, reentry occurred after less than one orbit May 23, 1977 (therefore not strictly a satellite, but included because of number allocated by USSR).
	DOD spacecraft U.S. 1977-38A Atlas-Agena D ETR	Total weight: 700kg. Objective: "Develop spaceflight techniques and technology." Description: Unavailable.	35 855	35 679	1435.1	0.2	Still in orbit.

Launch Date	Spacecraft, Country, Int'l Designation, Vehicle, Launch Site	Payload Data	Apogee (km)	Perigee (km)	Period (min)	Inclination (degrees)	Remarks
May 25	Cosmos 911 USSR 1977-39A C-1 Plesetsk	Total weight: Unavailable. Objective: "Investigation of upper atmosphere and outer space." Description: Unavailable.	1002	969	104.8	82.9	Probable military comsat; still in orbit.
May 26	Cosmos 912 USSR 1977-40A A-2 Plesetsk	Total weight: Unavailable. Objective: "Investigation of upper atmosphere and outer space." Description: Unavailable.	225	213	88.9	81.4	Probable military photo-reconnaissance satellite; recovered June 8, 1977.
	Intelsat IV-A F4 U.S. 1977-41A Atlas-Centaur ETR	Total weight: 1500kg. Objective: Launch into synchronous orbit of improved s/c with 2/3 greater capacity than Intelsat IV. Description: Rotating cylinder 6.99m high, 2.38m in diameter, covered with solar cells; carries earth-oriented platform with 20 communications repeaters, new antenna reflectors, telemetry & command subsystems.	35 755	35 346	1425.0	0.3	Third of a series of improved comsats launched by NASA for ComSatCorp., mgr. of Intelsat system. Apogee kick motor fired May 27 put s/c on station at 34.5°W over Atlantic Ocean; still in orbit.
May 30	Cosmos 913 USSR 1977-42A C-1 Plesetsk	Total weight: Unavailable. Objective: "Investigation of upper atmosphere and outer space." Description: Unavailable.	519	471	94.5	74.0	Probable radar-calibration satellite, possible explosion in orbit during September 1977; 21 objects identified as part of this launch; decayed from orbit between Feb. 78–Dec. 79.
May 31	Cosmos 914 USSR 1977-43A A-2 Baykonur-Tyuratam	Total weight: Unavailable. Objective: "Investigate upper atmosphere and outer space." Description: Unavailable.	305	202	89.6	65.0	Probable military photo-reconnaissance satellite; recovered June 13, 1977.
June 5	Ams 2 U.S. 1977-44A Thor-Burner 2 WTR	Total weight: 450kg. Objective: Support of Defense Meteorological Satellite Program. Description: Cylinder 6.4m long, 1.68m in diameter.	864	817	101.7	99.2	Advanced meteorological satellite, developed by USAF, similar to Tiros N planned for launch in 1978; still in orbit.

Date	Name/Designation	Description	Perigee (km)	Apogee (km)	Period (min)	Inclination (deg)	Remarks
June 8	Cosmos 915 USSR 1977-45A A-2 Plesetsk	Total weight: Unavailable. Objective: "Investigation of upper atmosphere and outer space." Description: Unavailable.	288	172	89.1	62.8	Probable military photo-reconnaissance satellite; recovered June 21, 1977.
June 10	Cosmos 916 USSR 1977-46A A-1 Plesetsk	Total weight: Unavailable. Objective: "Investigation of upper atmosphere and outer space." Description: Unavailable.	297	245	89.9	62.8	Probable military photo-reconnaissance satellite; recovered June 21, 1977.
June 16	Cosmos 917 USSR 1977-47A A-2e Plesetsk	Total weight: Unavailable. Objective: "Investigation of upper atmosphere and outer space." Description: Unavailable.	39 507	833	717.5	62.8	Probable military early-warning satellite; still in orbit.
	Goes 2 U.S. 1977-48A Thorad-Delta ETR	Total weight: 635kg. (fueled). Objective: Meteorological studies from geostationary orbit. Description: Spin-stabilized cylinder 3.4m long, 1.9m in diameter (incl. apogee boost motor); center tube supporting radiometer, telescope; scanner mirrors look through solar cells covering cylinder.	36 304	35 267	1436.0	0.9	Second in series of geostationary operational environmental satellites launched by NASA for NOAA; successful placement in transfer orbit; apogee boost motor fired June 16 put s/c on station at 75°W replacing Goes 1; turned over to NOAA July 29. Still in orbit.
June 17	Signe 3 USSR-France 1977-49A C-1 Kapustin Yar	Total weight: 102kg. Objective: Study of x-rays and gamma rays from cosmic sources. Description: Cylinder, 800mm high and 700mm in diameter, solar-pointed on vertical axis, with 4 solar panels.	519	459	94.4	50.7	French-built satellite launched under Franco-Soviet cooperative agreement; solar interplanetary gamma-neutron experiment designed to detect bursts of gamma rays and compare them with same events detected by other s/c. Decayed from orbit June 20, 1979.
	Cosmos 918 USSR 1977-50A F-1m Baykonur-Tyuratam	Total weight: Unavailable. Objective: "Investigation of upper atmosphere and outer space." Description: Unavailable.	243	128	88.1	65.1	Probable interceptor satellite; made close approach to Cosmos 909 (still in orbit), plunged into Pacific Ocean June 18, 1977.

Launch Date	Spacecraft, Country, Int'l Designation, Vehicle, Launch Site	Payload Data	Apogee (km)	Perigee (km)	Period (min)	Inclination (degrees)	Remarks
June 18	Cosmos 919 USSR 1977-51A B-1 Plesetsk	Total weight: Unavailable. Objective: "Investigation of upper atmosphere and outer space." Description: Unavailable.	819	267	95.5	71.0	Probable radar-calibration satellite; decayed from orbit Aug. 28, 1978.
June 22	Cosmos 920 USSR 1977-52A A-2 Baykonur-Tyuratam	Total weight: Unavailable. Objective: "Investigation of upper atmosphere and outer space." Description: Unavailable.	341	172	89.6	65.0	Probable military photo-reconnaissance satellite; recovered July 5, 1977.
June 23	Nts 2 U.S. 1977-53A Atlas F ETR	Total weight: 431 kg. Objective: Test the technology of the global positioning system Navstar. Description: Octagonal craft, 2 vanes, carrying cesium clocks for precise time-standard testing and solar array for Navy solar-cell experiments.	20 187	19 545	705.2	63.3	Navigation technology satellite, still in orbit.
June 24	Molniya 1-37 USSR 1977-54A A-2e Baykonur-Tyuratam	Total weight: Unavailable. Objective: Transmission of television programs and multichannel radio to stations throughout USSR. Description: Unavailable.	39 014	460	700.0	62.9	Still in orbit.
	Cosmos 921 USSR 1977-55A C-1 Plesetsk	Total weight: Unavailable. Objective: "Investigation of upper atmosphere and outer space." Description: Unavailable.	699	619	97.9	75.8	Probable launch-vehicle test, first of a new series; still in orbit.
June 27	DOD spacecraft U.S. 1977-56A Titan IIID WTR	Total weight: Unavailable. Objective: "Development of spaceflight techniques and technology." Description: Unavailable.	227	173	88.3	97.0	Probable USAF reconnaissance satellite; decayed from orbit Dec. 23, 1977.

Date	Satellite	Details					Remarks
June 29	Meteor 1-28 USSR 1977-57A A-1 Baykonur-Tyuratam	Total weight: Unavailable. Objective: Meteorology Description: Unavailable.	669	600	97.4	97.9	First Soviet vehicle to enter a retrograde orbit; possible start of a new Meteor system; still in orbit.
June 30	Cosmos 922 USSR 1977-58A A-2 Plesetsk	Total weight: Unavailable. Objective: "Investigation of upper atmosphere and outer space." Description: Unavailable.	293	204	89.5	62.8	Probable military photo-reconnaissance satellite; recovered July 13, 1977.
July 1	Cosmos 923 USSR 1977-59A A-2 Plesetsk	Total weight: Unavailable. Objective: "Investigation of upper atmosphere and outer space." Description: Unavailable.	815	797	101.0	74.1	Probable military comsat with data store-dump capability; still in orbit.
July 4	Cosmos 924 USSR 1977-60A C-1 Plesetsk	Total weight: Unavailable. Objective: "Investigation of upper atmosphere and outer space." Description: Unavailable.	548	512	95.2	74.0	Probable electronic ferret; decayed from orbit Feb. 10, 1981.
July 7	Cosmos 925 USSR 1977-61A C-1 Plesetsk	Total weight: Unavailable. Objective: "Investigation of upper atmosphere and outer space." Description: Unavailable.	631	608	97.1	81.2	Probable electronic ferret; still in orbit.
July 8	Cosmos 926 USSR 1977-62A C-1 Plesetsk	Total weight: Unavailable. Objective: "Investigation of upper atmosphere and outer space." Description: Unavailable.	1021	974	105.1	82.9	Probable navigation satellite; still in orbit.
July 12	Cosmos 927 USSR 1977-63A A-2 Plesetsk	Total weight: Unavailable. Objective: "Investigation of upper atmosphere and outer space." Description: Unavailable.	359	151	89.6	72.9	Probable military photo-reconnaissance satellite; recovered July 25, 1977.

Launch Date	Spacecraft, Country, Int'l Designation, Vehicle, Launch Site	Payload Data	Apogee (km)	Perigee (km)	Period (min)	Inclination (degrees)	Remarks
July 13	Cosmos 928 USSR 1977-64A C-1 Plesetsk	Total weight: Unavailable. Objective: "Investigation of upper atmosphere and outer space." Description: Unavailable.	1004	969	104.8	83.0	Probable navigation satellite, third-generation B type; still in orbit.
July 14	Himawari (Gms 1) Japan 1977-65A Thorad-Delta ETR	Total weight: 281kg. Objective: Launch into synchronous transfer orbit of s/c to provide nearly continuous meteorological observation, collection & transmission of data, and monitoring of solar activities. Description: Cylinder 3.1m long, 2.1m in diameter, covered with solar cells, spin-stabilized, carrying spin-scan radiometer.	35 779	35 531	1429.1	1.2	Part of World Weather Watch program, launched by NASA for Japan's Natl. Space Development Agency; apogee boost motor fired July 15 put s/c in synchronous orbit at 140°E; still in orbit.
July 17	Cosmos 929 USSR 1977-66A D-1 Baykonur-Tyuratam	Total weight: Unavailable. Objective: "Investigation of upper atmosphere and outer space." Description: Unavailable.	278	214	89.4	51.6	Probable manned-flight precursor; radio transmissions suggested 2-part vehicle; portion recovered Aug. 16, 1977; extensive maneuvers; decayed from orbit Feb. 2, 1978.
July 19	Cosmos 930 USSR 1977-67A C-1 Plesetsk	Total weight: Unavailable. Objective: "Investigation of upper atmosphere and outer space." Description: Unavailable.	513	480	94.5	74.0	Probable radar-calibration satellite and third C-vehicle launch in which satellite remained attached to upper stage (cf. 1977-11A, 1977-31A); decayed from orbit May 12, 1980.
July 20	Cosmos 931 USSR 1977-68A A-2e Plesetsk	Total weight: Unavailable. Objective: "Investigation of upper atmosphere and outer space." Description: Unavailable.	40 066	604	724.2	63.0	Probable early-warning satellite; still in orbit.
	Cosmos 932 USSR 1977-69A A-2 Baykonur-Tyuratam	Total weight: Unavailable. Objective: "Investigation of upper atmosphere and outer space." Description: Unavailable.	318	174	89.4	65.0	Probable military photo-reconnaissance satellite; recovered Aug. 2, 1977.

Date	Satellite			Objective/Description	Notes
July 22	Cosmos 933 USSR 1977-70A C-1 Plesetsk	407	383	92.5	65.8	Total weight: Unavailable. Objective: "Investigation of upper atmosphere and outer space." Description: Unavailable.	Probable radar-calibration satellite; decayed from orbit Nov. 1, 1978.
July 23	Raduga 3 USSR 1977-71A D-1e Baykonur-Tyuratam	35 943	35 626	1436.0	0.1	Total weight: Unavailable. Objective: Transmission (relay) of TV and multichannel radio communications. Description: Unavailable.	Operational Soviet geostationary comsat (Statsionar 2); still in orbit.
July 27	Cosmos 934 USSR 1977-72A A-2 Plesetsk	339	170	89.6	62.8	Total weight: Unavailable. Objective: "Investigation of upper atmosphere and outer space." Description: Unavailable.	Probable military photo-reconnaissance satellite; recovered Aug. 9, 1977.
July 29	Cosmos 935 USSR 1977-73A A-2 Plesetsk	246	213	89.1	81.3	Total weight: Unavailable. Objective: "Investigation of upper atmosphere and outer space." Description: Unavailable.	Probable military photo-reconnaissance satellite; recovered Aug. 11, 1977.
Aug. 3	Cosmos 936 USSR 1977-74A A-2 Plesetsk	395	217	90.6	62.8	Total weight: Unavailable. Objective: "Investigation of upper atmosphere and outer space." Description: Unavailable.	Intl. biological satellite carrying experiments & equipment from USSR, USA, Czechoslovakia, and France; an experiments package recovered Aug. 21; decayed from orbit Aug. 22, 1977.
Aug. 12	Heao 1 U.S. 1977-75A Atlas-Centaur ETR	455	435	93.5	22.8	Total weight: 2560kg. Objective: Surveying x-ray & gamma-ray sky from 150 ev to 10 Mev, to obtain high-resolution experimental data on astrophysical phenomena. Description: Irregular hexagon (offering maximum experiment-aperture area) 4.1m long, 2.4m in diameter, with solar cells covering external panels, & experiments weighint 1220kg.	Largest earth-oriented unmanned scientific s/c ever launched, first of 3 NASA high-energy astronomical observatories; located 15 previously unknown x-ray sources in first 100 days of operation; first 6 mo. to be spent in sky-mapping; decayed from orbit Mar. 15, 1979.

APPENDIX A   ASTRONAUTICS AND AERONAUTICS, 1977

Launch Date	Spacecraft, Country, Int'l Designation, Vehicle, Launch Site	Payload Data	Apogee (km)	Perigee (km)	Period (min)	Inclination (degrees)	Remarks
Aug. 20	*Voyager 2* U.S. 1977-76A Titan IIIE-Centaur ETR	Total weight: 822kg. (after separation) Objective: Investigation of Jupiter and Saturn planetary systems and the interplanetary medium. Description: Mission module, 10-sided framework with 10 compartments for electronic pkgs., 47cm high and 1.8m across from flat to flat; 13m-long deployable boom carrying 4 magnetometers & 3 radioisotope generators (RTGs); 3.6m-wide high-gain parabolic antenna supported by tubular trusswork; twin 10m whip antennas.	Heliocentric orbit, solar-system escape trajectory				First in twin Voyager series (named *Voyager 2* though launched first, as *Voyager 1* would overtake before reaching Jupiter); after Jupiter flyby, set for July 1979, s/c would use gravitational assist of planet to set course for Saturn flyby in 1981, possible Uranus flyby 1986.
Aug. 24	*Cosmos 937* USSR 1977-77A F-1m Plesetsk	Total weight: Unavailable. Objective: "Investigation of upper atmosphere and outer space." Description: Unavailable.	597	149	92.1	65.0	Probable electronic ferret for ocean surveillance; decayed from orbit Oct. 19, 1978.
	Cosmos 938 USSR 1977-78A A-2 Plesetsk	Total weight: Unavailable. Objective: "Investigation of upper atmosphere and outer space." Description: Unavailable.	305	154	89.1	62.8	Probable military photo-reconnaissance satellite; recovered Sept. 6, 1977.
	Cosmos 939 USSR 1977-79A C-1 Plesetsk	Total weight: Unavailable. Objective: "Investigation of upper atmosphere and outer space." Description: Unavailable.	1462	1434	114.8	74.0	Eight s/c on single carrier, probable technical comsats; still in orbit.
	Cosmos 940 USSR 1977-79B C-1 Plesetsk		1462	1395	114.4	74.0	
	Cosmos 941 USSR 1977-79C C-1 Plesetsk		1462	1415	114.6	74.0	

	Cosmos 942 USSR 1977-79D C-1 Plesetsk		1534	1463	115.9	74.0	
	Cosmos 943 USSR 1977-79E C-1 Plesetsk		1462	1452	115.0	74.0	
	Cosmos 944 USSR 1977-79F C-1 Plesetsk		1472	1462	115.3	74.0	
	Cosmos 945 USSR 1977-79G C-1 Plesetsk		1492	1463	115.5	74.0	
	Cosmos 946 USSR 1977-79H C-1 Plesetsk		1511	1463	115.7	74.0	
Aug. 25	Sirio Italy 1977-80A Thorad-Delta ETR	Total weight: 398kg. (at launch). Objective: Launch into synchronous transfer orbit of Italian comsat. Description: Spin-stabilized cylinder 0.9 long (2m including SHF antenna & apogee boost motor nozzle), 1.4m in diameter.	36 327	34 210	1409.7	0.2	Experimental comsat launched by NASA for Consiglio Nazionale delle Richerche (CNR)—Natl. Research Council of Italy—into successful transfer orbit; firing of apogee kick motor Aug. 27 put s/c into stationary orbit over South Atlantic at 15° W; turned over to Italy Sept. 24; still in orbit.

Launch Date	Spacecraft, Country, Int'l Designation, Vehicle, Launch Site	Payload Data	Apogee (km)	Perigee (km)	Period (min)	Inclination (degrees)	Remarks
Aug. 27	Cosmos 947 USSR 1977-81A A-2 Plesetsk	Total weight: Unavailable. Objective: "Investigation of upper atmosphere and outer space." Description: Unavailable.	318	201	89.7	72.8	Probable military photo-reconnaissance satellite; recovered Sept. 9, 1977.
Aug. 30	Molniya 1-38 USSR 1977-82A A-2e Plesetsk	Total weight: Unavailable. Objective: Transmission of television programs and multichannel radio to stations throughout USSR. Description: Unavailable.	39 777	571	717.6	62.9	Still in orbit.
Sept. 2	Cosmos 948 USSR 1977-83A A-2 Plesetsk	Total weight: Unavailable. Objective: "Investigation of upper atmosphere and outer space." Description: Unavailable.	226	211	88.9	81.4	Probable military photo-reconnaissance satellite; recovered Sept. 15, 1977.
Sept. 5	Voyager 1 U.S. 1977-84A Titan IIIE-Centaur ETR	Total weight: 822kg. (after separation). Objective: Investigation of Jupiter and Saturn planetary systems and the interplanetary medium. Description: Mission module, 10-sided framework with 10 compartments for electronic pkgs., 47cm high and 1.8m across from flat to flat; 13m-long deployable boom carrying 4 magnetometers & 3 radioisotope generators (RTGs); 3.6m-wide high-grain parabolic antenna supported by tubular trusswork; twin 10m whip antennas.	Heliocentric orbit, solar-system escape trajectory				Second launch of dual series; after overtaking Voyager 2, and a March 1979 Jupiter flyby, s/c would alter course for Saturn flyby in Nov. 1980; on a faster trajectory, Voyager 1 passed Voyager 2 Dec. 15, 1977.
Sept. 6	Cosmos 949 USSR 1977-85A A-2 Plesetsk	Total weight: Unavailable. Objective: "Investigation of upper atmosphere and outer space." Description: Unavailable.	324	176	89.5	62.8	Probable military photo-reconnaissance satellite; recovered Oct. 6, 1977.
Sept. 13	Cosmos 950 USSR 1977-86A A-2 Plesetsk	Total weight: Unavailable. Objective: "Investigation of upper atmosphere and outer space." Description: Unavailable.	281	205	89.4	62.8	Probable military photo-reconnaissance satellite; recovered Sept. 27, 1977.

ASTRONAUTICS AND AERONAUTICS, 1977 APPENDIX A

Date	Name/ID	Description	Perigee	Apogee	Period	Incl.	Remarks
	Cosmos 951 USSR 1977-87A C-1 Plesetsk	Total weight: Unavailable. Objective: "Investigation of upper atmosphere and outer space." Description: Unavailable.	1017	965	104.9	83.0	Probable navigation satellite, third-generation B type; still in orbit.
Sept. 16	*Cosmos 952* USSR 1977-88A F-lm Baykonur-Tyuratam	Total weight: Unavailable. Objective: "Investigation of upper atmosphere and outer space." Description: Unavailable.	1000	907	104.1	64.9	One of a pair of ocean-surveillance s/c (see *Cosmos 954*); orbit raised Oct. 8, probably to prevent nuclear generator from reentry disintegration while still dangerously radioactive; still in orbit.
	Cosmos 953 USSR 1977-89A A-2 Plesetsk	Total weight: Unavailable. Objective: "Investigation of upper atmosphere and outer space." Description: Unavailable.	299	151	89.0	62.8	Probable military photo-reconnaissance satellite; recovered Sept. 29, 1977.
Sept. 18	*Cosmos 954* USSR 1977-90A F-lm Baykonur-Tyuratam	Total weight: Unavailable. Objective: "Investigation of upper atmosphere and outer space." Description: Unavailable.	263	250	89.6	65.0	One of a pair of ocean-surveillance s/c (see *Cosmos 952*); difficulties in control after Nov. 1, 1977; reentered Jan. 24, 1978, over Canada Northwest Territory with nuclear reactor.
Sept. 20	*Cosmos 955* USSR 1977-91A A-1 Plesetsk	Total weight: Unavailable. Objective: "Investigation of upper atmosphere and outer space." Description: Unavailable.	640	628	97.4	81.2	Probable electronic ferret; still in orbit.
	Ekran 2 USSR 1977-92A D-le Baykonur-Tyuratam	Total weight: Unavailable. Objective: Broadcasting (relay) of TV programs to small communities in Siberia and the far-northern USSR. Description: Unavailable.	35 622	35 580	1426.5	0.4	Geostationary comsat (Statsionar T); still in orbit.
Sept. 22	*Prognoz 6* USSR 1977-93A A-2e Baykonur-Tyuratam	Total weight: Unavailable. Objective: Investigation of magnetosphere and solar wind. Description: Unavailable.	197 867	488	5688.0	65.0	Part of a series of s/c investigating solar wind, solar radiation, magnetic fields in circumterrestrial space; still in orbit.

Launch Date	Spacecraft, Country, Int'l Designation, Vehicle, Launch Site	Payload Data	Apogee (km)	Perigee (km)	Period (min)	Inclination (degrees)	Remarks
Sept. 23	DOD spacecraft U.S. 1977-94A Titan IIIB-Agena D WTR	Total weight: 3000kg. (?) Objective: "Development of spaceflight techniques and technology." Description: Unavailable.	335	134	89.1	96.5	USAF reconnaissance s/c, part of satellite & missile observation system; decayed from orbit Dec. 8, 1977.
Sept. 24	Cosmos 956 USSR 1977-95A F-2(?) Plesetsk	Total weight: Unavailable. Objective: "Investigation of upper atmosphere and outer space." Description: Unavailable.	862	353	96.8	75.8	Probable vehicle test; decayed from orbit June 27, 1982.
	Intercosmos 17 USSR 1977-96A C-1 Plesetsk	Total weight: Unavailable. Objective: Investigation of charged particles in near-earth space. Description: Unavailable.	512	465	94.4	83.0	Intl. cooperative satellite, carrying experimental equipment built in USSR, Hungary, Rumania, & Czechoslovakia; decayed from orbit Nov. 8, 1979.
Sept. 29	Salyut 6 USSR 1977-97A D-1 Baykonur-Tyuratam	Total weight: 18 900kg. (?) Objective: Launch of a civilian-configuration manned space station for long-duration flight through resupply. Description: Stepped cylinder 14m long, 4.15m max. diameter	349	229	90.3	51.6	S/c similar to Salyut 4 but with 2 docking units; decayed from orbit July 29, 1982; 105 objects associated with launch also decayed by that time.
Sept. 30	Cosmos 957 USSR 1977-98A A-2 Baykonur-Tyuratam	Total weight: Unavailable. Objective: "Investigation of upper atmosphere and outer space." Description: Unavailable.	331	149	89.3	65.0	Probable military photo-reconnaissance satellite; recovered Oct. 13, 1977.
Oct. 9	Soyuz 25 USSR 1977-99A A-2 Baykonur-Tyuratam	Total weight: 6570kg. (?) Objective: Place crew aboard Salyut 6. Description: Unavailable.	329 (max)	353	91.3	51.6	First attempt to put a crew aboard Salyut 6, failed because of "deviations in planned docking routine"; crew (Lt. Col. Vladimir Kovalenok & civilian flight engr. Valery Ryumin) reentered safely Oct. 11, 1977; possible damage to docking apparatus on Soyuz.

Date	Satellite	Description	Weight/Orbit data			Remarks
Oct. 11	Cosmos 958 USSR 1977-100A A-2 Plesetsk	Total weight: Unavailable. Objective: "Investigation of upper atmosphere and outer space." Description: Unavailable.	423	317	91.9	62.8	Probable military photo-reconnaissance satellite; recovered Oct. 24, 1977.
Oct. 21	Cosmos 959 USSR 1977-101A C-1 Plesetsk	Total weight: Unavailable. Objective: "Investigation of upper atmosphere and outer space." Description: Unavailable.	850	146	94.6	65.8	Probable target for satellite-interception tests by Cosmos 961 (which reentered Oct. 26); decayed from orbit Nov. 30, 1977.
Oct. 22	Isee 1 U.S. 1977-102A Thorad-Delta ETR	Total weight: 329kg. (at launch). Objective: Measurement of structure of magnetosphere boundaries and fluctuations; near-earth measurement of solar wind by 2 s/c. Description: Cylinder, 16 sides, 1.6m long and 1.7m across flats; aluminum equipment shelf on a 0.8m-long thrust tube; solar array; spin-stabilized.	138 124	280	3446.8	28.7	Twin payload successfully launched and separated; Isee 1 built by GSFC; Isee 2 by STAR consortium for European Space Agency, launch by NASA as secondary payload; both s/c still in orbit.
	Isee 2 U.S.-ESA 1977-102B Thorad-Delta ETR	Total weight: 158kg. Objective: Joint measurement by s/c separated by some distance of effect of sun on near-earth space, to detect origins of disturbances in plasmapause and magnetopause. Description: Cylinder 1.1m high, 1.3m in diameter, covered with solar cells; single equipment platform, 2 radial 15m booms, for experiment antennas, S-band antenna on top; spin-stabilized.	138 330	279	3453.8	28.7	
Oct. 25	Cosmos 960 USSR 1977-103A C-1 Plesetsk	Total weight: Unavailable. Objective: "Investigation of upper atmosphere and outer space." Description: Unavailable.	547	498	95.0	74.0	Probable electronic ferret; decayed from orbit Oct. 22, 1980.

APPENDIX A   ASTRONAUTICS AND AERONAUTICS, 1977

Launch Date	Spacecraft, Country, Int'l Designation, Vehicle, Launch Site	Payload Data	Apogee (km)	Perigee (km)	Period (min)	Inclination (degrees)	Remarks
Oct. 26	Cosmos 961 USSR 1977-104A F-1m Baykonur-Tyuratam	Total weight: Unavailable. Objective: "Investigation of upper atmosphere and outer space." Description: Unavailable.	1421 (max.)	269	101.8	66.4	Satellite interceptor launched to use Cosmos 959 as target; decayed from orbit Oct. 26 after 0.78 hr.
Oct. 28	Molniya 3-8 USSR 1977-105A A-2e Plesetsk	Total weight: Unavailable. Objective: Transmission of television programs and multichannel radio to stations throughout USSR. Description: Unavailable.	40 762	422	735.3	62.8	Still in orbit.
[Oct. 27]	Transat U.S. 1977-106A Scout WTR	Total weight: 94kg. Objective: Launch into appropriate orbit of s/c to provide Navy with worldwide 2-dimensional system for position-fixing. Description: Cylinder, 8-sided with 4 deployable solar panels, 2 special translators (transponders) for radio relay.	1106	1067	107.0	89.9	Launched by NASA for U.S. Navy, s/c to test Trident missile-tracking system and calibrate range-safety equipment at ground stations; part of Navy Transit system; still in orbit.
Cosmos 962 USSR 1977-107A C-1 Plesetsk		Total weight: Unavailable. Objective: "Investigation of upper atmosphere and outer space." Description: Unavailable.	1010	967	104.9	83.0	Probable navigation satellite, second-generation A type; still in orbit.
Nov. 22	Meteosat 1 U.S.-ESA 1977-108A Thorad-Delta ETR	Total weight: 697kg. (at launch). Objective: Launch into synchronous transfer orbit of s/c representing ESA contribution to global atmospheric research program (GARP). Description: Cylinder 4.3m high, 2.1m in diameter, spin-stabilized, carrying data-relay system and telescope radiometer.	35 692	34 913	1411.5	0.7	Launched by NASA for ESA into transfer orbit; apogee motor fired Nov. 23 put s/c into geostationary orbit at 0° above Gulf of Guinea; experiments scheduled to be operational in May 1978. Still in orbit.

ASTRONAUTICS AND AERONAUTICS, 1977 APPENDIX A

Date	Designation	Details	Apogee	Perigee	Period	Incl.	Remarks
Nov. 24	Cosmos 963 USSR 1977-109A C-1 Plesetsk	Total weight: Unavailable. Objective: "Investigation of upper atmosphere and outer space." Description: Unavailable.	1208	1180	109.3	82.9	Probable geodetic satellite; still in orbit.
Dec. 4	Cosmos 964 USSR 1977-110A A-2 Plesetsk	Total weight: Unavailable. Objective: "Investigation of upper atmosphere and outer space." Description: Unavailable.	357	168	89.8	72.9	Probable military photo-reconnaissance satellite; recovered Dec. 12, 1977.
Dec. 8	Cosmos 965 USSR 1977-111A C-1 Plesetsk	Total weight: Unavailable. Objective: "Investigation of upper atmosphere and outer space." Description: Unavailable.	515	464	94.4	74.0	Probable radar-calibration satellite; apparently broke up or exploded Dec. 16, 1979.
	Noss 2 U.S. 1977-112A Atlas WTR	Total weight: Unavailable. Objective: "Development of spaceflight techniques and technology." Description: Unavailable.	1169	1054	107.5	63.4	U.S. Navy ocean-survey satellite, part of Whitecloud system; other objects, possibly payloads, in similar orbits; still in orbit.
	U.S. 1977-112D Atlas WTR		1169	1054	107.5	63.4	
	U.S. 1977-112E Atlas WTR		1168	1055	107.5	63.4	
Dec. 10	Soyuz 26 USSR 1977-113A A-2 Baykonur-Tyuratam	Total weight: 6570kg. (?) Objective: Second attempt to put crew aboard Salyut 6. Description: Unavailable.	235	195	88.7	51.6	Soyuz 26 successfully docked with second docking unit on Salyut 6 on Dec. 11; crew (Maj. Yuri Romanenko & civilian flight engr. Georgy Grechko) would occupy station into Jan. 1978; Soyuz 26 used to return crew of Soyuz 27 to earth Jan. 16, 1978.

257

APPENDIX A ASTRONAUTICS AND AERONAUTICS, 1977

Launch Date	Spacecraft, Country, Int'l Designation, Vehicle, Launch Site	Payload Data	Apogee (km)	Perigee (km)	Period (min)	Inclination (degrees)	Remarks
Dec. 11	DOD spacecraft U.S. 1977-114A Atlas-Agena D ETR	Total weight: 700kg. (?) Objective: "Development of spaceflight techniques and technology." Description: Unavailable.	35 855	35 679	1435.1	0.2	Still in orbit.
Dec. 12	Cosmos 966 USSR 1977-115A A-2 Baykonur-Tyuratam	Total weight: Unavailable. Objective: "Investigation of upper atmosphere and outer space." Description: Unavailable.	294	203	89.5	65.0	Probable military photo-reconnaissance satellite; recovered Dec. 24, 1977.
Dec. 13	Cosmos 967 USSR 1977-116A C-1 Plesetsk	Total weight: Unavailable. Objective: "Investigation of upper atmosphere and outer space." Description: Unavailable.	1003	963	104.8	65.8	Probable target vehicle for satellite-interception tests; still in orbit.
Dec. 14	Meteor 2-3 USSR 1977-117A A-1 Plesetsk	Total weight: Unavailable. Objective: Meteorology. Description: Unavailable.	893	854	102.4	81.2	Third of second-generation Soviet weather satellites scanning earth in infrared and visible-light wavelengths; still in orbit.
Dec. 15	Sakura Japan 1977-118A Thorad-Delta ETR	Total weight: 677kg. (at launch). Objective: Launch into synchronous transfer orbit of domestic comsat to provide Japan with telephone and TV links. Description: Spin-stabilized cylinder 3.5m high, 2.2m in diameter, solar cells on exterior; antenna horn reflector on despun section of drive motor assembly; 6 K-band channels, 2 C-band channels for communications.	36 157	35 568	1440.0	0.1	Launched by NASA for Japan's Natl. Space Development Agency (NASDA) into synchronous transfer orbit; apogee boost motor fired Dec. 16 put s/c into synchronous orbit at 135° E due south of Japan, over New Guinea; still in orbit.
Dec. 16	Cosmos 968 USSR 1977-119A A-2 Plesetsk	Total weight: Unavailable. Objective: "Investigation of upper atmosphere and outer space." Description: Unavailable.	809	781	100.8	74.0	Probable military comsat with data store-dump capability; still in orbit.

Date	Satellite	Details				Remarks
Dec. 20	Cosmos 969 USSR 1977-120A A-2 Plesetsk	Total weight: Unavailable. Objective: "Investigation of upper atmosphere and outer space." Description: Unavailable.	316	179	89.5	62.8	Probable military photo-reconnaissance satellite; recovered Jan. 3, 1978.
Dec. 21	Cosmos 970 USSR 1977-121A F-1m Baykonur-Tyuratam	Total weight: Unavailable. Objective: "Investigation of upper atmosphere and outer space." Description: Unavailable.	1139	946	106.0	65.8	Probable antisatellite test; made close pass at Cosmos 967 (1977-116A) and apparently exploded (disintegrated in orbit); 57 objects identified as connected with this launch, some still in orbit.
Dec. 23	Cosmos 971 USSR 1977-122A C-1 Plesetsk	Total weight: Unavailable. Objective: "Investigation of upper atmosphere and outer space." Description: Unavailable.	1008	979	105.0	82.9	Probable navigation satellite, third-generation B type; still in orbit.
Dec. 27	Cosmos 972 USSR 1977-123A C-1 (?) Plesetsk	Total weight: Unavailable. Objective: "Investigation of upper atmosphere and outer space." Description: Unavailable.	1170	714	103.9	75.8	Probable vehicle test; still in orbit.
	Cosmos 973 USSR 1977-124A A-2 Baykonur-Tyuratam	Total weight: Unavailable. Objective: "Investigation of upper atmosphere and outer space." Description: Unavailable.	324	202	89.8	71.4	Probable military photo-reconnaissance satellite; recovered Jan. 9, 1978.

Appendix B

MAJOR NASA LAUNCHES, 1977

The following table of major NASA launches includes payloads carried by all rocket vehicles larger than sounding rockets launched in 1977 by NASA or under NASA direction.

During 1977, the U.S. had 26 launches, of which 24 were successful. Of these, 10 were launches by DOD. Of the remaining 14, 3½ were launches by NASA for its own programs: *Heao 1*, the two Voyager spacecraft launched in August and September, and *Isee 1* (part of a twin payload). Launches for others included 2½ for ESA (*Geos*, *Meteosat*, and *Isee 2*); 2 for Japan (*Himawari* and *Sakura*); *Nato IIIB* for NATO, *Palapa 2* for Indonesia, *Intelsat IVA F-4* for ComSatCorp, *Goes 2* for NOAA, *Sirio* for Italy, and a *Transat* payload for the U.S. Navy. The unsuccessful attempts, both in September, were OTS, a comsat for ESA destroyed shortly after launch, and an Intelsat IVA launched for ComSatCorp that exploded 1 minute into its flight.

The table includes vehicle and payload performance under categories S for successful, P for partially successful, and U for unsuccessful. A fourth category, Unk, indicates payloads that did not operate because of vehicle failure. These unofficial categories do not take into account that U missions may produce valuable information, or that payloads with a long-life design may fail to meet design requirements and become officially unsuccessful at a later date. Further information on these launches appears in Appendix A and in the indexed entries in the text.

MAJOR NASA LAUNCHES, 1977

Date	Name (NASA Code)	General Mission	Launch Vehicle; Site	Performance Vehicle	Performance Payload	Remarks
Jan. 28	*Nato IIIB*	Communications	Thorad-Delta ETR	S	S	Second of 3 planned comsats to be launched by NASA for NATO.
Mar. 10	*Palapa 2*	Communications	Thorad-Delta ETR	S	S	Launched by NASA for Indonesia; operational Apr. 15.
Apr. 20	*Geos*	To provide data on magnetic and electrical fields at 36 000km altitude.	Thorad-Delta ETR	U	P	Launched by NASA for ESA; booster malfunction put s/c in unsatisfactory transfer orbit; mission judged by NASA unsuccessful, though all experiments sending useful data.
May 26	*Intelsat IVA F-4*	To provide 2-way voice circuits & TV channels simultaneously, or telephone, TV, and other forms of communications combined; to serve as backup for *Intelsat IVA F-1 & F-2*.	Atlas-Centaur ETR	S	S	Third in series of improved comsats, launched by NASA for ComSat Corp.
June 16	*Goes 2*	Continuous observation of the earth's atmosphere on an operational basis.	Thorad-Delta ETR	S	S	Second operational s/c launched by NASA for NOAA; turned over to NOAA July 29.
July 14	*Himawari (Gms)*	Improvement of Japanese and international meteorological service through observation on almost continuous basis and monitoring of solar activity.	Thorad-Delta ETR	S	S	Launched by NASA for Japan's Natl. Space Development Agency.
Aug. 12	*Heao 1*	Survey of x-ray and gamma-ray sky for high-resolution astronomical data.	Atlas-Centaur ETR	S	S	First of series of 3 planned high-energy astronomical observatories; largest earth-oriented unmanned scientific s/c launched to date.
Aug. 20	*Voyager 2*	Investigation of Jupiter and Saturn planetary systems and interplanetary medium.	Titan IIIE-Centaur ETR	S	S	First of twin Voyager s/c, called 2 because it would be overtaken by other s/c before reaching Jupiter.
Aug. 25	*Sirio*	Communications	Thorad-Delta ETR	S	S	Launched by NASA for Italy's Consiglio Nazionale delle Richerche; turned over to Italy Sept. 24.

Sept. 5	*Voyager 1*	Investigation of Jupiter and Saturn planetary systems and interplanetary medium.	Titan IIIE-Centaur ETR	S	Second of Voyager series, overtook *Voyager 2* on Dec. 15.
	Isee 1, Isee 2	Scientific measurements from 2 points in space; study of magnetosphere and sampling of solar wind.	Thorad-Delta ETR	S	Secondary payload built by European STAR consortium; twin s/c launched by NASA, *Isee 2* for ESA.
Oct. 27	*Transat*	Testing Trident missile-tracking system, checking out range-safety equipment, serving as position-fixer	Scout WTR	S	Launched by NASA for the U.S. Navy.
Nov. 22	*Meteosat 1*	ESA contribution to World Weather Watch program, earth observation from geostationary orbit.	Thorad-Delta ETR	S	Launched by NASA for ESA; scheduled to be fully operational by May 1978.
Dec. 14	*Sakura (Cs)*	Development of domestic comsat system for Japan; experiments in K-band and C-band.	Thorad-Delta ETR	S	Launched by NASA for Japan's Natl. Space Development Agency.

Appendix C
MANNED SPACE FLIGHTS, 1977

Manned spaceflight in 1977 was limited to 3 Soviet launches, *Soyuz 24* (to dock with *Salyut 5*) and *Soyuz 25* and *26* to dock with *Salyut 6*, launched Sept. 29, 1977.

The space station *Salyut 5*, which had been launched June 22, 1976, decayed from orbit Aug. 8, 1977 after being used for docking by Soyuz flights *21*, *22*, and *23*. The first attempt at docking with *Salyut 6* was unsuccessful, but the crew of *Soyuz 26* occupied the station without difficulty and let their visitors in *Soyuz 27* take their vehicle home the following January; the men from *Soyuz 26* remained in orbit, returning in *Soyuz 27* the following March.

At the end of 1977, the U.S. had made 31 manned spaceflights: 2 suborbital, 20 in earth orbit, 3 in lunar orbit, and 6 lunar landings, using 43 different crewmen. The USSR had made 30 manned flights, all in earth orbit, with 38 cosmonauts. Total hours of manned flight were, for the U.S., 7681hr 10min; for the USSR, 6997hr 18min. Total cumulative man-hours in space were, for the U.S., 22 503hr 39min; for the USSR, 13 519hr 18min.

MANNED SPACE FLIGHT, 1977

Date Launched	Date Recovered	Designation; Crew	Weight (kg)	Duration (revolutions)	Remarks
Feb. 7	Feb. 25	*Soyuz 24* Victor Gorbatko Yuri Glazkov	6570 (est.)	425hr 17min	Docked with *Salyut 5* on Feb. 8, transferred the following day; undocked early on Feb. 25. Continued program started by *Soyuz 21* crew, probably earth-resources photography and military reconnaissance.
Oct. 9	Oct. 11	*Soyuz 25* Vladimir Kovalenok Valery Ryumin	6570 (est.)	48hr 43min	Failed to dock with *Salyut 6* because of "deviations in the planned docking routine"; no attempt to use other docking port; possible damage to Soyuz docking apparatus.
Dec. 10	Jan. 16, 1978	*Soyuz 26* Yuri Romanenko Georgy Grechko	6570 (est.)	898hr 19min	Docked with second docking unit on *Salyut 6* on Dec. 11; Grechko performed EVA Dec. 19 to inspect forward docking unit in preparation for *Soyuz 27* launch; *Soyuz 26* returned to earth in Jan. 1978 with crew of *Soyuz 27*, leaving aft unit open for supply vessel. *Soyuz 26* crew returned to earth in *Soyuz 27* on Mar. 16, 1978.

Appendix D
NASA SOUNDING ROCKET LAUNCHES, 1977

The following table lists the 53 sounding rockets of the Arcas class and above launched by NASA in 1977. The launches took place in Australia, Canada, Norway, and Sweden, as well as in the United States. Payloads were carried for the Naval Research Laboratory, the Smithsonian Astrophysical Observatory, the Uppsala Ionospheric Observatory, NOAA, and 3 NASA centers; 14 universities; 1 domestic institute, M.I.T.; and 3 foreign, the Univ. of Adelaide, the Univ. of Bern, and the Max-Planck Institut. Onboard experiments included 18 physics; 12 astronomy; 10 aeronomy; 9 astrophysics; 3 vehicle-system tests; and 1 space-processing test.

Information in the table came from Goddard Space Flight Center's Quick-Look Sounding Rocket Data sheets, issued after launches, with additional information from some of the experimenters concerned. Launch dates in the table are local time, with the date by Greenwich Mean Time (z) if different.

NASA-LAUNCHED SOUNDING ROCKETS, 1977

Launch Date	Rocket NASA Designation, Launch Site	Apogee (km)	Remarks
Jan. 8	Nike Tomahawk 18.1008UE Wallops Island, Va	138	Univ. of Bern/Univ. of Ill. magnetospheric physics; Bern experiment functioned properly; all Illinois experiments worked except ultraviolet.
Jan. 8	Astrobee D 23.008UE Wallops Island, Va.	83.19	Penn. State Univ. plasma physics.
Jan. 8 (Jan. 9 Z)	Super Arcas 15.152UE Wallops Island, Va.	86.23	Penn. State Univ./Univ. of Texas, Dallas. Experiment in plasma physics.
Jan. 12	Terrier Malemute 29.003AE Poker Flat Range, Al.	— (predicted 500km)	NOAA auroral studies; all experiments operated; radar beacon failed, actual altitude unavailable.
Jan. 14	Black Brant VC 21.049UL White Sands, N.M.	266.9	Univ. of Colo. lunar and planetary astronomy experiment in stellar ultraviolet.
Jan. 15	Aerobee 170 A 13.122UG White Sands, N.M.	197.7	Univ. of Wisc. high-energy astrophysics; soft x-ray sky survey.
Jan. 18	Terrier Malemute 29.001UE Poker Flat Range, Al.	11.4 F	Univ. of Wisc. auroral studies. Rocket motor failed at takeoff, no opportunity for equipment to function.
Jan. 23	Nike Tomahawk 18.1005UE Andoya, Norway	221	Univ. of Minn./Univ. of N.H. auroral studies.
Feb. 6	Nike Tomahawk 18.1004UE Andoya, Norway	224.3	Univ. of Minn./Univ. of N.H. experiment in plasma physics.
Feb. 7	Nike Tomahawk 18.1011UE	444	Univ. of Alaska experiment in plasma physics.

Feb. 11	Nike Tomahawk 18.1012UE Fairbanks, Ak.	447	Univ. of Alaska experiment in plasma physics.
Feb. 11	Super Arcas 15.154UE Kiruna, Sweden	—	Univ. of Houston experiment in plasma physics.
Feb. 11	Nike Tomahawk 18.213IE Kiruna, Sweden	—	Uppsala Ionospheric Observatory/Cornell Univ. auroral studies; reported successful, data unavailable.
Feb. 15	Nike Apache 14.500UA Fort Churchill, Canada	136.4	Univ. of Pittsburgh aeronomy experiment.
Feb. 15 (Feb. 16 Z)	Aerobee 200 26.054UH Woomera, Australia	175.6	Columbia Univ. x-ray astronomy experiment.
Feb. 15 (Feb. 16 Z)	Aerobee 200 26.052DH Woomera, Australia	— (predicted 189.1km)	NRL x-ray astronomy experiment; radar track lost; close to predicted trajectory.
Feb. 15	Super Arcas 15.155UE Kiruna, Sweden	—	Univ. of Houston plasma-physics experiment; problem with parachute resulted in partial success.
Feb. 16	Astrobee F 25.015GG White Sands, N.M.	197.6	GSFC stellar-ultraviolet experiment.
Feb. 16	Nike Tomahawk 18.212GE/IE Andoya, Norway	247	Norway/GSFC experiment, plasma physics composition of auroral F region.
Feb. 17 (Feb. 18 Z)	Aerobee 170 13.124UG Woomera, Australia	191.9	Johns Hopkins Univ. galactic-ultraviolet astronomy experiment.
Feb. 17 (Feb. 18 Z)	Aerobee 200 26.057GG Woomera, Australia	196.6	GSFC galactic-ultraviolet astronomy experiment.
Feb. 21 (Feb. 22 Z)	Aerobee 200 26.051UG Woomera, Australia	173.7	Johns Hopkins Univ. galactic astronomy experiment.

APPENDIX D

Launch Date	Rocket NASA Designation, Launch Site	Apogee (km)	Remarks
Feb. 21 (Feb. 22 Z)	Aerobee 170 13.123 IS Woomera, Australia	177.3	Univ. of Adelaide solar-physics experiment; some experiments damaged in parachute landing.
Feb. 23 (Feb. 24 Z)	Aerobee 200 26.053DH Woomera, Australia	183.8	NRL x-ray astronomy experiment.
Feb. 25 (Feb. 26 Z)	Nike Orion 31.001UA Fort Churchill, Canada	142.3	Univ. of Pittsburgh aeronomy experiment.
Mar. 8 (Mar. 9 Z)	Nike Orion 31.002UA Fort Churchill, Canada	137.7	Univ. of Pittsburgh aeronomy experiment; dayglow phenomena.
Mar. 9	Black Brant VC 21.041US White Sands, N.M.	191.4	Univ. of Colo. solar-ultraviolet physics experiment.
Mar. 12	Astrobee F 25.020UG White Sands, N.M.	224.7	Univ. of Colo. galactic-astronomy stellar-ultraviolet experiment; no data obtained during flight, but instrument operated properly after recovery.
Mar. 19	Astrobee F 25.025UE Fort Churchill, Canada	— (predicted 195km)	Univ. of Mich. plasma-physics (aurora) experiment; radar track lost; peak estimated close to prediction; data satisfactory.
Mar. 20	Aries 24.0021E Kiruna, Sweden	460	Max-Planck Institut plasma-physics experiment.
Mar. 22	Aerobee 200 A 26.059UG White Sands, N.M.	188.5	Univ. of Wisc. galactic-astronomy experiment.
Apr. 8	Nike Tomahawk 18.1013GA Fort Churchill, Canada	158.3	GSFC aeronomy experiment.
Apr. 16 (Apr. 17 Z)	Black Brant VC 21.054 UG White Sands, N.M.	135.6	Johns Hopkins Univ. galactic-astronomy experiment.

ASTRONAUTICS AND AERONAUTICS, 1977 — APPENDIX D

Date	Vehicle	Altitude	Experiment
May 20 (May 21 Z)	Aerobee 170 A 13.131DA White Sands, N.M.	23 F	NRL aeronomy experiment. Vehicle locked in roll, no opportunity for equipment to operate.
May 25	Aerobee 200 A 26.061UH White Sands, N.M.	191.9	Univ. of Wisc. high-energy astrophysics experiment.
June 9	Nike Black Brant VC 27.008UH White Sands, N.M.	204.5	Smithsonian Astrophysical Observatory high-energy astrophysics; failed to obtain proper data because of improper pointing.
June 21	Black Brant VC 21.045NP White Sands, N.M.	111	MSFC space-processing experiment.
July 12	Nike Orion 31.005UA White Sands, N.M.	113.4	Univ. of Colo. aeronomy experiment.
July 14	Orion 12.027GT White Sands, N.M.	59.3	GSFC vehicle-systems test; recovery successful.
July 19	Taurus Orion 12.1001WT Wallops Island, Va.	240	WFC flight test; successful.
July 21	Astrobee F 25.016GG White Sands, N.M.	205.8	GSFC galactic astronomy; successful.
July 27	Nike Black Brant VC 27.031UH White Sands, N.M.	202.7	Smithsonian Astrophysical Observatory high-energy astrophysics experiment.
July 27	Astrobee F 25.021UH White Sands, N.M.	189.5	Massachusetts Institute of Technology high-energy astrophysics experiment.
Aug. 10	Nike Apache 14.543UE Wallops Island, Va.	199.5	Univ. of Ill. plasma-physics experiment.
Aug. 17 (Aug. 18 Z)	Nike Tomahawk 18.182GA White Sands, N.M.	198 F	GSFC aeronomy experiment; did not operate properly, investigation is under way.

Launch Date	Rocket NASA Designation, Launch Site	Apogee (km)	Remarks
Aug. 17	Nike Tomahawk 18.1009UE White Sands, N.M.	266.7	Univ. of Texas plasma-physics experiment. Instrument failure.
Sept. 29 (Sept. 30 Z)	Astrobee F 25.022UA/US White Sands, N.M.	257	Univ. of S. Calif. lunar and planetary astronomy experiment; successful.
Oct. 2	Aerobee 200A 26.065UH/IH White Sands, N.M.	198	Smithsonian Astrophysical Observatory high-energy astrophysics experiment.
Dec. 2	Taurus Orion 12.1002WT Wallops Island, Va.	246.8	WFC vehicle test; satisfactory.
Dec. 11	Aerobee 200 A 26.060GG White Sands, N.M.	110	GSFC galactic-astronomy experiment; successful.
Dec. 11	Astrobee F 25.017GG White Sands, N.M.	140.47	GSFC galactic-astronomy experiment; successful.
Dec. 11	Nike Tomahawk 18.183GA White Sands, N.M.	127.5	GSFC aeronomy experiment; successful.
Dec. 14	Nike Orion 31.003UA White Sands, N.M.	138.9	Univ. of Pittsburgh solar-physics experiment; successful.

Appendix E

ABBREVIATIONS OF REFERENCES

Listed here are the abbreviations used for citing sources in the text. Not all the sources are listed, only those that are abbreviated.

AAAS Bull	American Association for the Advancement of Science's *AAAS Bulletin*
A&A	American Institute of Aeronautics and Astronautics' magazine, *Astronautics & Aeronautics*
A&A 1977	NASA's *Astronautics and Aeronautics, 1977: A Chronology* (this publication)
ABC	American Broadcasting Company
AEC Release	Atomic Energy Commission news release
Aero Daily	*Aerospace Daily* newsletter
Aero Med	*Aerospace Medicine* magazine
AF Mag	Air Force Association's *Air Force Magazine*
AFHF Newsletter	*Air Force Historical Foundation Newsletter*
AFJ	*Armed Forces Journal* magazine
AFSC *Newsreview*	Air Force Systems Command's *Newsreview*
AFSC Release	Air Force Systems Command news release
AIA Release	Aerospace Industries Association of America news release
AIAA *Facts*	American Institute of Aeronautics and Astronautics' *Facts*
AIAA Release	American Institute of Aeronautics and Astronautics news release
AIP *Newsletter*	American Institute of Physics *Newsletter*
AP	Associated Press news service
ARC *Astrogram*	NASA Ames Research Center's *Astrogram*
Astro Journ	American Astronomical Society's *Astrophysical Journal*
Atlanta JC	*Atlanta Journal Constitution* newspaper
AvWk	*Aviation Week & Space Technology* magazine
B News	*Birmingham News* newspaper

B Sun	Baltimore *Sun* newspaper
Bull Atom Sci	Education Foundation for Nuclear Science's *Bulletin of the Atomic Scientists*
Bus Wk	*Business Week* magazine
C Daily News	*Chicago Daily News* newspaper
C Trib	*Chicago Tribune* newspaper
Can Press	Canadian Press news service
CBS	Columbia Broadcasting System
C&E News	*Chemical & Engineering News* magazine
Cl PD	Cleveland *Plain Dealer* newspaper
Cl Press	*Cleveland Press* newspaper
Columbia J Rev	*Columbia Journalism Review* magazine
ComSatCorp Release	Communications Satellite Corporation news release
CQ	*Congressional Quarterly*
CR	*Congressional Record*
CSM	*Christian Science Monitor* newspaper
CTNS	Chicago Tribune News Service
D News	*Detroit News* newspaper
D Post	*Denver Post* newspaper
DASA Release	Defense Atomic Support Agency news release
DFRC	See FRC.
DJ	Dow Jones news service
DOC PIO	Department of Commerce Public Information Office
DOD Release	Department of Defense news release
DOT Release	Department of Transportation news release
EOP Release	Executive Office of the President news release
ESA Release	European Space Agency news release, use dated (not numbered)
FAA Release	Federal Aviation Administration news release
FBIS—Sov	Foreign Broadcast Information Service, Soviet number
FonF	*Facts on File*
FRC Release	Flight Research Center news release, after 8 Jan. 1976, became Dryden Flight Research Center (DFRC) news release
FRC *X-Press*	NASA Flight Research Center's *X-Press*
GE Forum	*General Electric Forum* magazine
Goddard News	NASA Goddard Space Flight Center's *Goddard News*
GSFC Release	NASA Goddard Space Flight Center news release

GSFC *SSR*	NASA Goddard Space Flight Center's *Satellite Situation Report*
GT&E Release	General Telephone & Electronics news release
H Chron	*Houston Chronicle* newspaper
H Post	*Houston Post* newspaper
INTELSAT Release	Intl. Telecommunications Satellite Org. news release
JA	*Journal of Aircraft* magazine
JPL *Lab-Oratory*	Jet Propulsion Laboratory's *Lab-Oratory*
JPL Release	Jet Propulsion Laboratory news release
JPRS	Department of Commerce Joint Publications Research Service
JSC Release	NASA Lyndon B. Johnson Space Center (Manned Spacecraft Center until 17 Feb. 1973) news release
JSC *Roundup*	NASA Lyndon B. Johnson Space Center's *Space News Roundup*
JSR	American Institute of Aeronautics and Astronautics' *Journal of Spacecraft and Rockets* magazine
KC Star	*Kansas City Star* newspaper
KC Times	*Kansas City Times* newspaper
KSC Release	NASA John F. Kennedy Space Center news release
LA *Her-Exam*	Los Angeles *Herald-Examiner* newspaper
LA Times	*Los Angeles Times* newspaper
Langley Researcher	NASA Langley Research Center's *Langley Researcher*
LaRC Release	NASA Langley Research Center news release
LATNS	Los Angeles Times News Service
LeRC Release	NASA Lewis Research Center news release
Lewis News	NASA Lewis Research Center's *Lewis News*
M HER	*Miami Herald* newspaper
M News	*Miami News* newspaper
M Trib	*Minneapolis Tribune* newspaper
Marshall Star	NASA George C. Marshall Space Flight Center's *Marshall Star*
MJ	*Milwaukee Journal* newspaper
MSFC Release	NASA George C. Marshall Space Flight Center news release
N Hav Reg	*New Haven Register* newspaper
N News	*Newark News* newspaper
N Va Sun	*Northern Virginia Sun* newspaper

NAA *News*	National Aeronautic Association *News*
NAA Record Book	National Aeronautic Association's *World and U.S.A. National World Aviation—Space Records*
NAC Release	National Aviation Club news release
NAE Release	National Academy of Engineering news release
NANA	North American Newspaper Alliance
NAS Release	National Academy of Sciences news release
NAS—NRC Release	National Academy of Sciences—National Research Council news release
NAS—NRC—NAE *News Rpt*	National Academy of Sciences—National Research Council—National Academy of Engineering *News Report*
NASA Actv	*NASA Activities*
NASA anno	NASA announcement
NASA GMR	NASA Headquarters "General Management Review Report"
NASA HHR—39	NASA Historical Report No. 39
NASA Hist Off	NASA History Office
NASA Hq *WB*	NASA Headquarters *Weekly Bulletin*
NASA Int Aff	NASA Office of International Affairs
NASA *LAR,* XIII/8	NASA *Legislative Activities Report,* Vol. XIII, No. 8
NASA Leg Off	NASA Office of Legislative Affairs
NASA MOR	NASA Headquarters Mission Operations Report, preliminary prelaunch and postlaunch report series (information may be revised and refined before publication)
NASA prog off	NASA program office (for the program reported)
NASA proj off	NASA project office (for the project reported)
NASA Release	NASA Headquarters news release
NASA Rpt SRL	NASA report of sounding rocket launching
NASA SP-4019	NASA Special Publication No. 4019
Natl Obs	*National Observer* magazine
Nature	*Nature Physical Science* magazine
NBC	National Broadcasting Company
NGS Release	National Geographic Society news release
NMI	NASA Management Instruction
NN	NASA Notice
NOAA Release	National Oceanic and Atmospheric Administration news release

NRL Release	Naval Research Laboratory news release
NSC Release	National Space Club news release
NSC *News*	National Space Club *News*
NSC *Letter*	National Space Club *Letter*
NSF *Highlights*	National Science Foundation's *Science Resources Studies Highlights*
NSF Release	National Science Foundation news release
NSTL Release	NASA National Space Technology Laboratories news release
NY News	*New York Daily News* newspaper
NYT	*New York Times* newspaper
NYTNS	New York Times News Service
O Sen Star	*Orlando Sentinel Star* newspaper
Oakland Trib	*Oakland Tribune* newspaper
Omaha W-H	*Omaha World-Herald* newspaper
ONR *Rev*	Navy's Office of Naval Research *Reviews*
P *Bull*	Philadelphia *Evening* and *Sunday Bulletin* newspaper
P Inq	*Philadelphia Inquirer* newspaper
PAO	Public Affairs Office
PD	National Archives and Records Service's *Weekly Compilation of Presidential Documents*
PIO	Public Information Office
PMR *Missile*	USN Pacific Missile Range's *Missile*
PMR Release	USN Pacific Missile Range news release
Pres Rpt 74	*Aeronautics and Space Report of the President: 1974 Activities*
SAO Release	Smithsonian Astrophysical Observatory news release
SBD	*Defense/Space Business Daily* newspaper
Sci Amer	*Scientific American* magazine
Sci & Govt Rpt	*Science & Government Report*, independent bulletin of science policy
SciServ	Science Service News service
SD	*Space Digest* magazine
SD Union	*San Diego Union* newspaper
SET Manpower Comments	Scientific Manpower Commission's *Scientific, Engineering, Technical Manpower Comments*
SF	British Interplanetary Society's *Spaceflight* magazine
SF Chron	*San Francisco Chronicle* newspaper
SF Exam	*San Franciso Examiner* newspaper

Sov Aero	*Soviet Aerospace* newsletter
Sov Rpt	Center for Foreign Technology's *Soviet Report* (translations)
SP	*Space Propulsion* newsletter
Spaceport News	NASA John F Kennedy Space Center's *Spaceport News*
Spacewarn	IUWDS World Data Center A for Rockets and Satellites' *Spacewarn Bulletin*
SR list	NASA compendium of sounding rocket launches
SSN	*Soviet Sciences in the News,* publication of Electro-Optical Systems, Inc.
St Louis G-D	*St. Louis Globe-Democrat* newspaper
St Louis P-D	*St. Louis Post-Dispatch* newspaper
T-Picayune	New Orleans *Times-Picayune* newspaper
Tech Rev	Massachusetts Institute of Technology's *Technology Review*
Today	*Today* newspaper
testimony	Congressional testimony, prepared statement
text	Prepared report or speech text
transcript	Official transcript of news conference or congressional hearing
UN Reg	United Nations Public Registry of Space Flight
UPI	United Press International news service
USGS Release	U.S. Geological Survey news release
USPS Release	U.S. Postal Service news release
W Post	*Washington Post* newspaper
W Star-News	*Washington Star-News* newspaper
WFC Release	NASA Wallops Flight Center news release
WH Release	White House news release
WJT	*World Journal Tribune* newspaper
WSJ	*Wall Street Journal* newspaper

Index

A-2, 187
A-10, 105
AAAS. See Am. Assn. for the Advancement of Science.
abort, 123, 155-156, 189, 190
Abruzzo, Ben, 172, 173
absorption, 145
Academy of Sciences, PRC. See China, Peoples Republic of.
Academy of Sciences, USSR. See Soviet Academy of Sciences.
acceleration, accelerator; accelerometer 117, 151, 215, 230
accessibility, 224
accidents, 19, 94, 105, 154-155, 224, 235
accuracy, 4, 9, 30, 55, 58, 79, 96, 121, 158, 196
acoustics, 32, 61, 71, 105-106, 118
ACPL. See atmospheric cloud physics laboratory.
Acurio, John, 232
Adams, Brock, 62
additives, 1
Adelaide, Univ. of, 37
Administracion National de Telecomunicaciones, Paraguay, 135
Administration of Long Distance Telecommunications, PRC, 154
advanced civilizations, 143
advanced research and technology, 147, 159
Advanced Research Projects Agency (ARPA), 5, 8, 71
Advanced X-Ray Astrophysics Facility (AXAF), 228
AEC. Atomic Energy Commission. See Energy Research and Development Administration.
AE-D, AE-E. See Atmosphere Explorers.
AERE Hartwell Co., 2
aerial survey. See monitoring; surveillance.
Aero Club of Washington, 172
Aerobee (sounding rocket), 37
aerodynamics, 16, 28, 108, 118, 122-123, 150-152, 154, 217-218, 219
Aeroflot (USSR passenger plane service), 29, 176-177
aeronautical, 28, 61, 182, 195, 207, 231-232, 235

Aeronautics and Space Research Faculty Program, 152
aeronautical and space technology (AST). See NASA.
Aeronutronic Ford Corp. See Ford Aerospace & Communications Corp.
Aeroparque George Newbery, Buenos Aires, 222
Aerosat (ESA project), 16, 122, 174-175
aerosols. See atmosphere pollution.
Aerospace Corp., 48
aerospace environment. See environment.
aerospace industry, 62, 107, 112, 134, 148, 172, 182, 193, 213, 219
aerospace technology. See technology applications.
Aetna Casualty & Surety Co., 72
affirmative action. See equal employment opportunity.
Africa, 10, 110, 139, 157, 179, 198, 210, 225, 234
Afro-American Historical & Cultural Museum, 35
age, 39, 214-215
agricultural production; agriculture. See crops.
agro-environmental programs, 5, 106-107, 119, 147
AIL (division of Cutler Hammer), 112
ailerons, 177
aiming, 121, 149-150, 150-152
air cargo. See air freight.
air carriers. See airlines.
Air Force, U.S., 8, 13, 15, 36, 72, 73, 105, 109, 111, 124-125, 132, 140 145, 147, 148, 154-155, 159, 167, 169, 172, 173, 177, 179, 181, 188, 194, 201, 203, 205, 235
—contractors, 7, 8, 18, 22, 73, 112, 192, 235
—Flight Dynamics Laboratory, U.S., 214
—Geophysics Laboratory (AFGL), 18, 93
—Global Weather Center, 58
—Satellite Communications System (AFSATCOM), 22
—Scientific Advisory Board, 147
—Systems Command (AFSC) 2, 18, 203, 214
air freight, 112
air pollution. See atmosphere pollution.
air sampling. See GASP.

air show. See Paris, France.
air traffic, 43, 128, 173, 213
airborne monitoring, 55, 72, 204
airborne warning and control system (AWACS), 53-54
airconditioning, 29, 128, 179, 203, 213, 232
aircraft
 —civilian, 25, 60, 107, 130, 134, 192, 222
 —communications, 13
 —design. See design.
 —energy efficiency (ACEE) program, 79, 110, 128, 177, 180, 218
 —guidance, 7
 —industry. See aerospace industry.
 —manufacturers. See aerospace industry.
 —military, 73, 74, 107, 147, 154-155, 199
 —power systems, 7
 —safety. See safety.
 —targets, 71
 —testing, 61, 217-218
 —transport, 25, 35, 64, 71, 79, 128, 218
 —use in research, 58, 60, 61, 84, 96, 134, 138-139, 161, 171, 180, 192, 193-194, 199, 204, 208-209, 216, 218, 221-222, 235
aircushion (landing system). See landing.
airdrop. See drop.
AiResearch Mfg. Co., 7
airflow. See aerodynamics.
airfoils. See design, aircraft.
airlift, 111
airlines, 62, 122, 176, 177, 180, 190-191, 193, 218, 220-221
airlock
 Dec 10-31 (*Soyuz 26* launch), 224
airplanes. See aircraft.
airports, 62, 66, 112, 152, 191, 213, 218, 222
airspace, 155
airworthiness. See handling.
Akademik Korolev (USSR ship), 127, 164, 198
Alabama, 127, 196, 211
Alabama A&M Univ., 152
Alabama Society of Professional Engineers, 71
Alabama Space & Rocket Center, 6, 71
Alabama, Univ. of, 74, 230
alarm. See safety devices.

Alaska, 47, 134
Alberta, Canada, 38
Aldebaran (star), 3
Aleksandrov, Anatoly P., 89, 183
Aleksandrov, L.A., 167
alert. See safety devices.
Alexander, Dr. Carter, 85
algae, 123
algebra. See mathematics.
Algernon Blair Industrial Contractors Inc., 139
alien. See extraterrestrial.
"All-American Boys" (book), 204
Allen, Bryan, 182
Allen, H. Julian, 16
alloys, 106, 170, 192
Alpha Centauri (star), 37
alterations. See modifications.
altitude (altimeter), 12, 58, 61, 63, 67, 76, 93, 98, 135, 141, 149, 150, 181, 191, 192, 193, 196, 198, 202, 209, 211, 215, 221, 235
aluminum, 45, 108, 119, 143, 170, 211, 215
American Airlines, 45
American Assn. for the Advancement of Science (AAAS), 39, 49
American Bar Assn., 13
American Council on Education, 148
American Geophysical Union, 22, 102, 106, 234
American Inst. of Aeronautics and Astronautics (AIAA), 19, 235
American Machine & Foundry Co., 175
American Meteorological Society, 22
American Physical Society, 75
American Society for Public Administration, 229, 230
American Society of Engineering Education, 107, 152
American Society of Mechanical Engineers (ASME), 24
American Univ., 220
Ames laboratory (NACA), 16
Ames Research Center (ARC)
 —contracts, 173, 216
 —missions, 103, 159
 —people, 16, 85, 107, 147, 152, 166, 167, 169, 190
 —programs, 15-16, 44, 47, 57, 61, 68, 72, 84, 85, 92, 107, 136, 137, 159, 166, 167, 186, 188, 190, 208, 215, 231, 235

amino acids. See organic.
ammonia. See organic.
ammunition. See weapons systems; explosions.
amplitude modulation (AM), 133
Ams 2 (USAF metesat), 244
analog. See computers.
analysis, 70, 167, 201, 211, 227
Anderson, Donald, 6
Anderson, Jack, 73
Anderson, Maxie, 172, 173
Anderson, Robert, 73
Andover, Maine, 10
Andrews AFB, Md., 203
Angola, Peoples Republic of, 179
Anik (Canadian comsat series). See Telesat Canada.
animals, 68, 143, 146, 167, 201
animation (equipment, exhibits, etc.), 172
anniversaries. See observances.
anomaly. See failure; malfunction; problem.
Antarctic, 31, 39, 66
antennas, 33, 235
— airborne, spaceborne, 55, 235
— ground-based, 4, 154–155
antisatellite measures. See interceptor satellite.
Antonov, O.K., 29
Apollo, 28, 31, 44, 52, 63, 72, 75, 101, 113, 115, 156, 191, 195, 220, 229, 235
Apollo 4, 215
Apollo 7, 24
Apollo 9, 162, 191
Apollo 10, 128
Apollo 11, 3, 132, 165, 181
Apollo 12, 101, 161, 165
Apollo 13, 101
Apollo 14, 165, 195
Apollo 15, 5, 101, 165, 191, 195
Apollo 16, 165
Apollo 17, 51, 101, 165
Apollo lunar-surface experiment packages (ALSEP), 156, 161, 165
Apollo-Soyuz Test Project (ASTP), 11, 23, 52, 89–90, 121, 128, 132, 187
applications, 56, 59, 176, 162, 168, 169, 175, 203
Applications, NASA Office of (OA). See NASA.
applications sounding-rocket program (ARSP), 138

applications technology satellite. See *Ats 1*.
Applied Physics Laboratory, Johns Hopkins Univ. (JHU/APL), 197
applied research, 118
approach and landing tests (ALT). See also Space Shuttle, testing, 143, 145, 212, 235
appropriations. See Congress; NASA budget.
Arab nations. (See also Middle East), 14, 103
Araks (French experiment), 117
Arava (Israeli transport aircraft), 71
archives, 195
arcing. See insulation.
ARCO Inc., 219
Arctic, 140, 234
Argentina, 222
Ariane (ESA launch vehicle), 16, 168, 225
Ariel (U.K. satellite), 174
Arizona, 47, 48, 80, 94, 193, 219
Arizona State Univ., 192
Arizona, Univ. of, 21, 107, 137
Arkalyk, USSR, 27
Arktika (USSR ship), 158
Armed Forces Radio and TV Service, 10
Arms Control and Disarmament Agency, 31
Army, U.S. See also Defense, U.S. Dept. of, 115, 134, 172, 194, 201, 216, 227, 231–232
— Air Corps, U.S., 131
— Ballistic Missile Agency (ABMA), 215
— Chemical Corps, 175
— (Corps of) Engineers, 33
— Propulsion Laboratory, LeRC, 231
Arnold, Dr. James, 146
artifacts, 124
artificial environment. See environment.
Aryabhata (India domestic comsat), 174
Ascension Island, 8
Asia, 10, 154
Asimov, Isaac, 57
Asinyarov, Gennady, 124
Assateague (Wildlife Refuge; Coast Guard sta.), Va., 69, 164
assembling; assembly, 45, 50, 64, 106, 113, 133, 158, 160, 192, 208, 211, 220, 228, 230, 233, 235
ASSESS (airborne science/Spacelab experiment-system simulation), 72, 84

assessment. See evaluation.
asteroids, 157, 106, 124, 145, 207
Astrain, Santiago, 65
astrometrics, 165
astronauts (See also individual names.), 5, 6, 10, 11, 18, 84, 85, 121, 132, 140, 144, 157, 162, 172, 179, 181, 185, 191, 198, 201, 204, 205-206, 210, 211, 214, 222, 228
astronomers, 9, 10, 18, 25, 31, 32, 47, 48, 56, 63, 66, 69, 72, 84, 108, 121, 137, 152, 165, 167, 189, 204, 206-207, 210, 216, 227, 233, 235
astrophysics, 137, 224, 228
Atchafalaya River, 33
Atkinson, Richard C., 117-118
Atlantic Ocean, 8, 10, 27, 30, 127, 139, 167, 171, 172, 180, 198, 210, 227
Atlas (NASA launch vehicle), 229
Atlas Centaur, 11, 37, 149, 180, 202, 205, 209, 221, 229
atmosphere
 —interference, 10, 186, 233
 —ionosphere, 58, 85
 —magnetosphere, 34, 102, 168-169, 176
 —planetary, 39, 49, 55, 59-60, 66, 68, 69, 92-93, 102, 119, 121, 136, 156, 161, 166, 184, 187, 207, 221-222
 —pollution, 72, 84, 138, 141, 165, 204
 —research, 21-22, 54, 56, 59, 66, 68, 127, 131, 137-138, 148, 166, 175, 197-198, 234
 —stratosphere, 138, 192
 —upper atmosphere, 19, 54, 193, 208-209
Atmosphere Explorer (AE) missions, 136
atmospheric cloud physics laboratory (ACPL), NOAA, 25, 130, 175
atmospheric physics laboratory. See atmospheric cloud physics laboratory.
atmospheric variability experiments (AVE), 127
atomic energy. See power sources, nuclear.
Atomic Energy Commission (AEC). See Energy Research and Development Administration (ERDA).
Ats 1, 64
attitude control (See control systems.), 182
auroras, 38, 40, 85
Australia, 10, 37, 48, 69, 81-82, 103, 133, 162, 174, 212

Austria, 32, 103
automated systems, 45, 47, 74, 76, 106, 109-110, 113, 119, 131, 151-152, 153, 170, 178, 192, 193, 196, 198, 216, 219, 220, 221-222, 224, 225, 229, 235
automobiles. See vehicles.
autopilot, 96
Avco Corp., 177
Avery, Robert M., 25
aviation industry. See aerospace industry; airlines.
aviation research, 60, 213
aviation safety reporting system (ASRS), 43, 128
avionics, 61, 112, 202, 207
Avions Marcel Dassault-Breguet, 107
awards, 11, 19, 20, 44, 49, 71, 87, 93, 101, 124, 133, 145, 152, 175, 181, 182, 188, 203, 205, 212, 220, 235
Azores Islands, 181

B-1 (bomber), 7, 112, 155
B-52 (U.S. aircraft), 119, 211
B-58, 235
B-70 (planned U.S. bomber), 13
B-707. See Boeing 707.
B-727. See Boeing 727.
B-737 (LaRC flying laboratory). See Boeing 737.
B-747. See Boeing 747.
background, 208-209, 210
backup. See schedule; support systems.
bacteria, 63, 201
Bahamas, 162
Bahcall, Dr. John, 10
Baker, James L., 64
Ball, George W., 40
Ball Brothers Research Corp., 136, 181, 186
ballistic missile. See weapons systems.
balloons, 76, 93, 98, 119, 127, 138, 172, 173, 181, 188, 215, 227
Baltimore, Md., 12, 164
Banana River, 28-29, 57
Bane, Don, 101
Barbon, Jill, 192
Barents Sea, 57
barge, 160, 211
barium, 36
Baron, Bob, 218
Barrere, Marcel, 184, 185
basic research, 117-118
Battelle Institute, Columbus div., 43
Battelle Memorial Institute, 17, 231

batteries, 15, 84, 94, 120, 149, 216
battery power. See power sources.
Baykonur, USSR, 222
beacons, 6, 132-133, 172, 173
beam fabrication. See space construction.
Bedford, Mass., 172
Belem, Brazil, 222
Belenko, Lt. Viktor, 26
Belgium, 32, 103, 154
Belgrade, Yugoslavia, 130
Bell Helicopter Textron Co., 216
Bell Laboratories, 62, 166, 177
Belton, Dr. Michael J.S., 38-39
Bendix Corp., 11, 63, 74, 150, 156
"bends." See hyperbaric research.
Bennich, Jerry D., 230
Benton, Dr. Eugene, 167
Bering Strait, 178
Berlin, Germany, 235
Berlin, Univ. of, 115
Bermuda Triangle, 64, 173
beryllium, 30
Betelgeuse (star), 104
bicentennial, U.S., 182
big-bang theory. See theory.
Big Bird (USAF reconsat), 22
Billiu, David S., 72, 84
binary stars. See stars.
biochemistry, 123, 190
bioengineering. See biology.
biology, 4, 32, 62, 65, 68, 71, 85, 101, 125, 159-160, 161, 167, 174, 184, 190, 201, 206, 215-216
biomedical applications office, MSFC. See medical research.
biophysics, 146
Bios 3 (Soviet experiment), 123
Biosatellite 3 (U.S. mission), 160
biosats, 159-160, 167
black history, 18, 35
black holes, 37, 149, 210
blackout (communications), 4
blades, propeller. See propellers.
Blamont, Jacques, 119
Bledsoe, Maj. Adolphus H., 181
Bleyle, Wilhelm, KG, 70
blood, 133
boats. See ships.
Boeing Commercial Airplane Co., 110, 128, 163, 192
Boeing Company, 8, 22, 42, 53, 112, 173
Boeing Engineering and Construction Co., 129

Boeing Services Intl., 102
Boeing 707, 66, 182
Boeing 727, 128, 162
Boeing 737, 132, 163, 212, 222
Boeing 747, 27, 45, 55, 64, 67, 107-108, 110, 113, 118, 122, 132, 135, 141, 150-151, 171, 175, 179, 182, 191, 196, 202, 207, 211, 228
Boese, Dr. R.W., 166
Boland, Rep. Edward P., 108, 135
Boldt, Dr. Elihu, 212
bolides. See meteorites.
Bolivia, 196
bombs. See explosives.
Bond, Langhorne M., 62
bones. See medical research.
Bonn Institute for Radio Astronomy, 10
Bonneville Power Administration, U.S., 80
book. See publications.
boom (aircraft; spacecraft), 75, 155-156, 167, 186, 195, 218
boosters. See rockets.
borosilicates, 233
Boston Univ., 147
botany, 32, 201
Bovay Engineers, Inc., 121
Bradt, Dr. Hale, 212
brain, 146
brakes; braking, 122-123, 150-152, 171, 196
Brantley, Whitt Jr., 128
Braufman, Adam I., 4
Brayton engine, 159
Brazil, 134, 222
breeder reactor. See power sources, nuclear.
Brezhnev, Leonid, 76, 143
briefing. See media.
Brigham City, Utah, 133
Bristol Scout (WWI U.K. plane), 141
Britain. See United Kingdom.
British Isles (See also United Kingdom.), 40
British Royal Society. See Royal.
broadcasting. See frequencies, broadcast; media.
Brooksbank, William A. Jr., 215
Broome, C. Calvin, 125
Brown, Gov. Edmund G., Jr., 148
Brown, Harold, 74, 147
Brussels, Belgium, 63
Bryson Construction Co., 64

BSE, 229
budget
—federal, 35-36
—military, 22, 52, 54
—NASA. See NASA, budget.
Buenos Aires, Argentina, 212, 222
Buffalo, 213
Bugreyev, Nikolai, 124
buildings; building trades (see also construction of facilities.), 213
Bulgaria, 20, 159, 174
buoy, 63, 79
buoyancy, 118
Bureau of Land Management, DOI, 47
burn; burner; burnout, 188, 199, 211, 233
business. See aerospace industry; commercial applications; industrialization.
Buzzard, Wallace, 214

C-5 galaxy, 111, 122, 222
C-47, 12
C-band communications, 139
cabling, 98, 113
calculations; calculators. See computations.
calibration, 37, 42, 85, 113, 155, 195, 210
California, 36, 40, 47, 80, 81, 105, 148, 162, 202, 228
California Institute of Technology (CalTech), 44, 137, 150, 190, 203, 230
California, University of
—Berkeley, 147, 167, 209, 230
—San Diego, 145-146, 150, 212
Calio, Dr. Anthony J., 175, 220
Callisto, 166
calorimeter. See instrumentation.
Calspan Corp., 216
CalTech. See California Institute of Technology.
CAMAC (electronic assembly system), 50
Cambridge, Mass., 40
cameras (See also photography; imaging), 216
—earth-based, 167
—satellite, 37, 176, 189, 210
—television, 12, 33, 104, 121, 160, 198, 224, 234
camouflage, 177
Canada, 11, 19, 32, 38, 56, 103, 110, 122, 134, 138, 153, 163, 175, 208, 213, 229

Canadian Department of Communications, 19, 47
Canberra, Australia, 4, 82
cancer, 146, 216
capacity, 206
Cape Canaveral, Fla., 8, 52, 62, 102, 103, 132, 167, 168, 195, 202
Cape Town, South Africa, 182
capital punishment. See law enforcement; public policy.
captive flight, 17, 103, 107-108, 118, 122-123, 135
Carassa, Prof. Francesco, 157
carbon 14 dating, 39-40
carbon compounds, 206
careers, 41
cargo-integration test equipment (CITE), 185-186
cargo logistics airlift-systems study (CLASS), 112
cargo transfer (see also payloads), 98
cargoes (See also air freight; payloads; shipping.), 162, 214
Caribbean Sea, 10, 171, 176
Carr, Gerald A., 121
carrier aircraft. See Boeing 747.
Carruthers, Dr. John R., 62
cars. See vehicles.
Carter, President Jimmy, 1, 10, 23, 30, 38, 48, 52, 56, 62, 64, 65, 67, 71, 72, 93, 102, 111, 114, 131-132, 185, 218
Caruso, Andrea, 65
Casani, John, 137, 167
castings, 106
Cataldo, Charles E., 11
catalog, 44, 117, 129, 205
catalyst, 221
catastrophe, 149
celebrations. See observances.
celestial, 69, 188
celestial sphere. See galaxy.
cells, 146, 152
censorship, 130
Centaur, 180, 220, 221
centers, NASA. See individual center listings.
Centre National d'Etudes Spatiales (France), 66, 119
centrifugal force, 57, 119
Cerenkov radiation, 31
certification, 182
Chad, Republic of, 110
changes. See modifications.

channels (communications). See frequencies, broadcast.
Chapman, Maj. Gen. Kenneth R., 53
Chappell, Dr. C.R., 67
charge-coupled devices, 22, 71
charged particles. See energetic particles.
Charyk, Dr. Joseph V., 72
chase planes (See also T-38.), 155
Chaudiere River, 110
checkout, 11, 211, 223-225
chemicals, 18, 73, 85, 101, 136, 146, 161-162, 188, 201, 230
chemistry, 17, 207, 222
Chesapeake Bay, 63
Chicago, Ill., 111
Chicago, University of, 102
Chien, Hsueh-shen, 131
Chile, 103
China, People's Republic of, 131, 154, 182, 229
Chincoteague, Va., 69, 164
Chirivella, Dr. Jose, 221
Chiron, 204
chlorella, 123
chlorine compounds, 146
Christmas Islands, 41
chrome; chromium, 175, 230
Chryse Planitia (region of Mars), 70, 125
Chrysler Corporation, 11
cinema. See films.
circuits, 133, 154
city-federal cooperation, 8, 188
civil aviation, 60, 110, 122, 178, 192
Civil Service Commission, U.S. (See also employment; NASA management.), 104, 161, 215
civilians, 21, 25, 121, 140, 148, 163, 173, 179, 195, 201, 205, 231, 235
Clark, Ray, 24
Clarksburg, Md., 138
clay, 190
cleaning, 218
Cleveland, Ohio, 12
clocks, 51
closed universe. See theory.
clothing, 59, 201
clouds (cloudcover), 18, 25, 36-37, 39, 85, 119, 127, 130, 210
coal (see also fuel.), 65, 73, 178
Coast Guard, U.S., 79, 206
coastal zone color scanner, 194
Cohen, Aaron, 143
cold, 227

Colin, Dr. Lawrence, 166
College of William and Mary, 18
Colli, Jean-Claude, 14
collimator. See instrumentation.
Collins, Michael, 116, 181
collision (See also crash; accident.), 155, 156, 177
colloquium. See meetings.
Colman, Everett R., 133
Cologne, W. Germany, 84
Colombia, 134
color, 142, 165, 194
Colorado, 80, 161, 176
Colorado, University of, 137
Columbia University, 37, 93, 114
Columbus, Ohio, 17
Columbus medal, 188
Coma cluster (stars), 37
comets, 204, 207
commands, 133, 186
commemoration. See observances.
Commerce, U.S. Dept. of (See also NOAA.), 67, 114, 141, 193
commercial airlines. See airlines.
commercial applications (See also industrialization.), 59, 139, 163-164, 169, 195, 199-200
commercialization. See industrialization.
committees, congressional. See Congress.
communications
 —global systems, 38, 124-125, 198, 220
 —lightwave, 62
 —problems, 118, 131, 134, 157
 —radio, 19, 74, 139
 —relay. See relay.
 —satellites. See comsats.
 —spacecraft, 55, 136-137, 226
 —systems, 74, 118, 122, 180
Communications Satellite Corporation (ComSatCorp), 10, 65, 72, 120, 138, 169-170, 175, 206, 213, 217, 228, 229
Communications Technology Satellite (Cts), Canada, 13, 47, 138, 208
communist bloc. See Soviet bloc.
community relations (See also public relations.) 3, 74, 192
Compagnia Industriale Aerospaziale, 157
comparisons, 13, 26, 121, 212
compass. See instrumentation.
compatibility, 185
competition (See also contest.), 71
 —economic, 8, 62, 153-154, 187, 203

—national science, 133
Complex 39B, KSC, 128
components. See instrumentation.
composite.
   —fuels, 133
   —materials, 45, 76, 79, 128, 163-164, 177, 180
computations (See also measurements.), 27, 142, 227
computers, 70, 143, 227
   —airborne, 26
   —assembly, 50
   —enhancement. See data processing.
   —maintenance, 74, 221
   —programs, 21, 128, 202-203
   —software, 38
   —spacecraft, 33-34, 55, 66, 150-152, 155-156, 192, 217
   —systems, 118, 155-156, 198, 213
computerization. See automated systems.
ComSatCorp. See Communications Satellite Corporation.
Comsat General (subsidiary of ComSatCorp), 72, 124, 139
comsats (See also names of individual communications satellites.), 4, 6, 7-14, 22, 32, 36, 38, 51-52, 54, 72, 122, 135, 153-154, 156-157, 161, 168, 173, 175, 181, 182, 202, 207, 208, 209, 221, 226, 229
Comstar system (domestic comsats), 72
Comstar-C, 229
Comstock, Bruce, 181
Conant, William W., 79
concept, 98
Concorde, 53, 66, 151, 213
concrete, 196
condensation, 9
conduction, 141
Cone, Dr. Clarence D., Jr., 146
conferences. See meetings.
configuration, 17, 133, 134, 138, 216, 218 222, 225
confirmation, 188, 196
Congo, Peoples Republic of the, 198
Congress, U.S., 108, 121, 132
   —conference committee, 108, 135
   —House of Representatives, 38, 60, 64, 108, 124-125, 180
   —Senate, 56, 62, 111, 114, 195, 217
conjunction period (astronomical), 15
conservation, 11-12, 185

Consiglio Nazionale della Richerche (CNR), Italy. See Natl. Research Council.
console, medical communications, 74
constellations, 104
Constitution, U.S., 62
construction of facilities, 53, 62, 63, 64, 65, 70, 72, 80, 113, 131, 132, 139, 188, 208, 228-229, 233
consumables. See supplies.
containerless processing (See also new techniques.), 30, 105, 192
contamination (See also pollution.) 30, 110-111, 168, 204
contest, 162, 182
Continental (aircraft engine), 130
contingency. See emergencies.
contour mapping. See topography.
Contractors Cargo Co., 113-114
control; control centers, 14, 15, 101, 155-156, 171, 207, 228
   —systems, 23, 30, 55, 63, 74, 96, 110, 118, 121, 125, 133, 134, 135, 136-137, 146, 149-150, 150-152, 157, 160, 161, 178-179, 186, 192, 199, 202-203, 207, 216, 221, 223-225
control, weather. See meteorology.
Controlled Biosynthesis Laboratory, USSR, 123
controllers. See controls.
controversy, 40-41
Convair Division. See General Dynamics.
Convair 990. See Galileo.
conventional, 8, 96, 129, 134, 199-200, 213-214, 216
conversion, energy/power. See power sources, conversion; solar conversion devices.
cooking, 124
cooling (See also cryogenics.), 132, 153, 213, 231
Cooper, Henry S.F., 49
Cooper, Robert S., 161
cooperation (See also city-federal; industry-government; international; interservice; NASA research; state-federal.), 59
Coordinating Committee of Nations, 141
coordination, 33, 61, 84, 159, 162
copper, 143
Cornell University, 48, 59, 69, 124, 143, 177
Corning Glass Works, 232-233

corrosion, 11
Cortright, Dr. Edgar M., 203
Cosmic Background Explorer, 50
cosmic radiation, rays, 31-32, 39-40, 149-150, 153, 188, 190, 208, 212, 215
Cosmic Research Institute, USSR, 66
cosmology, 56, 228
cosmonauts (See also the names of individuals.), 10, 20, 23, 26-27, 189-190, 223-225
Cosmos 782, 159-160
Cosmos 888, 238
Cosmos 889, 238
Cosmos 890, 238
Cosmos 891, 239
Cosmos 892, 239
Cosmos 893, 239
Cosmos 894, 240
Cosmos 895, 240
Cosmos 896, 240
Cosmos 897, 240
Cosmos 898, 241
Cosmos 899, 241
Cosmos 900, 241
Cosmos 901, 241
Cosmos 902, 241
Cosmos 903, 242
Cosmos 904, 242
Cosmos 905, 242
Cosmos 906, 242
Cosmos 907, 243
Cosmos 908, 243
Cosmos 909, 243
Cosmos 910, 243
Cosmos 911, 244
Cosmos 912, 244
Cosmos 913, 244
Cosmos 914, 244
Cosmos 915, 245
Cosmos 916, 245
Cosmos 917, 245
Cosmos 918, 245
Cosmos 919, 246
Cosmos 920, 246
Cosmos 921, 246
Cosmos 922, 247
Cosmos 923, 247
Cosmos 924, 247
Cosmos 925, 247
Cosmos 926, 247
Cosmos 927, 247
Cosmos 928, 248
Cosmos 929, 158, 248
Cosmos 930, 248
Cosmos 931, 248
Cosmos 932, 248
Cosmos 933, 249
Cosmos 934, 249
Cosmos 935, 249
Cosmos 936, 159, 167, 206, 249
Cosmos 937, 250
Cosmos 938, 250
Cosmos 939, 250
Cosmos 940, 250
Cosmos 941, 250
Cosmos 942, 251
Cosmos 943, 251
Cosmos 944, 251
Cosmos 945, 251
Cosmos 946, 251
Cosmos 947, 252
Cosmos 948, 252
Cosmos 949, 252
Cosmos 950, 252
Cosmos 951, 253
Cosmos 952, 253
Cosmos 953, 253
Cosmos 954, 253
Cosmos 955, 253
Cosmos 956, 254
Cosmos 957, 254
Cosmos 958, 255
Cosmos 959, 255
Cosmos 960, 255
Cosmos 961, 256
Cosmos 962, 256
Cosmos 963, 257
Cosmos 964, 257
Cosmos 965, 257
Cosmos 966, 258
Cosmos 967, 258
Cosmos 968, 258
Cosmos 969, 259
Cosmos 970, 259
Cosmos 971, 259
Cosmos 972, 259
Cosmos 973, 259
cost reduction, 12, 15, 35, 47, 50, 55, 57, 58, 61, 65, 66, 73, 80, 94, 104, 108-109, 111, 118, 131, 145, 156, 161, 162-163, 175, 178, 179, 188, 208, 218, 219, 220, 228
cotton, 59
counseling, 148
countdown. See launch control.
counting, 189, 193

court trials. See judicial proceedings.
Cousteau, Jacques-Yves, 149
crash, 105
craters, 76, 124
Craven Point, N.J., 63
creation of universe. See theory.
Creole, 103
crew (See also astronauts; cosmonauts; Space Shuttle, crew; pilots; payload or mission specialists; manned flight; individual names.), 193, 198, 223-225
criminal justice. See law enforcement; judicial proceedings.
Criswell, Dr. David, 157
criticism, 58-59, 105, 187, 204
crops, 5, 25, 59, 119, 141, 147, 165, 193, 218
cross-country Shuttle flights. See Space Shuttle testing.
crowding. See frequencies, broadcast.
cruise missiles. See weapons systems.
cryogenics (See also cooling.), 50, 58, 132, 136, 160, 186
crystal growth, 62, 192
Cs (*Sakura*, Japan comsat), 112, 226, 258, 261, 263
Cuba, 20, 174, 183, 196
cultural affairs, 130, 131, 143, 164
Cunningham, Walt, 204
Cygnus (constellation), 107
Czechoslovakia, 20, 159, 174, 196

damage, 66, 80, 119, 146, 154-155, 168, 180, 213-214
danger (See also safety.), 98, 127, 160, 184
Darmstadt, W. Germany, 74, 210
Daspit, Leo, 233
data
—acquisition, 4, 5, 14, 15, 62, 63, 72-73, 112-113, 145, 147, 171, 186, 189, 192, 194, 207, 219-220
—analysis, 15, 171
—collection. See data acquisition.
—distribution, 111, 134, 186-187, 189
—management, 169, 174, 185
—processing, 7, 56, 71, 136-137, 142, 143, 147, 234
—relay. See relay.
—transmission, 63, 125, 132-133, 139, 154, 155, 157, 158, 163, 171, 214
—users, 58, 111, 143, 186-187, 210
Davidsen, Dr. Arthur, 121

daylight; daytime, 158, 195
DC-10, 67, 79, 132
death, 16, 19, 29, 74, 105, 115, 235
debris, 98
decompression sickness ("the bends"). See hyperbaric research.
deep space network, 154, 165-166
defectors, 13, 26
defects. See malfunctions; problems.
Defense Communications Agency, 65
Defense, U.S. Dept. of (DOD) (See also Advanced Research Projects Agency; Air Force, U.S.; Army, U.S.; Marine Corps, U.S.; Navy, U.S.), 6, 7, 38, 52, 53, 54, 59, 65, 74, 105, 109, 112, 114, 122, 132, 140, 141-142, 154-155, 162, 163, 177, 197, 203, 211, 236
—contractors, 7, 8, 22, 109
—unidentified spacecraft, 83, 177, 197, 239, 241, 243, 246, 254, 258
defense satellite-communications system (DSCS), 83-84, 124, 243
defense spending, 134, 194
defense systems. See weapons systems.
dehumidifying, 16
Deimos (moon of Mars), 124, 152
delay (*See also* problems.), 103, 131, 137, 149-150, 157, 168, 171, 193, 209, 217
delivery, 206, 221
Delta (NASA launch vehicle), 14, 37, 52, 73, 106, 111, 112, 145, 156, 168, 171, 180, 193, 195, 202, 205, 210, 229
demonstration (See also exhibits.), 45, 113-114, 130-131, 148, 173, 179, 186, 196, 206, 212, 217-218, 222, 228, 235
Denmark, 1, 154
density, 93, 98, 124, 132, 168, 222
Denver, Colorado, 13
deployment, 75, 93, 155, 198, 211, 221, 235-236
depreciation, 163
Desert Research Institute, 25
deserts, 150
design
—aircraft, 28, 29, 61, 71, 110, 112, 130, 138, 151, 152, 180, 192, 199, 217-218, 219
—engineering, 24, 34, 107, 147, 153, 154, 160, 169, 173, 207
—instrumentation, 42, 47, 64, 68, 69, 71, 135, 165, 208-209, 211
—ships, 2

—spacecraft, 43, 48-49, 117, 121, 137, 156-157, 160, 162-163, 235-236
destruction; destruct system (See also interceptor satellites.), 177, 213, 234
detection; detectors, 22, 31-32, 58, 63, 71, 80, 101, 127, 140, 149, 160-161, 176, 188, 189, 196, 204, 210, 215, 216, 233
detente, 144
Detroit Diesel Allison, 134
deuterium, 109
developing nations, 130
deviations (See also malfunctions.), 96, 190, 223
devices. See equipment; instrumentation.
diagnosis. See medical research.
diesel, 179
diet. See food.
difficulty. See problems.
digital. See computers.
dimensions. See size.
diplomacy. See international relations.
direct broadcast, 129-130, 154
direct readout. See real-time monitoring.
direction. See aiming; orientation; management.
disaster. See emergencies.
discoveries (See also new products; new techniques.), 48, 59-60, 69, 124, 129, 130, 152, 158, 201, 203-204, 208-209
Discovery, 33
disease. See medical research.
dispersion, 192, 209
display. See exhibits.
disposable, 160, 170
distance, 10, 33-34, 57, 76, 176-177, 182-183, 195-196, 226-227, 233
distortion, 25-26
divers; diving, 160, 208
DNA. See genetics.
docking, 26-27, 89-90, 189-190, 223-224, 231
documentation, 89, 169, 182, 192-193
domestic comsats. See comsats; names of individual satellites.
"doomsday plane", 38
Doreen, 165
Douglas Aircraft Company (McDonnell Douglas Corp.), 112, 172
Downey, James E. III, 19
drafting; draftsmen, 194, 213
drag, aerodynamic/atmospheric, 72, 119-120, 191, 197, 211
Drexel University, 106
drogue. See parachutes.

drop, 93, 101, 119-120, 211
drought, 39-40, 80-81
drugs. See pharmaceuticals.
Dryden Flight Research Center
—contracts, 45
—people, 74, 101, 144-145, 169, 175, 191, 217-218, 221, 235
—programs, 15, 23-24, 42, 55, 61, 69, 96, 101, 103, 107-108, 113-114, 118, 135, 143, 168, 170, 171, 193, 202, 218, 221-222
Dscs II-7, II-8. See defense satellite-communications system.
Dubna, USSR, 230
Duke, Dr. Michael G., 28
Duke University, 216
Duki, USSR, 178
dummy. See simulations.
durability, 34, 71, 143, 159, 175, 211
duration, 71, 98, 121, 145, 189, 198, 206
Durham, N.C., 216
dust, 94, 107, 178, 219
Dutch. See Netherlands.
Duxbury, Thomas C., 176
dwarf stars. See stars.
dynamic, 55, 203, 211

Eaker, Gen. Ira C., 172
early-warning
—satellites, 22
—aircraft, 43
earth, 103, 143, 183, 209, 235
—resources monitoring, 47, 48, 54, 63, 84, 108, 111, 130, 134, 141, 147, 148, 168, 225, 229
earth station; earth terminal. See ground station.
earthquakes. See seismology; seismometer.
East Germany. See German Democratic Republic (GDR).
Easterling, Mahlon F., 175
Eastern Europe. See Soviet bloc.
Eastern Shore, 164
Eastern Test Range (ETR) (See also Cape Canaveral; KSC Spaceport.), 2, 52, 73, 102, 112, 155, 210, 226
Eberhart, Jonathan, 49
ecliptic, 195
ecology, 47, 63, 165
economic policy, 2, 3, 75, 76, 208
Economic Stimulus Bill, 118
Eddy, Dr. John A., 39-40
education, 115, 117, 157, 165, 177, 208

Edwards Air Force Base (AFB), Calif., 17, 150, 171, 178, 179, 191, 196, 219, 235
Effelsberg radiotelescope (Max Planck Inst.), 9
efficiency, 13, 79, 112, 145, 159, 200, 214
Efik, 103
Egypt, 103, 196
EIC Corp., 2
Eichman, Wilbert F., 156
Einstein theory of relativity, 4, 51
Eisele, Donn F., 24
Eisenhower, President Dwight D., 161
Ekran 2, 253
elasticity. See design, aircraft.
ELDO, (former) European launch-vehicle development organization; see ESA.
electric power. See power sources.
electrical systems; electricity, 21, 71, 75, 94, 111, 127, 168, 170, 190, 194, 202, 213, 227, 234
electrocardiograms; electrocardiograph (EKG), 74, 214
electrochemicals, 63
electroencephalograph, 214
electrolytes, 192
electromagnetism, 23, 63, 72, 102, 106
electronics, 26, 50, 71, 74, 112, 153, 155, 158, 177, 194, 202, 214
Electronics Research Center, Mass., 175
Electronics Society, PRC, 182
electrons (See also energetic particles.), 75, 85, 102, 109, 117
electrostatics. See insulation.
elevators; elevons, 128, 164
Elliot, Dr. James L., 48, 59-60, 69-70
elliptical orbit. See orbit, planetary; orbit, spacecraft.
El Toro, Calif., 155
embargo, 141-142
embolism. See medical research.
emergencies, 36, 50, 74, 138, 155, 160, 208, 214, 216, 217, 227
emissions. See radiation.
employment (See also Civil Service Commission, U.S.; NASA management.), 35, 41-42, 61, 75, 104, 107, 109, 113, 118, 125, 161, 169, 175, 214-215
Empresa Publica de Telecomunicacoes (EPTEL), 179
Encke, comet, 207
encounter. See impact; trajectories.
endangered species, 57
endurance. See duration; lifetimes.

energetic particles (See also ions; electrons; neutrons; protons; radiation.), 19, 25, 31-32, 75, 85, 109, 166, 176, 215
energy
—conservation, 12, 14, 28-29, 35-36, 60, 61, 73, 79, 94, 117, 128, 130, 147, 148, 159, 162, 177-178, 180, 185, 186-187, 188, 199-200, 218
—conversion. See power sources, conversion; solar conversion devices.
—crisis. See energy conservation.
—efficiency. See energy conservation.
—heating and cooling, 28-29, 128-129, 179, 186, 188, 203, 215
—levels, 208
—research, 53, 117-118, 134, 185, 194
—shortage. See energy conservation.
—sources. See power sources.
—storage, 80, 120, 179
—technology, 22, 71, 80, 145
—transfer (transmission) (See also power sources, conversion.), 61, 98, 136, 141, 146, 165
Energy Agreement of 1974, U.S.-USSR, 111
Energy Research and Development Administration (ERDA), U.S., 5, 29, 36, 53, 70, 94, 109, 110, 111, 121, 129, 140, 162, 172, 176, 177, 179, 186, 199-200
Energy, U.S. Dept of (DOE), 162, 185, 195, 209, 219, 220, 228
engineering, 70, 71, 80, 124, 136, 143, 151, 182, 196, 199, 213, 219, 230, 231, 235
engines. See power sources.
England. See United Kingdom.
Engle, Joe, 122, 135, 170, 171, 191
English, 103
Enterprise, 17, 23-24, 27, 107, 113, 118, 122, 135, 141, 150, 151, 168, 170, 171, 178, 190-191, 196, 202, 211, 230
entry (atmosphere). See reentry.
environment, 8, 21-22, 32, 63, 64, 68, 71, 73, 93, 123, 132, 137, 147, 159, 167, 168, 174, 181, 187, 194, 218, 222, 225
environmental monitoring, 5, 55, 62, 63, 110, 148, 194, 229
Environmental Protection Agency (EPA), U.S., 63, 192
epitaxial growth, 30
equal opportunity, 2, 21, 192

equator, 18, 67, 113, 136, 138, 153, 156, 210
equipment (See also instrumentation.), 54, 58, 59, 63, 80, 84, 106, 112, 113, 149, 163, 208, 212, 214, 223, 224, 231
equivalency principle, 51
Erickson, Dr. Edwin, 107
ERNO (Raumfahrttechnik GmbH, VFW-Fokker subsidiary), 48
error, 127, 155, 190
Esperanto, 103
ESRO, (former) European Space Research Organization. See ESA.
estimates, 196
Etam, W.Va., 10
Ets II (Japan engineering test satellite, *Kiku 2*), 41
Europa, 166
Europe, 19, 107, 157, 188, 210, 228, 234
European Broadcasting Union (EBU), 10
European Space Agency (ESA), 16, 25, 32, 53, 56, 65, 67-68, 72, 73, 74-75, 85, 112, 122, 153, 154, 163, 166, 168, 170, 171, 174-175, 180, 182, 183, 189, 193, 199, 202, 210, 213, 225, 229, 230
European Space Operations Center (ESOC), 75
European Space Technology Center (ESTEC), 189
Eurovision, 168
EVA. See extravehicular activity (spacewalk).
evaluation, 138, 197
Evans, Ronald E., 51
Evans, Dr. Stanley, L., 152
Evans, Gen. William J., 203
Everette, Bob, 231
Everglades Towing Co., 231
evolution, 201
exercises. See practice.
exhaust, 186, 192-193
exhibits, 24-25, 61, 69, 71, 132, 148, 172
exobiology. See biology.
expansion (See also theory.), 199, 209, 233
expendable, 65, 104, 105, 229,
expense, 105, 145, 162-163, 168, 170, 173
experiment specialists. See mission specialists.
Experimental Aircraft Association, 130
Experimental Test Pilots, Society of. See Society.

experiments,
—biology, 4, 65, 101
—communications, 5, 138, 208
—materials research, 59, 130
—scientific, 65, 67, 75, 156, 157, 159, 210
—systems, 72, 75, 84, 133, 149, 150, 159, 165, 175, 224, 225
exploitation, 130, 145-146
exploration (planetary; terrestrial), 6, 49, 52, 70, 106, 149, 156, 158, 161, 162, 169, 198-199, 207, 221-222
Explorer 1, 115
Explorer 42. See *Uhuru*.
Explorer 54, 55. See Atmosphere Explorer (AE) missions.
Explorers Club, 70
explosives, 36
external tank. See Space Shuttle.
extraterrestrial, 9-10
extravehicular activity (EVA), 18, 51
extravehicular mobility unit (EMU), 45
Exxon Corp., 73
eyesight. See vision.

F-8, 55
fabrication (See also construction.), 25, 42, 45, 47, 49, 56, 59, 128, 136, 158, 163, 177, 211, 216, 220
facilities, 61, 65, 113, 132, 161-162, 163, 169, 173, 177, 194, 211, 228-229, 231
facilities support. See support services.
failures (See also malfunctions.), 58, 70, 80, 102, 118, 136, 151, 160, 165-166, 168, 171, 180, 189-190, 202, 203, 205, 209, 223, 224, 227, 234
Fairbank, Dr. William, 75
Fairchild Industries, 105, 116
fares. See fees.
farming. See crops.
Fastie, Prof. William G., 121
fatality. See death.
fear, 40
feasibility, 64, 81, 213, 216
Federal Aviation Administration (FAA), 43, 62, 72, 112, 122, 128, 130, 163, 173, 174-175, 192, 193, 212, 220, 222
Federal Aviation Agency. See Federal Aviation Administration.
Federal Communications Commission (FCC), 139, 169-170, 206, 213, 217
federal employees. See employment.

federal funding. See funding, government.
Federal Republic of Germany (West Germany), 1, 4, 33, 48, 70, 74–75, 84, 154, 162, 166, 187-188
Federation Aeronautique Internationale (FAI). See Intl. Aeronautical Federation.
fees, 65, 105, 143, 162-163
Fein, Juergen, 84
Felixstowe (WWI U.K. flying boat), 141
females. See women.
Feoktistov, Dr. Konstantin, 224
ferry test flights. See Space Shuttle, testing.
Feuerbacher, Dr. Bernt, 67
fiberglass, 130
Field Museum, Chicago, 39
fighter planes. See aircraft, military.
films, 61, 64, 87, 149, 172
financing. See funding.
Finch, Dr. William A., 14
Fink, Daniel, 174
Finland, 1
FIRES (firefighters' integrated response equipment system), 59
Fischer, Craig L., M.D., 230
fish, 165
Fish and Wildlife Service, 57
fission. See power sources, nuclear.
Five Towns, 66
fixed-wing aircraft, 134
flare. See maneuvering.
fleet satellite-communications system. See FltSatCom.
Fleet Weather Facility, USN, 66
Flemish, 103
Flemming, Arthur S., award, 93
Flerov, Georgy, 230
Fletcher, Dr. James C., 44, 49, 52, 93, 105, 116
flight, 182
 —control. See control.
 —dynamics laboratory. See Air Force, U.S.
 —operations, 8
 —path, 96
 —quality. See handling.
 —test
  aircraft. See aircraft testing.
  instruments, 96
  missiles, 8
Flight Medicine Center, JSC, 222

flightworthiness. See handling.
flooding, 1, 33, 110, 119, 138
Florida, 75, 105, 128, 196, 202
Florida State University, 131, 148
Florida, University of, 146
flotation, 231
FltSatCom, 1, 124, 202, 229
fluorocarbons. See atmosphere pollution.
fluoroscopy. See x-rays.
flutter. See aerodynamics.
flux, 166
flyby, 119, 226–227
fly-by-wire controls. See control systems.
flying laboratory. See Galileo II.
flying quality. See handling.
"flying wind tunnel", 134
FM. See frequency modulation.
fog, 158
Fokker F-7, 182
food, 1, 123-124, 224
Food and Drug Administration (FDA), 169
force (See also pressure.), 153, 189
Ford, President Gerald M., 1, 23, 35, 59, 72
Ford, Sen. Wendell H. (D-Ky), 111
Ford Aerospace Co., 7, 15, 207-208, 226
Ford Motor Co., 199-200
forecasting. See future plans; meteorology; predictions.
foreign, 62, 213
forests, 193, 196
Fort Belvoir, Va., 216
Fort Eustis, Va., 231
Fort Hood, Tex., 70
Fort Monroe, Va., 8, 188
Fort Worth, Tex., 199
Foxbat. See MiG-25.
France, 1, 13, 32, 53, 66, 103, 107, 117, 119, 154, 159, 172, 174, 184, 185, 198
Freden, Dr. Stanley C., 14
free flight. See aircraft, use in research; Space Shuttle testing.
freedom of inquiry, 40
freight. See air freight; cargo; shipping.
French Guiana, 127
frequencies, broadcast, 129-130, 139, 153, 157, 159, 168, 207, 208
frequency modulation (FM), 133
Fri, Robert, 70
friction, 151
Friedman, Dr. Herbert, 149, 174, 212
Friendship 7, 27

Frosch, Dr. Robert A., 93, 106, 114, 131–132, 146, 175, 177, 178, 186, 187, 189, 212, 217, 228, 232
fruit flies, 167
Fucino, Italy, 157
fuel (See also power sources.), 14, 25, 36, 58, 60, 61, 65, 110, 113, 128, 133, 146, 159, 160, 176, 199–200, 213, 231
—conservation. See energy conservation.
—shortage. See energy conservation.
—storage, 106
Fullerton, Lt.Col. Charles Gordon, 63, 118, 135, 150, 151, 178, 193, 196
Fulton, Fitzhugh L., Jr., 36, 150, 171, 191, 202, 235
funding, government, 58–59, 64, 65, 67, 68, 72, 108–109, 111, 117, 124–125, 134, 141, 156, 174, 180–181, 188, 194, 199, 206–207, 217
fuselage, 118, 120, 217
fusion. See power sources, nuclear.
future plans, 40–41, 50, 58, 60, 64–65, 72, 89–90, 115, 167, 188, 189, 197, 198, 206–207, 229, 233, 235–236

g forces. See gravity.
Gagarin, Yuri A., 29, 67, 123
Gagarin award (USSR), 11
galactic energy. See power sources.
galaxies, 9, 10, 37, 54, 76, 117, 121, 129, 136, 149, 153, 176, 210, 228, 233
Galaxy. See C-5.
Galileo I, 235
Galileo II, 25, 72, 84
Galileo scholarships, 235
gallium compounds, 145
gamma radiation, 48, 117, 149–150, 178, 188, 212
Garmire, Dr. Gordon, 212
Gartrell, Harold E., 232
Garrett AiResearch, 134
gas; gasoline. See fuel.
gas-chromatograph mass spectrometer, 55
gas-exchange experiment, 70–71
gaseous bodies, 34, 39
gases, 15, 54, 65, 107, 129, 140, 146, 204, 228
gasoline. See fuels.
GASP (global air-sampling program), 204
Gates Learjet. See Learjet.
geese, 11–12

Gehrz, Robert, 109
Gelles, Dr. S.H.; Gelles Associates, 192
Gemini program, 101, 191
General Accounting Office (GAO), 104–105, 109
General Assembly, UN. See United Nations.
General Aviation Design & Analysis, 28
general-aviation turbine engine (GATE), 134
General Dynamics Corp., 8, 75, 199, 205, 235
General Electric Co., 68, 106, 124, 130, 154, 174, 175
generators. See power sources.
genetics, 40, 68, 118, 201
Genoa, Italy, 188
geochemistry. See chemistry.
geodesy, 82, 229
geography, 14, 176
Geological Survey, 80
geological resources. See earth-resources monitoring.
geology, 14, 39, 82
geophysics, 31, 82, 145
George AFB, Calif., 155
Georgia, 7, 196
Georgia Institute of Technology. See Georgia Tech.
Georgia Power Co., 70
Georgia Tech (Georgia Institute of Technology), 7, 57, 130, 146
Geos 1, 73, 74–75, 242, 245, 261, 262
GEOS-B, 229
geostationary meteorological satellite. See GMS.
geostationary operational environmental satellite (GOES), 33, 111, 112–113, 134, 148, 157, 171, 176, 210, 229
geostationary satellites, 4, 10, 33, 168
geosynchronous satellites, 73, 74–75
geothermal sources. See power sources.
German, 103
German Democratic Republic (East Germany), 20, 159–160, 174, 196
Germany, World War I, 115
Germany, Federal Republic of; West Germany. See Federal Republic of Germany.
getaway specials (GAS), Shuttle payloads, 4, 162
Giacconi, Prof. R., 228
Gibson, Dr. Edward G., 48, 121

Gibson, Roy, 170, 189
Gillam, Isacc T. IV, 144-145, 191
Ginter, R.D., 11
glass, 62, 175, 192, 232-233
Glazkov, Lt.Col. Yuri, 26, 27, 189
Glenn, Sen. John H. (D-Ohio), 27-28
glider, 151
global air-sampling program. See GASP.
global atmospheric research project. See GARP.
global communications (See also radio; television.), 10, 135, 207
global positioning. See navigation.
global TV transmission. See television.
Gms 1, 111, 131, 140, 248, 261, 262
Goddard award, 20
Goddard Space Flight Center
—contracts, 42, 148
—missions, 56
—people, 14, 55, 79, 166, 169, 210, 216
—programs, 7, 42, 54, 64, 66, 67, 129, 132-133, 137, 138, 148, 155, 161, 168, 172, 189, 215
Goes 1, 67
Goes 2, 112-113, 245, 261, 262
Gold, Gus, 232
Goldstone, Calif., 4, 154, 155
Goodyear Tire & Rubber Co., 65
Gorbatko, Col. Viktor V., 26, 189
Gorenstein, Marc, 209
Gossamer Condor, 182
government employees. See Civil Service Commission, U.S.; employment.
government funding. See funding.
government-industry cooperation. See industry.
Graphic Films Corp., 64, 87
Graves, Curtis M., 3
Gravitational Probe A. See gravity probe.
gravity, 4, 68, 98, 106, 129, 145, 156, 165, 167, 199, 211, 222, 226
Gravity Probe-A, 30, 51
Great Britain. See United Kingdom.
Grechko, Georgy, 23, 223-225
Greenberg, Daniel S., 141-142
Greenland, 40
Gregorek, Dr. G.M., 28
Griffin, Gerald D., 101, 145
Groo, E.S. (Todd), 123
ground control. See control systems.
—crews, 205
—data; ground truth, 196

—stations (See also individual locations and systems.), 1, 10, 55, 56, 154, 157, 174, 207, 210
—testing, 80, 103, 178, 202
—transporter. See vehicles.
growth, 192
Grumman Aerospace Corp., 47, 59, 71, 106
GTE Automatic Electric Co., 221
Gubarev, Aleksey, 23
guest observer, 150
guidance (See also control systems.), 22, 30, 107-108, 118, 122-123, 178
Guidry, Louis E. Jr. (Skip), 36, 202
Gulf of Mexico, 33, 36, 176, 194
GUMPS (Goddard-University of Maryland particle study), 215
gunpowder, 165
Gurevich, M.I., 29
Gurovsky, Dr. Nikolai, 206
Guzhenko, Timofey B., 158
gyroscope, 63, 137, 150, 155
gyrostabilization, 14

habitat, 107, 137
Hachenberg, Otto, 9
Hackwell, John, 109
Haise, Fred, 118, 135, 150-151, 178, 193, 196
Hale telescope, Mt. Palomar, 233
Haley, Alex, 18
Hall, Asaph, 152
Halley, Edmund, 40
Halley's comet, 50, 106, 119, 170, 207
Halliday, Ian, 38
Halo, high-altitude large optics, 58
Hamilton Standard Div., United Technologies, 45, 50
Hampton, Va., 8, 18, 188, 233
hamsters, 146
handbooks. See publications.
handicapped, 172
handling, 133, 150, 196, 199, 214, 215, 216, 218, 233
hangar, 113
Happe, Ralph A., 192
hardware, 44, 45, 50, 61, 72, 79, 141, 168, 173, 177, 187, 202-203, 207, 208, 211, 230-231
Harlow, John H., 121, 201
Harmon trophy, 235
harmonics, 19
Harris, Patricia Roberts, 47

Harris Electronic Systems Div., 1
Hartig, Prof. George, 121
Harvard-Smithsonian Center for Astrophysics. See Smithsonian.
Harvard University, 129, 150, 228
harvest. See crops.
hatch, crew, 118
Havana, Cuba, 183
Hawaii, 31, 32, 131
Hawaii, University of, 39
Hawaiian Institute of Geophysics, 31
Hawker Siddeley Ltd., 53
Hayes, Bethanne, 38
hazard. See danger.
health (See also medical research.), 85, 193, 206, 208
Heao 1, 63, 72, 137, 149–150, 156, 210, 212, 249, 261, 262
HEAO-B, 229
hearings. See Congress.
heart, 5
Hearth, Donald F., 132, 152, 194, 212
heat, 12, 62, 70, 71, 93, 109, 150, 156, 159, 161, 162, 177, 179, 188, 199, 215, 228, 232, 233
heat-capacity mapping mission (HCMM), 229
heating and cooling. See energy.
heatshield, 93
heavy hydrogen. See deuterium.
helicopters, 61, 134, 169, 172, 173, 205, 222
heliogyro. See solar sailer.
helium, 129, 166, 188
Heller, Jack A., 159
Hello, Bastian (Buz), 112
Helt, Capt. Robert C., 181
hemophilia, 133
Henize, Karl, 72, 84
Hercules, 188
heroes, 204
Hessberg, Dr. Rufus, 206
high-altitude atmosphere sounding. See atmosphere, upper.
high-energy astronomy observatory (HEAO), 47, 63, 137, 149–150, 156, 174, 188, 210, 212
high-speed aircraft (See also supersonics.), 96
highways, 94, 192
Himawari. See *Gms 1.*
Hinners, Dr. Noel W., 30, 137, 181, 206, 210, 212, 226

history, 6, 24, 35, 39–40, 61, 67, 128, 141, 164
Hitch, Charles J., 40
hodoscope, 215
Holland. See Netherlands.
Hollinshead, Charles, 104
Holman, Jonathan, 133
holography, 17
homing devices, 177
Honest John (rocket), 36
Honeywell, Inc., 112
Hong Kong, 131
honors. See awards.
"Hopkins." See Johns Hopkins University.
Horton, Victor W., 36, 202
hospitals (See also medical research.), 5, 8, 74, 214
hotels, 179
House of Representatives, U.S. See Congress.
housing, 148
Housing and Urban Development (HUD), U.S. Dept. of, 12, 47
Houston, Texas, 3, 5
Hovis, Dr. Warren, 194
Howard University, 145, 177, 216
Hudson Laboratories, Columbia University, 93, 114
Hughes Aircraft Co., 22, 42, 68, 125, 148, 153–154, 181
Hughes Tool Co., 172
human rights, 130, 144
humane groups, 146
humans, 103, 143, 146, 165, 178, 185
Hungary, 20, 159, 174, 196
hunter/killer satellite. See interceptor satellite.
hunting, 11–12
hurricane. See meteorology; storms.
Hydac, 18, 36
Hydra, 209
hydraulic, 153
hydrazine, 113, 221
hydrodynamics, 25, 80
hydrogen (See also liquid hydrogen.), 109, 129, 159, 199
Hydromet Service, USSR, 127
hydrophones, 31–32
hydroponics, 123–124
hydrostatic tests, 64
hygiene, 202
hypersonics, 219

hypotheses. See theory.
ice, 18, 39, 66, 79, 110, 158, 227
Iceland, 141, 172
Idaho, 47
identification, 43, 188, 196, 205, 213
IL-62, 176
ILC Industries, Inc., 201
Illinois, 127
Illinois Institute of Technology, 56
Illinois, University of, 201
illness. See medical research.
Ilyushin, Sergei V., 29
Ilyushin, Vladimir, 29
image processing. See data processing.
imaging, 7, 10, 11, 58, 80, 110, 131, 137, 148, 175, 176, 186, 189, 210, 216, 224, 225
IMEWS-7. See integrated missile/early-warning satellite.
impact, 22, 76, 178, 205, 218, 226
improvements. See new products; new techniques.
inauguration, 10
Independent Business Telephone Co., 221
India, 4, 32, 174, 196
Indian Ocean, 4, 41, 48, 59, 69, 131, 139, 180
Indian River, Fla., 57
indicators. See instrumentation.
Indonesia, 4, 41, 51, 103
industrialization (See also space industrialization.), 57, 65, 71, 111, 180, 208
industry-government cooperation, 25, 37, 45, 57, 59, 65, 71, 119, 132, 145–146, 163, 174, 177, 178, 196, 199–200, 205, 213, 219, 220
inert flights, Shuttle orbiter. See Shuttle testing.
inertia. See moment of inertia.
inertial guidance. See control systems.
inertial navigation. See navigation.
inflation. See pressure.
 —economic, 22, 121, 163
information (See also data; media.), 74, 130, 193
infrared, 9, 22, 51, 54, 58, 72, 84, 109–110, 131, 161, 169, 189
 —scanning, 12, 148, 161
 —sources, 44
 —spectra, 10, 107, 210
 —systems, 22, 71, 155
inhibition. See prevention.
injection, 187

injuries. See accidents; safety.
INMARSAT: International Marine Satellite Organization, 46
innovation. See discoveries; new products; new techniques.
Insat, 4
insecticides, 5
insects, 5, 61, 68, 218
inspection. See checkout.
installations, 179, 211
Institute for Advanced Study, Princeton University, 10
Institute for Defense Analysis, U.S., 147
Institute for High Temperatures, USSR, 111
Institute of Noise Control and Engineering, 192
Institute of Space and Aeronautical Science, Japan, 23, 38
instrumentation (See also equipment; sensor systems.), 3, 10, 18, 21, 25, 33–34, 53, 56, 57–58, 63, 66, 71, 73, 74, 85, 93, 96, 101, 109, 112, 117, 119, 127, 132–133, 134, 137, 148, 154–155, 156, 158, 159, 161, 164, 165, 167, 168, 169, 171, 175, 177, 181, 188, 189, 194, 196, 198, 208–209, 211, 212, 214, 215, 216, 219, 220, 222, 234
insulation, 12, 76, 131, 168, 175
integrated missile/early-warning satellite (IMEWS-7), 239
integration, 74, 133, 136, 162, 168, 189, 198, 202, 207, 211, 232, 233
intelligence, military. See spy systems.
Intelsat program, 10, 97–98, 173, 180, 202, 205, 209, 229
Intelsat IVA F4, 97–98, 244, 261, 262
intensity, 149, 215, 216
interagency affairs, 53
Inter-American Telecommunications Conference, 222
interceptor satellites (See also weapons systems.), 6–7, 22, 52, 54, 59, 102, 177
intercontinental weapon. See weapons systems.
Intercosmos, 174, 196, 206
Intercosmos 16, 66
Intercosmos 17, 254
interference, 4, 19, 36, 113, 131, 132, 134, 155, 157, 159, 168, 207, 218, 233
interferometer, 155, 166
Intergovernmental Maritime Consultative Organization, UN, 46

Intergovernmental Personnel Act, 162
Interior, U.S. Dept of (DOI), 47, 111, 178
International Association against Painful Experiments on Animals, 146
international aviation. See international relations; International Civil Aviation Organization.
International Business Machines (IBM), 72, 145, 185, 186
International Civil Aviation Organization, 212
international cooperation, 24-25, 44-45, 46, 57, 65, 67, 81-82, 87, 115-117, 129-130, 139, 141-142, 159, 183, 208
—inter-European, 1-2, 56, 65, 85
—USSR and others, 4, 20, 28, 31-32, 56, 59, 111, 117, 121-122, 127, 137-138, 143-144, 159-160, 164, 167, 173-174, 184, 206
—U.S./Europe, 9, 32, 53-54, 79, 85, 136, 170, 174-175, 186, 187-188, 189, 193, 196, 198-199
International Council of Scientific Unions, 131
International Magnetospheric Project, 73
international relations, 62, 177
international sun-earth explorer (ISEE), 112, 193, 195
International Telecommunications Satellite Organization (INTELSAT), 2, 65-66, 106, 110, 135, 154, 179, 181, 198
International Telecommunications Union (ITU), 207
international ultraviolet explorer (IUE), 229
interplanetary, 68, 102-103, 119, 123, 170, 176
interpreters, 164
interservice cooperation, 7-8
Intersputnik, 173-174
interstate. See highways.
interstellar space, 143
intertank. See Space Shuttle external tank.
inventions. (See also new products; new techniques; patents.), 35, 63, 152
inventory. See resource management.
investigations, 38, 142, 170, 209
Investigators Working Group, 67, 183
investment, in space program. See funding, government.
Io, 102, 166
ion-propulsion devices, 43, 119, 170

ionization, 85
ionosphere. See atmosphere.
ions, 146, 230
Iowa, State University of, 111
Iran, 103
iron, 215
irrigation, 48, 219
Irwin, James B., 5
Isee 1, 2, 195, 255, 261, 263
ISEE-C, 229
isolation, 60, 72, 123
isotopes. See chemicals.
Israel Aircraft Industries, Inc., 71
Italy, 65, 111, 119, 139, 142, 154, 156-157, 195
ITT World Communications, Inc., 140
Itten, Dr. K.I., 55

Jackson, Dr. William D., 111, 121
Jaffe, Leonard B., 174
Jamesburg, Calif., 10
jamming, 13, 155, 160
Japan, 13, 23, 26, 29, 32, 38, 39, 40, 41, 45, 46, 58, 105, 106, 111, 131, 140, 141, 154, 202, 210, 226, 229
Jeffs, George W., 180
Jelm Mt., Wyoming, 109
jet
—aircraft, 2, 8, 67, 73, 150, 151, 196, 208, 218
—engines, 13, 73, 192
Jet Propulsion Laboratory
—contractors, 43, 136, 219, 220
—missions, 62, 108, 125, 154, 155, 159, 199
—people, 72, 84, 106, 137, 142, 175, 176, 190, 221, 230, 232
—programs, 44, 48-49, 55, 68, 101, 154, 155-156, 159, 166, 167, 170, 181, 186, 188, 221-222
jettison, 119, 156, 160, 167
JFK. See Kennedy International Airport.
Joersz, Capt. Eldon W., 181
Johannesburg, South Africa, 69
Johns Hopkins University, 37, 121, 197
Johnson, Katherine G., 35
Johnson Space Center
—contractors, 24, 44, 45, 70, 118, 147, 201, 207, 234
—people, 48, 56, 84, 101, 104, 109, 121, 143, 153, 169, 175, 215-216
—programs, 7, 16, 21, 28, 41-42, 50, 53, 54, 61, 71, 74, 85, 109, 118, 132,

140, 143, 144, 145-146, 148, 156, 157, 165, 168, 172-173, 179, 185, 188, 191, 194, 196, 201, 202, 205-206, 214, 215-216, 218, 222, 228, 232
Johnston, Bradford, 146
Johnstown, Pa., 138
Joiner, Dr. Bennett, 106
joint American-Soviet particle-intercalibration (JASPIC) program, 85
Jones, Robert T., 152
journals, scientific. See publications.
Joyce, Dr. Richard, 39
judicial proceedings, 12
Juno 1, 115
Jupiter, 3, 33, 39, 49, 68, 76, 102, 103, 106, 108, 115, 135-136, 137, 140, 145, 159, 165, 166, 167, 169, 186, 187-188, 204, 227

Kaman Aerospace Corp., 34
Kapustin Yar, USSR, 66
Kazakhstan, USSR, 27, 190
Kelly, Al, 56
Kelly, Roy, 156
Kennedy, President John F., 115
Kennedy (, John F.) International Airport, NYC, 66, 212
Kennedy Space Center
 —contractors, 21-22, 35, 57, 68-69, 102, 118, 128, 139, 185-186, 221
 —missions, 33-34, 47
 —people, 101, 104, 130, 145, 169, 229
 —programs, 2, 15, 16, 21-22, 24, 28-29, 54, 61, 62, 68, 75, 113, 127, 168, 202, 215, 218, 234-235
 —Spaceport, 21, 75, 113, 132, 133, 179, 205, 207, 209, 211, 228-229, 231, 234
Kentucky State University, 68
Kerguelen Island, 69
Kerwin, Joseph P., M.D., 48
Kettering Group, 158
Key West, Fla., 183
Kezios, Dr. Stothe P., 24
Khabarov Krai, USSR, 178
Kiku 2, 41, 237, 240
Killeen, Tex., 70
killer satellites. See interceptor satellites.
Kincheloe, Iven P., 235
Kingsbury, James E., 50
Kitt Peak National Observatory, 38-39, 51, 66
Klate Holt Co., 212
Kleinknecht, Kenneth S., 153

Klett, Reiner, 4
Klimuk, Pyotr, 23
Kline, Raymond A., 123, 209
Koran, 103
Korean War, 235
Korolev medal (USSR), 11
Kourou, Fr. Guiana, 127, 198
Kovalenok, Vladimir, 189, 223
Kowal, Charles, 203-204
Kraft, Dr. Christopher C., 56, 109, 144, 228
Kramer, Dr. James J., 194
Kramp, Klaus, 84
Krasnoyarsk Institute of Physics, USSR, 123
Kremer prize, 182
Krider, Dr. Phillip, 21
Kuhn, Dr. Peter M., 161
Kuiper airborne observatory (NASA), 48, 59, 69, 107
Kunkel, Bruce A., 18
Kurchatov Institute, USSR, 109

L-5 Society, 12
L-1011, 177
L-band communications, 139
Labor, U.S. Dept. of, 194
labor relations, 164
laboratory, 72, 235
Lakeland, Fla., 76
laminar flow. See aerodynamics.
Lampton, Dr. Michael, 230
Lancaster, Calif., 17
Lance, Bert, 64
land lines, 138
land use, 7, 48, 63
lander, Viking. See Viking.
landing, 69, 113, 122-123, 150-152, 165, 196-197, 212, 213-214, 221, 222, 229, 234
landing rights, 53
Landsat program, 7, 14, 25, 42, 47, 48, 52, 54, 55, 63, 80, 108, 110, 111, 135, 175, 193, 196, 229
Landsat 1, 141
Langley AFB, Va., 8, 188, 233
Langley aeronautical laboratory, NACA, 16, 219
Langley Research Center
 —contractors, 8, 28, 30, 35, 79, 110, 128, 145, 158, 163, 177-178, 192, 212-213, 219, 233

—people, 35, 63, 71, 132, 152, 169, 212, 215–216, 227, 233
—programs, 8, 18, 61, 112, 130–131, 132, 138–139, 140–141, 143, 146, 147, 148, 172, 181, 188, 192, 194, 203, 212, 218, 222, 227, 233
Langmuir Laboratories, N.M., 235
languages, 143
lanthanum, 2
Lanzerotti, Dr. L.J., 166
Large Magellanic Cloud, 37
large orbiting telescope. See space telescope.
lasers, 2, 9, 17, 66, 82, 140–141, 146, 195
latex, 192
Latin America (See also individual countries in Central and South America.), 10, 176
latitudes, 85
launch control. See control.
—program, 64–65, 145, 189, 211, 229, 234
—reimbursable. See reimbursable.
—rockets. See rockets.
—support. See support services.
—vehicles. See name of individual rocket: Ariane, Atlas, Centaur, Delta, Saturn, Scout, Titan.
law enforcement, 189
Lawless, Dr. James, 190
Lawrence Laboratory, Calif., 209
Lazarev, Vasily, 23
lead (element), 230
leak, 180, 213
Learjet, 54, 132
Leary, Timothy, 12
leasing, 124–125, 139, 175
lecture, 189
Ledford, Harold, 11
Lee, Chester M. (Chet), 143, 163
Lee, Thomas J., 170
Lehigh University, 192
Lei Hung, 182
levitation, electromagnetic, 30, 106
Levy, Lillian, 192
Lewis Research Center
—contractors, 34, 129, 134, 199, 221
—people, 133, 134, 147, 159, 169, 194, 220
—programs, 12, 47, 53, 61, 67, 73, 94, 134–135, 159, 180, 204, 208, 231–232
Libra, 69
Lichtenberg, Byron, 230

life. See lifetimes.
life; life detection; life forms; life sciences. See biology.
life support, 45, 68, 85, 123, 137, 223–224
lifetimes, 80, 119, 120, 139, 143, 150, 159, 165, 186
lift. See aerodynamics.
lifting bodies, 217–218
light; lightwaves, 66, 107, 131, 178, 233
lightning, 21, 127, 207, 234
Lima, Ohio, 12, 13
limb darkening, 69
Lincoln Laboratory, MIT, 219
Lindbergh, Charles A., 24, 25
Lineas Aereos Nicaragua (LANICA), 176
liquid-fuel rocket (See also liquid hydrogen; liquid oxygen.), 106, 151, 153
liquid hydrogen (See also hydrogen.), 44, 153, 160, 231
liquid oxygen (See also oxygen.), 25, 44, 50, 64, 153, 160, 205, 231
liquid propellant. See liquid-fuel rocket; fuel.
Lisovsky, Dr. G., 123
lixiscope, 216
loads, 64, 110, 111, 119–120, 153, 205, 214
location, 67, 79, 176, 205
Locke, Capt. Walter (USN), 8
Lockheed Corp., 22, 35, 71, 177, 178
Lockheed Electronics Co., 44, 147
Lockheed-Georgia Co., 112
Lockheed Missiles and Space Co., 136, 220, 233
Lockheed SR-71, 181
log book, 152
London, England, 46, 124, 146, 182
Long Island, N.Y., 66
Lord, Dr. Arthur, 106
Lord Electric Co., 173
Lorre, Jean, 142
Los Angeles, Calif., 148, 218
Louisiana, 196
Louisville, Kentucky, 38
Lovelace, Dr. Alan M., 89, 123, 169, 183
Lowman, Dr. Paul D., 14
Lubarsky, Dr. Bernard, 147
Lucas, Dr. William R., 11, 170, 201, 214, 215
lumber. See forests.
Luna 24, 28
lunar curatorial facility, JSC, 24, 53
—exploitation, 145–146

—module. See moon landing.
Lunar Odyssey, 71
lunar-polar orbiter mission, 207
—research, 145-146, 156
—rover, 178
—samples, 53
Lunar Science Institute, 28, 57
Lundin, Dr. Bruce T., 147
Lunokhod, 117
Lynn, Donald, 142

M-100B, 137
MacCready, Dr. Paul, 182
McCulloch, James C., 52
McDonald, Dr. Frank B., 210, 212
McDonald observatory, N.M., 104
McDonnell Douglas Corp., 11, 68, 79, 106, 173, 205
McDonnell Douglas Technical Services, Inc., 37, 52
McGinnis, David F., 110
McLucas, Dr. John L., 72
McMurdo Sound, Antarctica, 39
McMurtry, Thomas C., 36, 150, 171, 191, 202
McTigue, John G., 175
Mach numbers, 132
Madras, India, 69
magazines. See media.
magnet; magnetic instruments; magnetometer, 111, 121, 158, 195
magnetic fields, 40, 73, 98, 134, 153, 165, 169, 215, 222
magnetodynamics, 54
magnetohydrodynamics, 111, 121
magnetosphere. See atmosphere.
magnification, 105, 106
Mahon, Rep. George (D-Tex), 125
Maine, 181
maintenance, 2, 13, 18, 26, 45, 66, 74, 134-135, 202, 207, 210, 227
Makarov, Oleg, 23
Malerba, Franco, 230
malfunction (See also failure.), 54, 55, 58, 62, 63, 70-71, 72, 73, 75, 103, 118, 122, 150, 155-156, 160, 168, 171, 213
management, 65, 66, 123, 178
manatees, 57
maneuvering; maneuvers, 71, 151, 177, 198, 199, 210, 211
manned flight
—international, 206

—Shuttle (See also Shuttle crews.), 17, 68, 107, 118, 160, 208, 222, 234
—Soviet, 10, 67, 158, 187
—U.S., 27-28, 45, 54, 121, 128
manpower. See employment.
manuals. See publications.
manufacturers; manufacturing. See aerospace industry; fabrication; space industrialization.
mapper, thematic, 42
mapping, 10, 14, 44, 80, 82, 98, 109, 138, 165
Marine Corps, U.S., 121, 140, 148, 155, 173, 179, 194, 201, 205
Mariner missions, 33-34, 76, 161
Marion Power Shovel Co., 24
Marisat system, 72, 124, 139
maritime (See also individual satellites and systems.), 1-2, 46, 72, 81, 207, 229
Mark, Dr. Hans, 147
Mark, Sir Robert, 189
Markey, Stanley, 175
Marots, 16
Mars, 3, 6, 15, 25, 28, 31, 39, 48, 50, 52, 62, 76, 101, 106, 124, 132, 140, 152, 155, 161, 165, 181, 206-207, 221-222
—rover, 50, 52, 106, 207
Marshall Space Flight Center
—contractors, 5, 7, 11, 18, 25, 35, 45, 51, 54, 56, 59, 64, 74, 75, 76, 79, 80, 113, 130, 137, 175, 181, 186, 188, 211
—missions, 18, 56, 132-133
—people, 11, 19, 50, 67, 84, 115-117, 128, 149, 152, 157, 169, 174, 179, 201, 209, 211, 215, 227, 229-230
—programs, 6, 15-16, 18, 24, 30, 44, 50, 51, 59, 64, 67, 69, 71, 80, 95, 98, 101, 105, 119, 127, 152, 153, 154, 157, 160, 168, 170, 174, 178, 179, 183, 198, 201, 202-203, 205, 208, 211, 228, 230, 231
marsquake, 6
Martin, James S. Jr., 20
Martin, Dr. John J., 132
Martin Marietta (Aerospace) Corp., 25, 53, 64, 80, 101, 160, 170, 181, 198, 211, 231, 235
Maryland Center for Public Broadcasting, 13
Maryland, University of, 12, 63, 215
masers, 51
mass spectrometers, 152, 166

Massachusetts Institute of Technology (MIT), 4, 31, 71, 106, 133, 147, 182, 212, 219, 230
materials, 57, 59, 65, 76, 79, 95, 108, 124, 129, 130, 131, 145, 218, 219, 227, 233
—applications, 11, 45, 59, 69
Materials Experiment Assembly, 192
—processing, 4, 30, 32, 59, 62, 71, 105–106, 137, 141, 192, 194
mathematics (See also data processing.), 143, 227, 235
mating, 80, 113
matter (See also theory.), 54, 75, 209, 215, 227
Matthoefer, Hans, 187
Maui, Hawaii, 31-32
Max Planck Institute of Astronomy, 9
M.B. Associates, 220
measurements, 4, 69, 80, 82, 85, 98, 127, 137, 154, 187, 192, 194, 195, 208–209
mechanics, science of; mechanisms, 131, 194
Medal of Freedom. See awards.
medals. See awards.
media (See also publications; television.), 10, 12, 49, 53, 64, 74, 87, 143, 157, 158, 164, 212
medical research, 23, 25, 32, 57, 74, 84, 85, 146, 157, 169, 174, 187, 193, 194, 206, 214, 216, 222
medicine. See medical research; pharmaceuticals.
Mediterranean Sea, 14
meetings, 2, 8, 28, 39-40, 41, 52, 57, 60, 63, 67, 71, 75, 89, 137, 142, 145–146, 147, 152, 154, 157, 174, 189, 192, 203, 206, 212, 220, 227, 228, 234
Megill, Dr. L.R., 4
Meharry Medical College, 152
Mellors, Col. W.J., 154
memorials. See observances.
Menzies, Dr. Robert T., 72, 84, 230
Merbold, Ulf, 230
mercury, 170
Mercury, 28, 161
Mercury program, 27, 187, 216
Mercury (support services firm), 233
metals, 108, 128, 190
Meteor 1-28 (USSR metesat), 247
Meteor 2-02, 238
Meteor 2-03, 258
Meteor 2-07, 241
meteorite; meteoroid, 38, 39, 76, 128, 132, 165, 224, 225

meteorology, 18, 19, 25, 39-40, 56, 58, 62, 64, 66, 67, 101, 119, 127, 134, 164, 165, 167, 171, 174, 176, 198, 199, 210, 212, 222, 234
Meteosat (ESA project), 16, 112, 202, 210, 213, 256, 261, 263
metesats (See also individual satellites.), 40, 112, 161, 229
methane, 51, 190, 201
Mexico, 40
MF-12 (USSR rocket), 85
Miami, Fla., 171, 176
Michigan, 18
Michoud Assembly Facility (NASA), 11, 25–26, 44, 51, 64, 121, 170
micrometeorites. See meteorites.
microphones, 31-32
microwaves. See power sources.
microwave landing system. See microwave scanning-beam landing system.
microwave radiation, 208
microwave scanning-beam landing system, 113, 122, 178, 191, 222, 229
Middle East (See also Arab nations; individual nations in that area.), 10, 154, 210
Middleton, Bob, 179
Midgette, Woodrow W., Jr., 194
MiG-25 (Foxbat; USSR aircraft), 13, 26
Mikoyan, A.A., 29
Milan (Italy) Polytechnic Institute, 157
Milford, Rep. Dale (D-Tex), 60
military, 154–155, 213
Military Command and Control System, U.S., 38
—flight operations, 18, 154-155
—personnel, 22
—use of space, 54, 122, 155
Milky Way, 54, 209, 215
Millis, Dr. Robert, 48
minerals, 134, 165
Mini Sniffer, 221
miniaturization, 70, 169
mining, 57, 178, 199
ministars. See stars.
minority business, 11
minority employment; recruitment (See also women.), 2, 21, 41–42, 133
Miosi, Antoine, 117
Mirage (French jet fighter aircraft), 107
mirrors, 66, 109, 233
miscibility, 192
missiles. See weapons systems.
mission control. See control.

mission specialists; payload specialists (See also individual astronaut.), 21, 67, 72, 84, 132, 140, 157, 167, 179, 183, 191, 194, 201, 205-206
mission support. See support services.
Mississippi, 196
Mississippi River, 33, 44
Missouri, 127
Mitchell, Capt. Edgar D., 195
Mitchell, James P., 194
Mitchell, Royce, 211
Miteq Inc., 106
Mitre Corp., 72
Mitsubishi Electric Co., 226
mobile equipment, 72, 84, 139, 167, 185-186, 207, 224, 231
mockup. See simulation.
models, 55, 59, 132, 136, 138, 163, 188, 194, 217-218
modifications, 55, 64, 71, 72, 95, 113, 118, 121, 150, 168, 173, 197, 199, 207-208, 216, 229, 231
Mojave Desert, 148-149, 150, 155, 159
molecules, 9, 146, 190
Molniya 1-36, 241
Molniya 1-37, 246
Molniya 1-38, 252
Molniya 2-17, 239
Molniya 3-7, 242
Molniya 3-8, 256
moment of inertia, 75
money. See budget; funding.
Mongolia, 20, 174
monitoring, 57, 63, 80, 85, 156, 160, 161, 186, 202, 214
—space-based, 1-2, 11-12, 18, 25, 33, 47, 55, 58, 84, 102, 112-113, 129-130, 134, 165-166
Montreal, Quebec, 110
Moody, Jewel W., 11
moon (earth's) (See also lunar research.), 28, 31, 57, 76, 109, 115, 145-146, 174, 175
—landings, 28, 115, 117, 132, 165, 181, 191, 220
moonquakes. See seismology.
moons. See planetary satellites.
Moore, R. Gilbert, 4
Morgens, Howard J., 65
Moscow, USSR, 111, 121, 167, 173, 206
Moscow, University of, 76
Moss, Frank, 56

mothballing. See deactivation.
motion pictures. See films.
motion sickness. See medical research.
motor cars; motorcycles. See vehicles.
Motorola, Inc., 219, 220
motors. See power sources, engines.
Mt. Palomar Observatory, 10, 66, 204, 233
mountains, 80-81
movies. See films.
MST-3 (Tansei, Japan experimental satellite), 38
MU-3C (Japan rocket launcher), 38
MU-3H, 23, 38
Muller, Dr. Richard, 209
multiplexers (MUX), 221
muon, 31-32
Murphy, John, 130
music, 143
Mustang (U.S. fighter plane), 16
Myers, Dale P., 195
Myrtle Beach AFB, S.C., 105

N rocket (Japan launch vehicle), 29, 40, 41, 106
Nantucket Sound, 79
NASA 905 (aircraft). See Boeing 747.
Nashville, Tenn., 177
National Academy of Engineering, U.S., 124
National Academy of Sciences, U.S., 144, 207
National Advisory Committee for Aeronautics (NACA), 16, 147
National Aeronautic Association (NAA), 19, 76, 162, 172, 181, 182, 213
National Aeronautics and Space Administration (NASA)
—advisory committees/council, 56, 128
—budget, 18, 23, 48, 50, 64-65, 67, 68, 72, 104, 108, 114, 135-136, 156, 161, 199, 206-207, 214-215, 217, 228
—centers, 123, 209
—contractors (Headquarters), 1, 18, 42, 55, 68
—future plans, 50, 64-65, 67, 106, 114-115, 116, 117
—Headquarters. See management.
—management, 3, 51, 56, 61, 65, 104, 115-117, 121, 123, 161, 169, 197, 203, 209, 214-215
—Office of Aeronautics and Space Technology (OAST), 195

— Office of (Space and Terrestrial) Applications (OA), 62, 146, 162, 168, 220
— Office of Space Flight, 181
— Office of Space Science, 32, 146, 150, 153, 220
— people, 3, 11, 51, 52, 56, 62, 93, 114, 115-117, 123, 169, 218, 220
— research, 55, 60
   — Defense, Dept. of, 134, 197
   — Energy, Dept. of/ERDA, 5, 15-16, 94, 199-200, 209, 219, 220, 228
   — ESA. See international cooperation.
   — Federal Aviation Administration, 222
   — Housing and Urban Development, Dept. of, 12
   — Interior, Dept. of, 57
National Air and Space Museum (Smithsonian), 3, 44, 50, 67, 181
National Center for Atmospheric Research, U.S., 39
National Civil Service League, 3
National Consortium for Black Professional Development, 133
National Environmental Satellite Service (NESS), DOC, 110
National Fire Prevention and Control Administration, 59
National Geographic Society, 203
National Hurricane Center, 171
National Institute of Dental Research, U.S., 216
National Institutes of Health, U.S., 152
National Oceanic and Atmospheric Administration (NOAA), U.S., 11, 25, 55, 67, 69, 80, 110, 112, 134, 148, 160-161, 171, 176, 194
National Parachute Test Range, U.S., 199, 211
National Physical Laboratory, U.K., 227
national policy, 1, 3, 48, 56, 58-59, 60, 106-107, 108, 147, 187, 213, 229
National Register of Historic Places, 6
National Research Council (CNR), Italy, 157
National Science Board, U.S., 31
National Science Foundation, U.S., 62, 110, 117, 134, 194, 201
National Space Club, 154, 220
National Space Development Agency, Japan, 29, 41, 58, 106, 131, 226

National Space Institute, 116
National Space Technology Laboratories (NSTL), NASA, 44, 55, 120, 160, 169, 170
National Transonic Facility, LaRC, 132
National Weather Service, U.S., 234
Nato II A, II B, 7
Nato III A, 7
Nato III B, 7, 14, 238, 261, 262
NATO-IIIC, 229
natural disasters. See emergencies.
natural gas. See gases.
natural resources, 130
Naugle, Dr. John E., 220
Navajo Tribal Council, 193
Naval Observatory, U.S., 152
Naval Ocean Systems Center, U.S., 231
Naval Research Laboratory, U.S., 120, 149, 174
navigation, 35, 53, 67, 122, 124-125, 155, 158, 159, 165, 190, 197, 224
Navigation Technology Satellite (NTS), U.S. Navy, 120
navsats (See also names of individual satellites.), 6, 70, 120, 158, 159, 197, 221
Navstar, 22, 54, 122
Navy, U.S., 6, 65, 66, 93, 114, 120, 124, 139, 140, 148, 157, 173, 178, 179, 184, 194, 197, 201, 202, 205, 207, 208, 229, 231
— contractors, 8, 192
Near East. See Middle East.
Nebraska, 48, 219
nebula IC 133, 9
Nelepov, Evgeny, 164
Nelson, Howard W., 105
nephelometer, 166
Neptune, 38-39, 49, 204
Netherlands, 9, 43, 44, 136, 154, 186, 200
network, 82, 133, 165, 207, 210
Neugebauer, Prof. Gerry, 44
neutral buoyance facility, MSFC, 64, 118, 160, 208, 224
neutrons, 75, 109, 149
Nevada, University of, 25
New Delhi, India, 182
New England, 63
New Guinea, 226
New Jersey, 63
New Mexico, 66
New Orleans, La., 44, 160, 170

new products (See also discoveries; inventions; patents.), 2-3, 42-43, 58, 59, 62, 71, 96, 107, 120, 130-131, 140-141, 145, 169, 175, 195-196, 208, 213-214, 224, 230
New Scotland Yard (London), 189
new techniques (See also discoveries; inventions; patents.), 25-26, 45, 56, 67, 69, 85, 96, 109, 127, 128, 132, 141, 142, 146, 159, 160, 177, 178, 196, 214, 216, 220, 224
New York, N.Y., 182
New York Bight, 63
New York Port Authority, 66
newspapers. See media.
Nicaragua, 176
Nichols, Nichelle, 41
nickel, 190
nickel-cadmium batteries. See batteries.
nickel-hydrogen power cells, 2, 120
Nicks, Oran W., 132
Nicollier, Claude, 72, 84, 230
Niemann, Dr. H.B., 166
Nigeria, 103
night, 196
Nimbus program, 194
Nimbus 3, 161
Nimbus 5, 66
Nimbus 6, 64, 79, 172
Nimbus-G, 229
Nimrod (British airborne radar craft), 53-54
Nissan Motor Co., Ltd., 106
nitrogen; nitrogen compounds, 132, 192-193
Nixon, President Richard M., 52, 143
Noaa 2, 55
Noaa 5, 11
NOAA-A, 229
noise, 2, 66, 118, 135, 143, 195, 200
Noise-Con 77, 192
nominations, 93
non-nuclear weapons. See weapons systems.
Norske Meteorological Institute, Oslo, 79
North American Rockwell. See Rockwell International Corp.
North American Weather Consultants, 18
North Atlantic Treaty Organization (NATO), 7, 14-15, 53-54, 72, 141, 213
North Dakota, 127
North Pole, 158, 182, 234
North Sea, 1, 79

northern lights. See auroras.
Northern Telecom, Inc., 221
Northrop Services, Inc., 24
Norway, 1, 32, 46, 79
Noss 2, 257
nova, 174
Nts 2, 120, 246
nuclear energy; nuclear power; nuclear reactors; nuclear research. See power sources, nuclear.
—testing, 71
—weapons (See also weapons systems.), 8, 40-41
Nuclear Regulatory Commission, 53
numbers; numerical analysis. See mathematics.
Nunelly, Hubert, 13
Nunn, Sen. Sam (D-Ga.), 74
nutrition. See food.
N.V. Phillips (Netherlands firm), 200

Oak Ridge Associated Universities, 172
Oberth, Hermann, 115
observances, 67, 69, 124, 132, 148-149, 152, 162, 165, 182, 184, 187, 189, 192, 215, 223, 234
observations, 48, 59, 69-70, 98, 127, 158, 182, 210
observatories (ground-based), 109-110, 156
obsolete, 180
obstacles. See problems.
occultation, 69
ocean, 149, 152, 184, 198, 212, 231
oceanography, 58, 93, 109, 193-194
Ockels, Wubbo, 230
Odenwald, W. Germany, 75
Odessa (Tex.) Medical Center, 74
odor, 27
Office des Postes et Telecommunications, Upper Volta, 198
Office of Management and Budget (OMB), White House, 64, 169
Office of Science and Technology Policy, White House, 71, 111
Ohio River, 44
Ohio State University, 28, 219
oil (See also fuel.), 1, 16, 65, 73, 79, 110-111, 130, 188
Okhotsk, Sea of, USSR, 178
Olsen, Dr. Edward J., 39
O'Neill, Gerard K., 12, 137, 149, 185
Ontario, Canada, 110

OOE. See out-of-ecliptic mission. See solar polar.
open universe. See theory.
Operation Paperclip. See Paperclip.
Ophiuchus (constellation), 174
optics, 58, 62, 175, 192, 193
orbit, planetary, 13, 124, 204
—spacecraft, 75, 151, 158, 163, 168, 198, 211, 223
orbital-test satellite (ESA). See OTS.
orbiter. See Space Shuttle orbiter; Viking project; individual spacecraft.
orbiter-processing facility. See Space Shuttle orbiter.
Orbiting Solar Observatory (OSO) program, 54
ordnance. See explosives; weapons systems.
Oregon, 80
organic matter, 39, 184, 190, 201
Organization of American States (OAS), 222
orientation, 98, 156, 158
origins. See theory.
Orillion, Alfred G., 19-20
Oriol (USSR spacecraft), 117
Orion (constellation), 104
Orroral Valley, Australia, 82, 133
Osborn, Howard J., 152
OTS (ESA's orbital test satellite), 16, 112, 168, 171, 180, 202, 205, 229
Ottawa, Ontario, 110
Ottawa River, 110
out-of-ecliptic (OOE) mission. See solar polar.
outer planets, 28, 34, 49, 140, 145
Owens-Illinois, Inc., 203
oxygen (See also liquid oxygen.), 50, 68, 123, 159
ozone, 138, 193, 204

Pacific Conference on Chemistry, 190
Pacific Ocean, 23, 31, 57, 58, 131, 134, 139, 140, 178, 226
package. See payload.
Page, George F., 229
paint, 154
Pakistan, 103, 131
Palapa 2, 51-52, 240, 261, 262
Palmdale, Calif., 17
Pan American World Airways, 67, 182, 204, 205
Pan American Union. See Organization of American States.

Panama Canal, 121
Papazian, Dr. John, 106
Paperclip, Operation, 115
parachutes, 53, 93, 95, 101, 119-120, 152, 159, 211-212, 222, 231
Paraguay, 135
Paris, France, 24, 52, 84, 105, 153
Parker, Loyd C., 5
Parker, Robert, 84
Parnell, Dr. Thomas A., 88
particle generator, 25
particles, energetic. See electrons; ions; muons; neutrons.
parts. See instrumentation.
passengers (See also airlines.), 19, 29, 130, 193, 199, 220
patents (See also inventions.), 152, 218, 232
pathfinder spacecraft, 33, 226
Patrick AFB, Fla., 205
patterns, 154
payload
 —missile; launch vehicle, 8, 221
 —Shuttle. See Space Shuttle payloads.
 —specialists. See mission specialists.
Peaceful Uses of Outer Space, UN Committee on, 44, 129-130
peaceful uses of space, 57, 143-144
Pearl River (NASA barge), 121
Peenemunde (German launch site WWII), 115
Pegasus 1, 132
Peking, China (PRC), 154
Pennsylvania, 138
Pennsylvania State University, 72
Pensacola, Fla., 36
Pentagon. See Defense, U.S. Dept. of.
People's Republic of China. See China, People's Republic of.
Pepin, Dr. Robert, 28
Percy, Sen. Charles (R-Ill.), 74
performance, 37, 42, 92-93, 96, 98, 108, 111, 156, 159, 168, 171, 186, 231
Perkin-Elmer Corp., 136, 192, 232
Perkins, Porter C., 204
Perry, William J., 177
Persian, 103
Persian Gulf, 14
personnel. See employment.
Perth, Australia, 69
perturbation, 15
pesticides, 1
Peterson, Dr. Hale, 212
petroleum. See oil.

Petrone, Rocco A., 56
Petrov, Dr. Boris, 174, 206
Phobos (moon of Mars), 31, 76, 124, 152, 176
pharmaceuticals, 25, 74, 214
Phoenix, Ariz., 94
phonograph record. See recording systems.
photodetectors. See detection.
photography (See also cameras; film.), 7, 14, 23, 27, 48, 59, 104, 105, 125, 131, 142, 165, 174, 176, 196, 222
photometry, 210
photons, 49, 119, 140, 189
physics, 23, 65, 75, 115, 230, 233
physiology. See biology.
piggyback flights, 118, 135, 141
Pilliod, Charles J., Jr., 65
pilotless vehicle, 17
pilots, 13, 26, 128, 132, 140, 148, 150, 172, 179, 181, 191, 196, 201, 202, 205; 235
Pioneer mission, 92, 143, 188, 221, 226-227
Pioneer 10, 11, 3, 33, 102, 103, 143, 161, 169, 226-227
Pioneer Parachute Co., 231
pitch rate, 154
Pittsburgh, University of, 138
planetary environment. See environment; atmosphere, planetary.
planetary exploration. See exploration.
planetary satellites, 48, 59, 69, 76, 102, 124, 136, 152, 187-188, 203-204, 207, 226-227
planetary science, 28, 48, 62, 66, 92-93, 102-103, 107, 108, 161, 174, 206-207
planning. See future plans.
plants. See botany.
plasma physics, 32, 155
plastic, 17, 93, 119
platforms, data collection, 113, 171
Plum Brook station, LeRC, 34
Pluto (planet), 3, 39, 51, 104
plutonium, 65, 161-162
Pocomoke City, Md., 164
Pogue, William R., 121
pointing. See aiming.
poisoning. See medical research.
Poland, 20, 115, 159, 174, 196
polar-orbiting satellite (See also names of individual spacecraft.), 11, 131, 136
polar research, 66, 117, 234

polar route, 173, 182, 204
Polaris (weapon), 8
police, 189
politics, 107, 129-130, 131, 143-144, 157, 172, 187, 195
poll. See public opinion.
pollution. See also contamination.
  —ash, 8
  —atmosphere. See atmosphere.
  —chemical, 5, 110-111
  —environmental, 25, 63, 79, 141, 165, 195, 200
population, 60
Port Everglades, Fla., 205, 231
portable earth terminal, 47, 138, 208
portable systems, 50, 138
Portugal, 227
Porz-Wahn, W. Germany, 33
Poseidon (weapon), 8
postage. See stamps.
postponement. See delay.
potassium, 146
Powder Puff Derby, 162
power levels. See energy levels.
power sources, 58, 98, 159, 165-166, 172, 199, 208, 216
  —atomic. See power sources, nuclear.
  —batteries, 2, 3, 94, 190
  —conversion (See also solar conversion devices.), 140-141, 216, 232
  —electric, 5, 34, 227
  —engines, 61, 73, 159, 199-200
  —galactic, 43
  —generators, 159
  —geothermal, 134
  —lasers, 146
  —magnetohydrodynamics, 111, 121
  —microwaves, 5, 56, 57, 66, 141, 153, 175
  —nuclear, 35-36, 65, 104, 109, 118, 140, 146, 158, 159, 161-162, 215, 216, 230
  —satellites. See solar power satellites.
  —solar (See also solar conversion devices; solar power satellites.), 5, 14, 15-16, 23, 43, 45, 53, 65, 75, 76, 94, 145, 159, 161, 168, 179, 189, 203
  —steam, 7-8, 14, 188, 233
  —wind, 34, 129, 176
power supply. See power sources.
power systems, spacecraft. See spacecraft.
powersats. See solar power satellites.

Pozinski, Norman, 220
practice, 150, 211, 230-231
Prague, Czechoslovakia, 184, 185
precipitation, 119
precision. See accuracy.
predictions (See also meteorology; future plans.), 25, 58, 59, 62, 80, 127, 165, 192
prefabrication, 69
preservation, 152
press; press conferences. See media.
Press, Dr. Frank, 30-31, 71, 111, 144
pressure, 2, 64, 93, 159, 171, 203, 213-214, 222
pressurization, 45, 118, 224
prevention, 146
prices. See fees; competition, economic.
prime mover. See vehicles.
Primer, Howard, 13
Prince of Wales, 193
Princeton University, 10, 12
printing (See also media; publications.), 165
prizes. See awards.
probe, 68, 92-93, 136, 159, 161-162
problems, 27, 62, 63, 111, 113, 118, 129-130, 137, 154-155, 156, 158, 159, 160, 162, 165, 167, 171, 176, 186, 189-190, 193, 196, 209, 213, 214, 221, 233
Procter & Gamble, 65
procurement, 18, 79, 219
—policy, 2, 37, 105, 124-125
productivity, 71, 219
Professor Vize (USSR research ship), 85
Prognoz (USSR spacecraft series), 117, 176
Prognoz 6, 253
programming, 55, 176
propaganda, 130
propagation, 139
propane, 94
propellant. See fuel.
propellers, 180, 221
propulsion, 61, 125, 133, 147, 167, 170, 225
prosthesis, 169
protection (See also insulation.), 207
protonosphere. See atmosphere.
protons, 75, 85
prototypes, 36, 42, 67, 74, 130, 165, 188, 230
Proxmire, Sen. William (D-Wis.), 64, 108, 217

psychology, 123
public opinion, 116, 156, 213
public policy. See national policy.
public relations (See also community relations.), 6, 63, 67, 74
publications (See also media.), 106-107, 128, 136, 204
publicity. See media.
Puerto Rico, 222
pulsars, 102, 210
pump, 199
pyranometer, 53

quality, 169
quarks, 75
quasars, 10, 121, 149, 210
Quebec, 110

racing. See contest.
radar, 8, 13, 26, 58, 74, 78, 224, 227
radiation; radioactive material (See also electromagnetism; energetic particles.), 39-40, 50, 102, 129, 137, 146, 167, 208, 216
radio (See also frequencies; broadcast.), 4, 19, 94, 118, 153, 155, 158, 159, 179, 209, 217, 224
radio-technical commission for aeronautics, U.S., 122
radioastronomy, 76, 137
radioisotope thermal generators (RTGs). See power sources, nuclear.
radiometer, 66, 131, 148, 161, 166, 210
radiotelescope, 9, 66, 76
Raduga 3 (Statsionar comsat, USSR), 249
Ragent, Dr. Boris, 166
railroads, 133, 192
rain, 69
random-access measurement system, 79
range, 57, 82, 195-196, 198
Rango, Dr. Albert, 55
Ransburgh, John L., 230
RATAN-600 (USSR radiotelescope), 76
rates, 213, 217
rats, 146, 167
RCA (formerly Radio Corporation of America), 104, 153-154, 220
RCA Astroelectronics Div., 153-154, 234
RCA Global Communications, Inc., 139
RCA Service Co., 158
reactivation, 132-133
reactors. See power sources, nuclear.

real-time monitoring, 57
receivers; reception, 208
Reclamation, U.S. Bureau of, 80
reconditioning. See rehabilitation.
reconnaissance (See also monitoring; surveillance.), 22, 54, 221
record setting, 23, 67, 76, 121, 128, 133, 149, 158, 172, 176, 181, 182, 198, 219, 227, 235
recording, 133, 143, 174
recording systems, 67, 143, 160, 214
RECOUP (refuse-consuming utility plant), 88, 233
recovery, equipment; materials, 38, 53, 91, 101, 119, 138, 205, 211-212, 231
recycling, 65, 110, 131, 188, 211, 229, 233
Red Cross, American, 138
Red Flag (PRC journal), 131
red giant. See stars.
Redstone (ballistic rocket), 6, 115
Redstone (range instrumentation vessel), 213
Redstone Arsenal, 64, 115, 215
reduction in force (RIF). See employment.
redundancy, 14, 151
Reeder, Jack, 222
reentry, from orbit, 154, 175, 198, 215
refinery; refining. See fuel.
reflectivity, 39, 119
reform, 62
refrigeration. See cooling; cryogenics.
refuse. See waste disposal.
regularity, 209
regulation, 62, 129, 185, 213
rehabilitation, 80, 137, 139, 224, 228
rehearsal. See practice.
Reid, Henry John Edward, 212
reimbursable, 54, 65, 148, 197, 226, 229
Reinhard, Charles E., 173
relativity, 4, 51
relay, 1, 67, 113, 134, 166, 197, 210, 222, 223-224, 226
reliability, 2, 50, 80, 120, 135, 137, 159, 161, 223-224
remote, 98, 121, 198, 201, 210, 211, 219, 221, 232, 234
  —sensing, 42, 56, 63, 110, 147, 174, 193
remotely powered vehicles, 221
rendezvous
  —orbital, 10, 207
  —planetary, 49, 207
  —spacecraft, 10, 89, 170
Rensselaer Polytechnic Institute, 192
reorganization, government, 114, 185, 197, 203
repair. See rehabilitation.
reporting systems, 43, 128, 130
reprimand, 195
reprocessing, 65
rescue, 172, 233-234
research, atmospheric. See atmosphere.
research and development (R&D), funding for, 65, 117, 134, 194
reservoirs, 81
resolution, optical, 42, 66, 76, 107, 137, 169, 189, 210, 222
resource management, 42, 47, 63, 161, 185, 193, 196
Resources for the Future, Inc., 40
respiration, 71
retirement, 195, 214-215, 219
retrieval (See also rehabilitation.), 98, 137, 163, 198, 205, 210, 211, 228, 231
retropropulsion, 187, 222
reusable (See also recycling.), 36, 53, 79, 95, 101, 118, 120, 133, 168, 198, 211-212
revival. See reactivation.
Reynolds, Smith & Hills, 128
Rhode Island, 66
Rice Institute, 188
Richardson, John, 169
rings, planetary. See planetary satellites.
Rio de Janeiro, Brazil, 222
Rio de la Plata, Argentina, 222
Rio Grande, 40
rivers, 110
RNA. See genetics.
Robinson Printing Co., 11
robots, 106
Rocket Propulsion Laboratory, USAF, 58
Rocketdyne Div., Rockwell Corp., 11, 51, 55, 79, 121
rockets (See also names of individual rockets.), 37, 98, 105, 121, 137, 138, 151, 164, 167, 197-198, 211, 215, 217, 267-272
rocks, 106
Rockwell, W.F., Jr., 70
Rockwell International Corp., 7, 17, 22, 24, 51, 55, 68-69, 70, 71, 73, 76, 79, 109, 112, 114, 118, 125, 175, 176, 177, 180, 192, 195, 202, 207

Romanenko, Lt. Col. Yuri, 223, 225
Romania, 20, 159, 174
Rome, Italy, 181
Rood, Robert W., 128
Ross, Miles, 63, 101
Rostovsev, Capt. Oleg, 164
rotation. See spin rate.
rotor-systems research aircraft (RSRA), 61, 134, 216
Royal Aeronautical Society, London, 16, 182
Royal Astronomical Society, U.K., 124
Royal Electrical and Mechanical Engineers, U.K., 154
Royal Observatory Greenwich, U.K., 69
Royal Society, U.K., 60
RTG, radioisotope thermal generators. See power sources, nuclear.
Rubel, Werner H., 230
runoff, water, 55
runway, 113, 154
Ryumin, Valery, 189, 223

Sacramento, Calif., 218
safety
 —automobile, 19
 —aviation, 19, 43, 128, 135, 138
 —devices, 94, 161, 224
 —fire, 47, 131, 159, 218
 —nuclear, 1
 —range, 180
 —systems, 47, 54, 64, 163, 178, 208, 224
Sagan, Carl, 143, 149
sailboats. See ships.
St. Lawrence River, 110
St. Louis, Mo., 33
St. Regis Paper Co., 196
St. Thomas, V.I., 179
Sakura. See *Cs.*
Salisbury, Md., 192
salvage. See recovery.
Salyut program (USSR space vehicle), 20, 90, 158, 161, 206
Salyut 4, 23
Salyut 5, 10, 26, 27, 189, 206
Salyut 6, 180, 189, 223, 254
sampling, 63, 106, 167, 193, 204, 207, 222
San Diego, Calif., 218
San Diego State University, 14
San Francisco, Calif., 182, 195, 218, 234
San Francisco, University of, 167
Sandford, Maxwell, 104

Sandia Laboratories, N.M., 109
sanitation, 63
Santa Barbara Research Center, 148
Santa Rose Island, Fla., 36
Satcom series (U.S. comsats), 154
Satellite Business Systems, Inc., 72, 154
Satellite Experiment Laboratory, NOAA, 193-194
satellite network. See network.
Satellite Power Systems (proposed mission), 5
satellites, planetary. See planetary.
Satrack (USN Trident support system), 197
Saturn (family of launch vehicles), 11, 24, 44, 55, 61, 71, 95, 113, 115, 127, 132, 139, 215, 228
Saturn (planet), 3, 33-34, 39, 49, 60, 102, 103, 106, 137, 145, 155, 161, 186, 204, 207, 226-227
Saudi Arabia, 14, 53
savings. See cost reduction.
Sawyer, Ralph (Buzz), 217
scanning, 17, 54, 149, 150, 155, 194, 212
SCATHA, satellite charging at high altitude, 168
schedule
 —launches, 37, 64-65, 162-163, 202, 229, 234
 —operations, 42, 217
scheduled airlines, 19
Scherer, Lee, 104
Schlesinger, James R., 185
Schmitt, Sen. Harrison M. (R-NM), 11, 111
Schneider, Stanley R., 110
scholarships. See awards.
Schuerer, Paul H., 11
Schwartz, Dr. Daniel A., 174
Schweikart, Russell L. (Rusty), 162
science adviser, White House, 31, 144
Science Applications, Inc., 22
science court, 1
science writing, 49
scientific community, 1, 18, 40, 132
scientists, 32, 35, 75, 84, 107, 131, 133, 143, 183, 220
Scorpius (constellation), 174
Scott, Dr. David R., 191
Scout (NASA launch vehicle), 11, 197
scrap, 211
sea. See oceanography; name of geographical area.

seals, 184
search and rescue, 64, 172, 173, 205
SeaSat-A, 1–2, 58, 82, 229
Seattle, Wash., 102
Securities and Exchange Commission, U.S., 22
security measures, 17, 38, 54, 122, 130
seismology, 31, 62, 71, 82, 156, 165, 222
seismometer, 3, 6, 140, 165
Selcuk, Dr. M. Kudret, 232
selenide (thermoelectric units), 162
semiconductors, 192
Senate, U.S. See Congress.
sensitivity, 188, 208
sensor systems
 —aircraft, 13, 96
 —ground-based, 109, 178
 —spacecraft, 22, 55, 58, 66, 71, 74, 137, 155, 174, 177, 215
Sensor Technology, Inc., 219, 220
separation, 73, 93, 107, 122–123, 135, 150, 151, 158, 160, 166, 168, 179, 180, 191, 195, 196, 211
SERT, space electric rocket tests (LeRC program), 170
Sevastyanov, Vitaly, 23
707; 737; 747. See Boeing.
sferics, 166
shadow, 176
Shanghai, China, 154
shapes, 124
Shatalov, Lt.Gen. Vladimir, 26–27, 223, 224
Shepard, Alan B., Jr., 115
shipping (See also air freight; cargo.), 44, 112
ships, 56, 64, 66, 85, 127, 137, 158, 164, 173, 178, 180, 194, 198, 205, 213, 223, 231
Shklovsky, I.S. (USSR astrophysicist), 124
shore (facilities; terminals), 207
Short, Dr. Nicholas M., 14
short-stack SRB, 95
short takeoff and landing (STOL), 72, 173
shortage. See fuel conservation.
Shroud of Turin, 142
Shulenko, Mariya, 123-124
Shuttle. See Space Shuttle.
Siberia, 123, 140, 159, 167, 178
Sieff, Dr. Alvin, 166
Sierra Leone, 103

Sierra Nevada, 80
signals, 132–133, 140, 155, 159, 168, 175, 198, 202, 213
Signe 3 (French/USSR scisat), 66, 117, 237, 245
Sikorsky Aircraft Co., 134
silicon, 42, 145, 175, 219, 220
silver compounds, 18–19
Simpson, Dr. John A., 102
simulations, 42, 44, 53, 55, 61, 64, 71, 72, 84, 85, 119, 120, 123, 132, 138, 153, 160, 185, 190, 191, 202–203, 206, 208, 211, 224, 230, 231
Sinclair Laboratories, 192
Sirio (Satellite Italiano Richerche Industriale Orientata), Italy comsat, 111, 139, 156–157, 251, 261, 262
size, 149, 158, 160
Skyjacker (DFRC experimental aircraft), 217–218
Skylab program, 6, 27, 50, 52, 64, 72, 121, 128, 160, 161, 165, 198, 211, 232, 235
Skylab 1, 80, 162
Skylab 4, 48, 121
Slayton, Donald K. (Deke), 42, 143, 196
Slidell Computer Complex (NASA), 147
slingshot effect. See gravity.
small astronomical satellites (SAS), 129
Small Business Act, 11
small craft. See ships.
Smith, Dr. Bradford A., 137
Smith, Richard B., 174
Smithsonian Astrophysical Observatory, 204
Smithsonian Center for Astrophysics, Harvard, 129, 174
Smithsonian Institution, 50
Smoot, Dr. George, 209
SNAP, system for auxiliary nuclear power. See power sources, nuclear.
SNAP 19, SNAP 27, 161
Snavely, Lt.Gen. William W., 53
snow, 55, 80–81, 172, 214
social goals, 60, 208
social policy. See national policy.
social problems (See also community relations.), 76, 147, 189
socialist bloc. See Soviet bloc.
Societe de Telecommunications Internationales du Chad, 110
Societe Europeenne de Propulsion (SEP), 225

Society for Space Travel (Verein fur Raumschiffart/VfR), German rocket society, 115
Society of Experimental Test Pilots, 235
Socorro, N.M., 235
sodium, 146
Soffen, Dr. Gerald A., 52
soft landing. See landing.
software, 130, 186, 196, 207
soil, 101, 106
Soil Conservation Service, U.S., 80
soil samples, 31, 70–71
sol (Mars day), 6
solar activity, 39, 117, 119, 148, 153, 176
solar cells. See solar conversion devices.
solar conversion devices
 —arrays, 43, 94, 128, 160, 170, 189, 228
 —cells, 43, 75, 145, 161, 220, 226
 —collectors, 14, 29, 71, 128, 130, 175, 179, 232
 —panels. See array.
solar corona, 15, 37
solar energy. See power sources, solar.
solar environment, 113
solar heating and cooling. See energy.
solar particles. See solar activity.
solar physics, 32, 199
solar polar mission (OOE, out-of-ecliptic), 68, 199
solar-power satellites (powersats), 5, 23, 57, 70, 76, 180
Solar Power, Inc., 219
solar radiation. See solar activity.
solar research. See sun.
solar sailer (JPL future project), 48–49, 119, 170
solar system, 3, 9, 48, 64, 103, 104, 107, 119, 199, 204, 207, 209, 226–227, 228
solar wind, 39–40, 48–49, 73, 155, 156
Solarex Corp., 43, 219, 220
solid-fuel rocket, 35, 53, 58, 74, 80, 95, 101, 119, 133, 151, 153, 154, 160, 205, 211, 228
solid rocket (booster). See solid-fuel rocket.
solvents, 110
sonic booms, 105
sounders; soundings, 148
sounding rockets. See rockets.
sounds, 31–32, 143, 184
sources of radiation. See power sources, nuclear; radiation.

South Africa, Union of, 182
South America, 138, 176
South Pole (See also Antarctica.), 182
Southern Bell Telephone and Telegraph Co., 221
Southern California, University of, 167
Souza, Kenneth, 167
Soviet Academy of Sciences, 76, 89, 180, 183, 206
Soviet Air Force, 13
Soviet bloc, 129–130, 141–142
Soviet Ministry
 —of Health, 206
 —of Power and Electrification; of Power Plant Construction, 76
Soviet space activity (See also Union of Soviet Socialist Republics.), 10, 22, 23, 26–27, 28, 52, 54, 56, 59, 66, 76, 89–90, 117, 127, 146, 158, 164, 167, 176, 180, 184, 187, 189–190, 195–196, 197–198, 206, 207, 210, 222–225
Soyuz program (USSR manned flight), 10, 20, 89, 187, 189–190, 222–225
Soyuz 7, 26
Soyuz 15, 190
Soyuz 17, 23
Soyuz 19, 206
Soyuz 20, 23
Soyuz 21, 27
Soyuz 23, 190
Soyuz 24, 26, 239, 265, 266
Soyuz 25, 189, 223, 254, 265, 266
Soyuz 26, 222–225, 257, 265, 266
Space Age, 148, 165
Space Biology and Medicine working group, U.S./USSR, 206
space
 —colonization, 12, 24–25, 68, 137, 141, 149, 185
 —communities. See space colonization.
 —construction; structure (See also space station.), 45, 47, 58, 64, 69, 76, 106, 235–236
Space Day, 148
space defense systems (See also interceptor satellites; weapons systems.), 22, 58
Space Defense Technology Program (DOD), 177
space
 —environment, 57, 85, 148
 —exploration. See exploration.
 —flight, NASA office of. See NASA.
 —industrialization, 25, 47, 57, 137, 180

—investment in. See funding, government.
—manufacturing. See materials processing; space industrialization.
—medicine. See medical research.
—observance, U.S., 132
—platform. See space station.
—powersat. See solar-power satellites.
—processing. See materials processing.
Space Research Institute (USSR Acad. of Science), 17
—science, 49-51, 53, 65, 71, 73, 134, 135-136
—settlement. See space colonization.
Space Shuttle
—astronauts. See Space Shuttle crews.
—boosters. See Space Shuttle propulsion.
—costs, 65, 108-109, 121
—crews, 21, 42, 45, 63, 107-108, 132, 140, 143, 144, 148, 163, 172-173, 185, 201, 202, 205-206, 208, 222, 228
—experiments. See Space Shuttle payloads.
—external tank, 25, 44, 50, 51, 55, 64, 72, 80, 113, 120, 151, 153, 160, 170, 202, 228, 231
—facilities, 64, 113, 201-202, 228-229
—fees, 65, 143, 163
—funding, 104-105, 108-109, 121, 217
—impact, 75
—launch, 50, 113, 128, 139, 211, 228, 234
—main engine (See also Space Shuttle propulsion.), 6, 51, 55, 79, 120, 121, 160, 201, 202
—management, 143, 153, 201
—military use, 54, 163
—operations, 1, 21, 23-24, 50, 62, 68-69, 106, 113-114, 143, 145, 150-152, 162-163, 165, 168, 175, 208, 234
—orbiter, 17, 23-24, 27-28, 36, 42, 45, 55, 63, 74, 80, 103, 107, 108, 113, 118, 135, 143, 153, 165, 168, 170, 178-179, 190-191, 193, 196-197, 202, 207, 211, 230
—passengers, 57
—payloads, 4, 6, 10, 18, 25, 48, 50, 52-53, 54, 58, 64-65, 68, 69, 70, 71, 76, 98, 104-105, 119, 143, 160, 162-163, 170, 185-186, 187-188, 192, 198, 199, 207, 215, 228, 234
—performance, 107-108

—policy, 48, 90, 104-105, 143, 162-163, 206
—prices. See Space Shuttle fees; Space Shuttle users.
—propulsion, 35, 44, 55, 74, 79, 95, 101, 113-114, 119-120, 151, 154, 202, 211
—solid-fuel rocket booster. See Space Shuttle propulsion.
—systems, 55, 68-69, 143, 178-179, 234
—testing, 6, 17, 23-24, 36, 42, 44, 45, 51, 55, 63, 64, 71, 74, 80, 95, 101, 103, 107-108, 113-114, 118, 120, 135, 143, 148, 160, 165, 168, 170, 178, 185, 190-191, 193, 196-197, 202, 205, 207, 211, 228, 230, 234, 235
—training. See Space Shuttle crews.
—upper stages, 37, 163
—users, 143, 163
space spider, 69
space station, 23, 50, 72, 89-90, 98, 106-107, 128, 158, 165, 187, 223
space structure. See space construction; space station.
space suits, 45, 160, 208, 210, 224
Space Task Group, 216
Space Telescope
—contractors, 18, 232-233
—planning/management, 10, 18, 52, 56, 66, 108, 135, 136-137, 189, 207, 210
Space Transportation System (STS) (See also Space Shuttle.), 59, 105, 143, 163, 173
space travel, 71, 115-116
spacewalk. See extravehicular activity (EVA).
spacecraft
—communications, 4-5
—power systems, 74-75, 136
—retrieval, 18
Spacelab
—crew, 84, 85, 166-167, 183, 230
—experiments, 32, 65, 67, 72, 85, 119, 152-153, 157
—management, 16, 48, 64-65, 67, 153, 215
—mission, 72, 73, 206
—operations, 1, 28, 65
—payloads, 25, 85, 130, 133, 175
—preparation, 25, 52-53, 70, 72-73, 84, 85, 133, 152-153, 166-167, 170, 215

Spaceport. See Kennedy Space Center, Spaceport.
Spain, 1, 154
Spanish (language), 103
SPAR, space-processing applications rockets, 30, 105
spare, 207
Spaw Glass, Inc., 53
specifications, 165
spectrograph; spectrometer, 155, 166, 168, 210
Spectrolab, Inc., 220
spectroscopics, 51
spectrum, 149
speed (See also high-speed aircraft.), 61, 67, 96, 150, 151, 152, 160, 168, 176–177, 179, 181, 186, 191, 196, 197, 202, 209, 211, 218
Speer, Dr. Fred, 149, 156
Sperry Rand Corp., 7, 54
Sperry Support Services, Inc., 54
Sperry Univac, 74
spin, aircraft, 134, 138
— motor; spacecraft, 119, 131, 195
— rate (planetary, galactic), 38–39, 124, 209
— stabilization, 73
— stages (spacecraft), 173
Spirit of St. Louis, 3
Spitzbergen (Norway), 79
Springfield, Ill., 12
Sprott, Kingswood, Jr., 76
Sputnik 1 (USSR satellite), 115, 117, 165, 184, 187, 189
spy systems. See surveillance.
SR-71 (U.S. reconnaissance aircraft). See Lockheed.
SSN-8 (USSR missiles), 57
stability, 36, 75, 151, 186, 218
stabilization; stabilizers, 79, 163
stalling, aircraft. See safety, aviation.
stamps, 195
stand-in. See Space Shuttle orbiter; simulation.
standards, 59, 213
Stanford University, 16, 75, 107, 147
Star Trek, 41
stars; stellar, 37, 104, 107, 117, 121, 137, 143, 149, 158, 188, 209, 210, 228
State, U.S. Department of, 141
state-federal cooperation, 80, 94, 162
state funding of projects, 110
static electricity. See insulation.

static testing, 44, 55, 95, 159, 160
stationkeeping (satellite), 113
statistics, 138
steam. See power sources.
steering. See guidance.
stellar object. See star.
Stephenson, Charles A., 173
Stevenson, Sen. Adlai E. (D-Ill), 114
Steward Observatory (Univ. of Ariz.), 107
Stickney crater (on Phobos), 124
Stirling engine, 199–200
Stofan, Andrew J., 220
Stone, Dr. Edward P., 137
storage (See also energy storage.), 51, 222, 229
Stormovik (Il-2. USSR attack plane), 29
storms, 94, 119, 127, 131, 165, 168, 171, 176, 234–235
Strack, Bill, 135
strap-on motors, 106
strategic weapons. See weapons systems.
stratosphere. See atmosphere.
Strecker, Dr. D.W., 107
strength (signal), 133
stress, 153
Strittmatter, Dr. Peter, 107
structures, 44, 61, 69, 79, 95, 120, 128, 149, 153, 156, 160, 163, 205, 211, 231
students, 4, 41–42, 133, 143, 152, 235
Stuhlinger, Dr. Ernst, 116
styrofoam, 130
submarines, 6, 8, 57
subsonic aircraft, 35, 66, 152
sun, 66, 68, 104, 133, 145, 149, 158, 161, 199
sunspots. See solar activity.
Super Loki (U.S. rocket), 37
superclusters, galactic, 129
superhigh frequencies (SHF), 139
supernovas, 32, 37, 215
supersonic (aircraft; transport), 29, 64, 66, 213
supersonics, 9, 64, 219
supplies, 21, 224
support (funding, services, systems, teams), 42, 44, 45, 55, 68–69, 102, 113, 139, 147, 156, 168, 169, 197, 207, 208, 212, 221, 222, 226, 228–229, 231–232, 234
surface, 145, 168, 175, 214, 226
surgery, 5

surveillance (See also monitoring.), 58, 63, 101-102, 122, 136, 149-150, 221-222
survival, 22, 68, 202, 226
Swain, Charles W., 148
Swainson, Wyn Kelly, 17
Sweden, 66, 103, 154, 174, 200
Switzerland, 84, 154
symposium. See meetings.
synchronous (orbits; satellites), 2, 124, 138, 139, 148, 210
synthesis, 65, 73, 230
system for auxiliary nuclear power (SNAP). See power sources, nuclear.
systems engineering. See design, engineering.
Syvertson, C.A., 147, 188

T-38 (U.S. trainer/chase aircraft), 132, 150
tactical weapons. See weapons systems.
tadpoles, 225
tail, airplane. See design, aircraft.
tailcone, 151, 197
takeoff, 66
Tallahassee, Fla., 148
tandem launch, 193, 195
Tanegashima (Japan launch site), 30, 41
Tansei 3 (Japan experimental satellite), 38, 239
tape recording. See recording systems.
target satellites. See interceptor satellites.
Taylor, Michael, 72, 84
Taylor, Quentin, 62
Taylor, T.H., Construction Co., 63
teaching, 14
Teal Ruby (DOD project), 71
teamwork, 231
Tech House (LaRC), 130-131, 148
technicians, 182, 194
techniques, improved. See new techniques.
technology
 —applications, 11, 56, 71, 141-142, 145-146
 —comparisons, 13
 —funding, 194
 —problems, 76
 —transfer, 71
 —utilization, 51, 130, 131, 208, 212
telecommunications, 60, 182
teleconference, 47
Teledyne, Inc., 30, 134

Teledyne Industries, Inc., 221
telegraph, 168
telemetry, 158
teleoperator retrieval system. See remote.
telephone, 74, 168, 214, 221, 226
telephotography, 101
teleportation, 16
Telesat Canada, 32, 153, 229
telescope, 18, 37, 38-39, 48, 50, 66, 70, 72, 107, 109, 136-137, 149, 152, 188
Telespazio (Italy comsat control), 157
television, 10, 12-13, 33, 52, 104, 121, 129-130, 154, 165, 178, 198, 208, 223-224, 226, 234
temperature, 39-40, 62, 67, 93, 127, 128, 148, 150, 159, 161, 165, 168, 171, 175, 177, 203, 232
Tennessee, 127
Tennessee River, 44
Tennessee State University, 145
Tennessee Valley Authority (TVA), 227
terminal configured vehicle (TCV). See microwave landing system.
terrain. See topography.
Terrile, Dr. Richard J., 230
terrorism, 189
Tesla, Nikola, 19
test pilots. See pilots
testing
 —aircraft, 132, 235
 —ballistic missiles, 8, 57
 —equipment, 58, 59, 80, 112, 128, 163
 —instrumentation, 93
 —Space Shuttle. See Space Shuttle testing.
tethered satellite, 98, 181
Texas, 3, 41, 127, 138
Texas Instruments, Inc., 220
Texas Southern University, 3
textbooks, 14
thematic mapper, 42
theory, 51, 55, 75, 117, 129, 190, 208-209, 217
thermal-infrared scanning. See scanning.
thermal insulation. See insulation.
thermodynamics, 32
thermoelectric materials, 140
thermosphere. See atmosphere.
Thiel, Dr. Charles, 62
Thiokol Corp., 35, 133
third world, 14
Thompson, James R., 121
Thompson, Dr. Rodger, 107

Thornton, Dr. William E., 85
thrust, 153, 160, 205
thunderstorm research international program (TRIP), 21-22, 127, 234-235
thunderstorms, 21, 127
Tiedemann, Trudy, 74
tidal forces, 124
tilt-rotor aircraft, 61, 216
timber. See forests.
time, 51, 82, 149, 208
timing, 175
TIROS-N, 229
Titan (moon of Saturn), 49
Titan (U.S. solid-fuel rocket), 211
Titan Centaur (rocket), 33, 155, 167, 220, 221
titanium, 108, 233
Tokyo, Japan, 131, 140, 182
Tokyo, University of, 23, 38
Tomahawk (USN cruise missile), 8
tomato can. See interceptor satellites.
tools. See instrumentation.
topography, 109, 176
Toulouse Space Research Center (France), 117
tourism, 6, 61, 67, 69
towing, 231
tracking, 57, 93, 155, 175, 189, 221, 222
 —ground stations, 1, 82, 180
 —radio beacons, 6, 172, 184
 —ships, 10, 64, 66, 180
 —space-based, 58, 71, 79
 —and data acquisition, NASA office of. See NASA.
 —and data-relay satellite system (TDRSS), 1
Tracor Marine, Inc., 231
tractor. See vehicles.
traffic. See air traffic; communications problems; orbit, spacecraft.
training, 193, 194, 231
 —astronauts (See also Space Shuttle, crews.), 21, 84, 107, 118, 132, 208
 —cosmonauts, 20, 223-224
trajectories, 35, 68, 145, 167, 186, 215, 223, 226
Transat (USN navsat), 197, 256, 261, 263
Transcaucasia, USSR, 59
Transit (USN global navigation system), 197
Transit 3A (USN navsat), 161
Transit improvement program (TIP), 197
transmission. See energy transfer.

transmission, 47, 179, 208, 210
transonic, 132
transponder, 15, 197
transport aircraft. See aircraft.
transportation, 60, 111, 113-114, 121, 122, 137, 141, 175, 192
Transportation, U.S. Department of, 112, 192, 212
trash. See wastes.
traveling-wave tube (TWT), 181
treaty. See international relations.
trees, 39-40
trials, law court. See judicial proceedings.
Trident (USN ballistic missile), 8, 157, 197
Tropex (USSR-sponsored international tropical weather experiment), 56
trophy. See awards.
TRS, teleoperator retrieval system. See remote.
truck. See vehicles.
Truly, Richard, 122, 135, 170, 171, 191
Truszynski, Gerald M., 219-220
TRW Systems, Inc., 1, 22, 47, 103
Tucson, Ariz., 94
Tulsa, Okla., 45
tumors. See medical research.
Tupolev (USSR SST jetliner), 105
Tupolev, A.N., 29
turbines, 134-135
turbulence, 61, 161, 191, 218
turnaround time, 186
Turner Construction Co., 173
typhoons, 131

U-2 (U.S. jet reconnaissance aircraft), 208
U-25 facility, USSR, 111
Uchinoura (Japan launch site), 38
UH-1H (U.S. helicopter), 132
Uhlman, Dr. Donald, 106
Uhuru (*Explorer 42*), 129
UK-6 (British radiation-measurement spacecraft), 229
Ukraine, region of USSR, 167
ultrahigh frequency (UHF), 139
ultrasonic, 233
ultraviolet, 10, 37, 117, 119, 121, 169, 189, 210
underground testing, 71, 140
undersea; underwater, 8, 208
unemployment, 75, 107, 108
unidentified flying objects, 218

uniformity, 185, 209
Union Baptist Bible College, 3
Union of South Africa. See South Africa.
Union of Soviet Socialist Republics (USSR), 146, 158, 176, 234
— international cooperation, 4, 46, 52, 85, 89-90, 111, 119, 121, 127, 143-144
— manned spaceflight, 10, 89, 123-124, 180, 189-190, 206, 222-225
— technology, 13, 111, 121, 195-196
— weapons systems, 13, 57, 101
United Bank of California, 148
United Kingdom, 1, 40, 44, 46, 53, 65, 66, 84, 115, 119, 141, 152, 154, 186, 193, 212, 227, 229
United Kingdom National Physical Laboratory, 227
United Kingdom Science Research Council, 9
United Nations, 44, 93, 103, 114, 129-130
United Space Boosters, Inc., 35
United States, 24, 46, 53
United States Information Agency (USIA), 10
United States Strategic Institute, 172
United Stirling (Sweden), 200
United Technologies Corp., 35, 50, 192
Univac 1108 (computer), 74
"Universe" (NASA film), 64, 87
universe, origins of. See theory.
University (of). See rest of name.
university community, 71, 146, 148, 172, 174, 188, 228
unmanned flight, missions, 36, 42, 72, 149
unpowered flight, 17, 36, 42, 150-151
upper atmosphere. See atmosphere.
upper stages. See Space Shuttle.
Upper Volta, 198
uranium, 109, 130, 230
Uranus (planet), 3, 34, 38, 48, 49, 59-60, 69, 103, 137, 161, 204, 226-227
Urban, Dr. Eugene W., 157
Urdu (language), 103
U.S. Booster, Inc (See also United Space Boosters.), 231
users; user community (See also data utilization.), 143
Users Requirements Committee, 59
Utah State University, 4
utilities, private, 15-16, 129
Utopia Planitia (region of Mars), 6, 125

V-2 (WW II German rocket weapon), 115
vacuum, space, 54, 59
— tubes, 13
Van Allen belt, 102
Vandenberg AFB, Calif., 109, 120, 211, 229
Vanderhoff, Dr. John W., 192
Vanguard (USN rocket), 115
vapor, 192
variable systems, 134, 138
vegetation (See also botany.), 47, 123
vehicle assembly building (VAB), KSC, 61, 113
vehicles (See also transportation.), 19, 63, 74, 111, 113-114, 122, 175, 192, 199-200
velocity. See speed.
Venus (planet), 28, 92-93, 106, 119, 161, 221
Venus orbiting imaging radar. See VOIR.
verification, 32, 43
Vernon, France, 225
Versailles, Treaty of, 115
Verschoore, Chuck, 153
vertical takeoff and landing (VTOL) craft, 173
vertical test stand, 80
very large array (VLA), 66
Vessot, Dr. Robert, 30, 51
Veterans Administration (VA), U.S., 8, 146, 188
veterinary medicine, 157
Veverka, Joseph, 124
VHF (very high frequency), 113
vibration, 55, 63, 110, 113, 118, 135, 170, 202, 205
videotape, 10
Vienna, Austria, 129
Viking: lander, orbiter. See Viking project.
Viking mission, 3, 15, 33, 49, 52, 125, 155, 161
Viking project, 3, 15, 20, 25, 48, 52, 70, 101, 125, 181
Viking 1, 31, 55, 70, 132, 176
Viking 2, 31, 62, 76
vinegar flies, 68
Vinson, Mack, 229-230
Virgin Islands, U.S., 179
Virginia, 5, 137, 164
Virginia Polytechnic Institute (VPI), 5
visibility, 69, 75, 94
visible, 174, 178, 189, 210, 216

vision, 216
vitamins, 123
Vogt, Dr. Rochus E., 190
voice communications, 158, 208
VOIR (Venus orbiting imaging radar), 207
volcanoes, 207
Volna program (USSR comsats), 207
von Braun, Dr. Wernher, 115-117, 141, 209
von Zahn, Dr. Ulf, 166
Vought Corp., 30, 177
Voyager mission, 33, 49, 102, 103, 108, 137, 140, 143, 145, 155-156, 161, 167, 169, 184, 186, 221, 226-227
Voyager 1, 167, 186, 227, 252, 263
Voyager 2, 155-156, 186, 227, 250, 262

Wagner, Dr. J. Bruce, 192
Waldheim, Dr. Kurt, 44, 103
Walker, Curtis L., 232
Wallops Flight Center
 —launches, 30, 127, 164
 —programs, 5, 69, 85, 127, 134, 138, 164, 169, 178, 192, 206
Wang, Dr. Taylor, 106
war, 74
warning. See safety devices; safety systems.
Warwick, Dr. James W., 137
Washington, University of, 167
wastes, 51, 111, 188, 233
Watanabe, Kazuo, 131
water, 106, 123, 160, 218, 224
 —pollution. See pollution, environmental.
 —survey, 33, 48, 80, 141
 —tank, 64
 —vapor, 10, 110, 161, 204
"waterhole." See frequencies, broadcast.
waves, 231
weapons systems, 162, 203
 —aircraft, 29
 —interceptors. See interceptor satellites.
 —intercontinental, 8
 —missiles, rockets, 8-9, 16, 57, 58, 115, 197
 —nuclear, 65, 131, 177
 —satellites, 6-7, 22, 101-102, 177
weather; weather control. See meteorology.
Weaver, Leon, 84

Webb, C. Horton, 227
Weber, Mrs. Elizabeth, 67
Webster, J.J., 81
weight, 45, 79, 114, 123, 128, 134, 145, 150, 162, 177, 182, 211, 214
weightlessness (See also zero gravity.), 30, 57, 59, 64, 146, 160, 167, 206, 208, 224
Weis, Judge Joseph F., Jr., 13
Weitz, Paul, 160
welding, 25
West Germany. See Federal Republic of Germany.
Western American Energy Corp., 51
Western Test Range (WTR), 9, 197, 229
Western Union Corp., 22
Western Union International, 140
Western Union Space Communications Co., Inc., 1
Westinghouse Electric Corp., 13, 22, 49, 175, 220
Westlake, Reginald, 65
whales, 184
wheel, 165
Whitaker, Ann, 230
Whitcomb, Richard T., 71, 152
White (Gen. Thomas D.) space trophy, 203
white dwarfs. See stars.
White Sands, N.M., 1
 —Missile Range, 30, 93, 121
 —Test Facility, 44, 71
Whitecloud (USN navigation system). See *Noss 2*.
Wiebe, CPO Richard, 208
Wiedemeier, Dr. Herbert, 192
wildlife, 47, 157, 193
Wilkins, Dr. Judd, 63
Wilkinson, Dr. J.H., 227
Williams, Dr. Bill A., 85
Williams Research Corp., 134
Williamson, Dave, 218
wind, 67, 211, 214, 222, 231
wind power; windmills. See power sources.
wind tunnels, 28, 42, 61, 110, 132, 138, 173
Windsor, Richard M., 37
Windworks, Inc., 176
winglets, 71-72, 110, 130
wings, aircraft. See design.
Winter, Dr. David, 206
wireless, 19
wiring, 177

Wisconsin, 130, 176
Witteborn, Dr. Fred, 107
Woese, Dr. Carl, 201
Wolff, A.B., 186
women, in air and space program (See also names of individuals.), 41, 57, 74, 79, 123, 132, 157, 162, 176, 185, 191, 194, 201, 205-206
Wood, H. William, 65
Woods Hole Oceanographic Institute, 93, 114
Woomera, Australia, 37
work force adjustment. See employment.
working groups. See meetings.
work load, 232
workshops. See meetings.
World Meteorological Organization, 131, 210
World War I, 115, 141, 172
World War II, 16, 115, 147
worldwide communications. See global communications.
Wouch, Jerry, 106
wreck, 205, 214
Wright Flyer (airplane), 3
Wright Patterson AFB, Ohio, 199, 214
Wright trophy, 172
Wyoming, University of, 109

X-1 (NASA experimental aircraft), 219
X-15 (U.S. research aircraft), 235
XB-70 (experimental aircraft), 235
XC-8A. See Buffalo aircraft.
XV-15. See tilt-rotor research aircraft.
x-ray, 37, 48, 54, 58, 72, 113, 129, 149-150, 153, 156, 174, 210, 212, 216, 228

Yakovlev, A.S., 29
Yardley, John F., 124, 171, 209, 217
YF-12 (U.S. aircraft), 235
Yin, Dr. Lo I., 216
Yost, Paul E. (Ed), 181, 227
Young, A. Thomas, 106, 207, 226
Young, John, 143

Zelenchuk, USSR, 76
Zeppelin, 141
zero gravity, 25, 152
zinc, 93, 190
Zuckerman, Lord (Baron) Solly, 40
Zurich, University of, 55

The NASA History Series

HISTORIES

Anderson, Frank W., Jr., *Orders of Magnitude: A History of NACA and NASA, 1915-1980* (NASA SP-4403, 2d ed., 1981).

Benson, Charles D., and William Barnaby Faherty, *Moonport: A History of Apollo Launch Facilities and Operations* (NASA SP-4204, 1978).

Bilstein, Roger E., *Stages to Saturn: A Technological History of the Apollo/Saturn Launch Vehicles* (NASA SP-4206, 1980).

Boone, W. Fred, *NASA Office of Defense Affairs: The First Five Years* (NASA HHR-32, 1970, multilith).

Brooks, Courtney G., James M. Grimwood and Loyd S. Swenson. Jr., *Chariots for Apollo: A History of Manned Lunar Spacecraft* (NASA SP-4205, 1979).

Byers, Bruce K., *Destination Moon: A History of the Lunar Orbiter Program* (NASA TM X-3487, 1977, multilith).

Compton, W. David, and Charles D. Benson, *Living and Working in Space: A History of Skylab* (NASA SP-4208, 1983).

Corliss, William R. *NASA Sounding Rockets, 1958-1968: A Historical Summary* (NASA SP-4401, 1971).

Ezell, Edward Clinton, and Linda Neuman Ezell, *On Mars: Exploration of the Red Planet, 1958-1978* (NASA SP-4212, 1984).

Ezell, Edward Clinton, and Linda Neuman Ezell, *The Partnership: A History of the Apollo-Soyuz Test Project* (NASA SP-4209, 1978).

Green, Constance McL., and Milton Lomask, *Vanguard: A History* (NASA SP-4202, 1970; also Washington: Smithsonian Institution Press, 1971).

Hacker, Barton C., and James W. Grimwood, *On the Shoulders of Titans: A History of Project Gemini* (NASA SP-4203, 1977).

Hall, R. Cargill, *Lunar Impact: A History of Project Ranger* (NASA SP-4210, 1977).

Hallion, Richard P., *On the Frontier: Flight Research at Dryden, 1946-1981* (NASA SP-4303, 1984).

Hartman, Edwin P., *Adventures in Research: A History of Ames Research Center, 1940-1965* (NASA SP-4302, 1970).

Levine, Arnold, *Managing NASA in the Apollo ERA* (NASA SP-4102, 1982).

Newell, Homer E., *Beyond the Atmosphere: Early Years of Space Science* (NASA SP-4211, 1980).

Pitts, John A., *The Human Factor: Biomedicine in the Manned Space Program to 1908* (NASA SP-4213, 1985).

Roland, Alex, *Model Research: The National Advisory Committee for Aeronautics, 1915-1958* (NASA SP-4103, 1985).

Rosenthal, Alfred, *Venture into Space: Early Years of Goddard Space Flight Center* (NASA SP-4301, 1968).

Rosholt, Robert L., *An Administrative History of NASA, 1958-1963* (NASA SP-4101, 1966).

Sloop, John L., *Liquid Hydrogen as a Propulsion Fuel, 1945-1959* (NASA SP-4404, 1978).

Swenson, Loyd S., Jr., James M. Grimwood, and Charles C. Alexander, *This New Ocean: A History of Project Mercury* (NASA SP-4201, 1966).

REFERENCE WORKS

Aeronautics and Space Report of the President, annual volumes for 1975-1982.
The Apollo Spacecraft: A Chronology (NASA SP-4009, vol. 1, 1969; vol. 2, 1973; vol. 3, 1976; vol. 4, 1978).
Astronautics and Aeronautics: A Chronology of Science, Technology, and Policy, annual volumes 1961-1976, with an earlier summary volume, *Aeronautics and Astronautics, 1951-1960*.
Dickson, Katherine M., ed., *History of Aeronautics and Astronautics: A Preliminary Bibliography* (NASA HHR-29, 1968, multilith).
Hall, R. Cargill, ed., *Essays on the History of Rocketry and Astronautics: Proceedings of the Third through the Sixth History Symposia of the International Academy of Astronautics* (NASA CP-2014, 2 vols., 1977).
Hall, R. Cargill, *Project Ranger: A Chronology* (JPL/HR-2, 1971, multilith).
Looney, John J., ed., *Bibliography of Space Books and Articles from Non-Aerospace Journals, 1957-1977* NASA HHR-51, 1979, multilith).
Roland, Alex F., *A Guide to Research in NASA History* (NASA HHR-50, 6th ed., 1982, available from NASA History Office).
Skylab: A Chronology (NASA SP-4011, 1977).
Van Nimmen, Jane, and Leonard C. Bruno, with Robert L. Rosholt, *NASA Historical Data Book, 1958-1968*, vol. 1, *NASA Resources* (NASA SP-4012, 1976).
Wells, Helen T., Susan H. Whiteley, and Carrie E. Karegeannes, *Origins of NASA Names* (NASA SP-4402, 1976).

Recent volumes are available from Superintendent of Documents, Government Printing Office, Washington, DC 20402; early volumes from National Technical Information Service, Springfield, VA 22161.

www.ingramcontent.com/pod-product-compliance
Lightning Source LLC
Chambersburg PA
CBHW081718170526
45167CB00009B/3621